TensorFlow 深度学习
第 2 版(影印版)
Deep Learning with TensorFlow, 2nd Edition

Giancarlo Zaccone,
Rezaul Karim 著

南京　东南大学出版社

图书在版编目(CIP)数据

TensorFlow 深度学习:第 2 版:英文/(意)吉安卡洛·扎克尼(Giancarlo Zaccone),(德)礼萨·卡里姆(Rezaul Karim)著. —影印本. —南京:东南大学出版社,2019.5

书名原文:Deep Learning with TensorFlow,2nd Edition

ISBN 978-7-5641-8326-4

Ⅰ.①T… Ⅱ.①吉… ②礼… Ⅲ.①人工智能-算法-英文 Ⅳ.①TP18

中国版本图书馆 CIP 数据核字(2019)第 046177 号

图字:10-2018-500 号

© 2018 by PACKT Publishing Ltd.

Reprint of the English Edition, jointly published by PACKT Publishing Ltd and Southeast University Press, 2019. Authorized reprint of the original English edition, 2018 PACKT Publishing Ltd, the owner of all rights to publish and sell the same.

All rights reserved including the rights of reproduction in whole or in part in any form.

英文原版由 PACKT Publishing Ltd 出版 2018。

英文影印版由东南大学出版社出版 2019。此影印版的出版和销售得到出版权和销售权的所有者——PACKT Publishing Ltd 的许可。

版权所有,未得书面许可,本书的任何部分和全部不得以任何形式重制。

TensorFlow 深度学习 第 2 版(影印版)

出版发行:	东南大学出版社
地　　址:	南京四牌楼 2 号　邮编:210096
出 版 人:	江建中
网　　址:	http://www.seupress.com
电子邮件:	press@seupress.com
印　　刷:	常州市武进第三印刷有限公司
开　　本:	787 毫米×980 毫米　16 开本
印　　张:	30.25
字　　数:	592 千字
版　　次:	2019 年 5 月第 1 版
印　　次:	2019 年 5 月第 1 次印刷
书　　号:	ISBN 978-7-5641-8326-4
定　　价:	108.00 元

本社图书若有印装质量问题,请直接与营销部联系。电话(传真):025-83791830

mapt.io

Mapt is an online digital library that gives you full access to over 5,000 books and videos, as well as industry leading tools to help you plan your personal development and advance your career. For more information, please visit our website.

Why subscribe?

- Spend less time learning and more time coding with practical eBooks and Videos from over 4,000 industry professionals
- Learn better with Skill Plans built especially for you
- Get a free eBook or video every month
- Mapt is fully searchable
- Copy and paste, print, and bookmark content

PacktPub.com

Did you know that Packt offers eBook versions of every book published, with PDF and ePub files available? You can upgrade to the eBook version at www.PacktPub.com and as a print book customer, you are entitled to a discount on the eBook copy. Get in touch with us at service@packtpub.com for more details.

At www.PacktPub.com, you can also read a collection of free technical articles, sign up for a range of free newsletters, and receive exclusive discounts and offers on Packt books and eBooks.

Contributors

About the authors

Giancarlo Zaccone has over ten years of experience in managing research projects in scientific and industrial areas.

Giancarlo worked as a researcher at the CNR, the National Research Council of Italy. As part of his data science and software engineering projects, he gained experience in numerical computing, parallel computing, and scientific visualization.

Currently, Giancarlo is a senior software and system engineer, based in the Netherlands. Here he tests and develops software systems for space and defense applications.

Giancarlo holds a master's degree in Physics from the Federico II of Naples and a 2nd level postgraduate master course in Scientific Computing from La Sapienza of Rome.

Giancarlo is the author of the following books: *Python Parallel Programminng Cookbook, Getting Started with TensorFlow, Deep Learning with TensorFlow*, all by Packt Publishing.

You can follow him at https://it.linkedin.com/in/giancarlozaccone.

Md. Rezaul Karim is a research scientist at Fraunhofer FIT, Germany. He is also pursuing his PhD at the RWTH Aachen University, Aachen, Germany. He holds BSc and MSc degrees in Computer Science. Before joining Fraunhofer FIT, Rezaul had been working as a researcher at Insight Centre for Data Analytics, Ireland. Previously, he worked as a Lead Engineer at Samsung Electronics. He also worked as a research assistant at Database Lab, Kyung Hee University, Korea and as an R&D engineer with BMTech21 Worldwide, Korea.

Rezaul has over 9 years of experience in research and development with a solid understanding of algorithms and data structures in C, C++, Java, Scala, R, and Python. He has published several research papers and technical articles concerning Bioinformatics, Semantic Web, Big Data, Machine Learning and Deep Learning using Spark, Kafka, Docker, Zeppelin, Hadoop, and MapReduce.

Rezaul is also equally competent with (deep) machine learning libraries such as Spark ML, Keras, Scikit-learn, TensorFlow, DeepLearning4j, MXNet, and H2O. Moreover, Rezaul is the author of the following books:

Large-Scale Machine Learning with Spark, Deep Learning with TensorFlow, Scala and Spark for Big Data Analytics, Predictive Analytics with TensorFlow, Scala Machine Learning Projects, all by *Packt Publishing*.

> Writing this book was made easier by amazing efforts by many open source communities and documentation about many projects. Further, I would like to thank a wonderful team at Packt for their sincere cooperation and coordination. Finally, I appreciate numerous efforts by the TensorFlow community and all those who have contributed to APIs, whose work ultimately brought the machine learning to the masses!

About the reviewers

Motaz Saad holds a PhD in Computer Science from the University of Lorraine. He loves data and likes to play with it. Motaz has over ten years of professional experience in NLP, computational linguistics, and data science machine learning. Motaz currently works as an assistant professor at the faculty of Information Technology, IUG.

Sefik Ilkin Serengil received his MSc in Computer Science from the Galatasaray University in 2011.

Sefik has been working as a software developer for a FinTech company since 2010. Currently, he is a member of the AI team as a data scientist in this company.

Sefik's current research interests are Machine Learning and Cryptography. He has published several research papers on these topics. Nowadays, he enjoys speaking to communities about these disciplines.

Sefik has also created several online courses on Machine Learning.

Vihan Jain has made several key contributions to the open-sourced TensorFlow project. He has been advocating for the adoption of TensorFlow since two years. Vihan has given tech-talks and has taught tutorials on TensorFlow at various conferences. His research interests include reinforcement learning, wide and deep learning, recommendation systems, and machine learning infrastructure. Vihan graduated from the Indian Institute of Technology, Roorkee, in 2013 with the President's gold medal.

> I express my deepest gratitude to my parents, brother, sister, and my good friend and mentor, Eugene Ie.

Packt is Searching for Authors Like You

If you're interested in becoming an author for Packt, please visit `authors.packtpub.com` and apply today. We have worked with thousands of developers and tech professionals, just like you, to help them share their insight with the global tech community. You can make a general application, apply for a specific hot topic that we are recruiting an author for, or submit your own idea.

Table of Contents

Preface	**ix**
Chapter 1: Getting Started with Deep Learning	**1**
A soft introduction to machine learning	**2**
Supervised learning	4
Unbalanced data	6
Unsupervised learning	6
Reinforcement learning	8
What is deep learning?	8
Artificial neural networks	**11**
The biological neurons	12
The artificial neuron	13
How does an ANN learn?	**15**
ANNs and the backpropagation algorithm	15
Weight optimization	16
Stochastic gradient descent	17
Neural network architectures	**18**
Deep Neural Networks (DNNs)	18
Multilayer perceptron	19
Deep Belief Networks (DBNs)	20
Convolutional Neural Networks (CNNs)	22
AutoEncoders	24
Recurrent Neural Networks (RNNs)	26
Emergent architectures	27
Deep learning frameworks	**27**
Summary	**30**

Table of Contents

Chapter 2: A First Look at TensorFlow — 31
A general overview of TensorFlow — 32
What's new in TensorFlow v1.6? — 33
- Nvidia GPU support optimized — 34
- Introducing TensorFlow Lite — 34
- Eager execution — 35
- Optimized Accelerated Linear Algebra (XLA) — 35

Installing and configuring TensorFlow — 36
TensorFlow computational graph — 37
TensorFlow code structure — 41
- Eager execution with TensorFlow — 44

Data model in TensorFlow — 46
- Tensor — 46
- Rank and shape — 48
- Data type — 50
- Variables — 53
- Fetches — 54
- Feeds and placeholders — 55

Visualizing computations through TensorBoard — 57
- How does TensorBoard work? — 57

Linear regression and beyond — 59
- Linear regression revisited for a real dataset — 67

Summary — 73

Chapter 3: Feed-Forward Neural Networks with TensorFlow — 75
Feed-forward neural networks (FFNNs) — 76
- Feed-forward and backpropagation — 77
- Weights and biases — 79
- Activation functions — 81
  - Using sigmoid — 84
  - Using tanh — 84
  - Using ReLU — 84
  - Using softmax — 85

Implementing a feed-forward neural network — 85
- Exploring the MNIST dataset — 86
  - Softmax classifier — 88

Implementing a multilayer perceptron (MLP) — 95
- Training an MLP — 96
- Using MLPs — 98
  - Dataset description — 99
  - Preprocessing — 101
  - A TensorFlow implementation of MLP for client-subscription assessment — 103

Deep Belief Networks (DBNs)	111
Restricted Boltzmann Machines (RBMs)	112
Construction of a simple DBN	115
Unsupervised pre-training	116
Supervised fine-tuning	117
Implementing a DBN with TensorFlow for client-subscription assessment	118
Tuning hyperparameters and advanced FFNNs	**126**
Tuning FFNN hyperparameters	126
Number of hidden layers	126
Number of neurons per hidden layer	127
Weight and biases initialization	128
Selecting the most suitable optimizer	129
GridSearch and randomized search for hyperparameters tuning	130
Regularization	130
Dropout optimization	133
Summary	**136**
Chapter 4: Convolutional Neural Networks	**139**
Main concepts of CNNs	**140**
CNNs in action	**142**
LeNet5	**143**
Implementing a LeNet-5 step by step	**144**
AlexNet	152
Transfer learning	154
Pretrained AlexNet	154
Dataset preparation	**156**
Fine-tuning implementation	**157**
VGG	160
Artistic style learning with VGG-19	162
Input images	163
Content extractor and loss	164
Style extractor and loss	167
Merger and total loss	168
Training	168
Inception-v3	**171**
Exploring Inception with TensorFlow	172
Emotion recognition with CNNs	**173**
Testing the model on your own image	185
Source code	187
Summary	**190**

Chapter 5: Optimizing TensorFlow Autoencoders — 191
How does an autoencoder work? — 192
Implementing autoencoders with TensorFlow — 195
Improving autoencoder robustness — 200
- Implementing a denoising autoencoder — 201
- Implementing a convolutional autoencoder — 207
  - Encoder — 207
  - Decoder — 208
Fraud analytics with autoencoders — 217
- Description of the dataset — 217
- Problem description — 218
- Exploratory data analysis — 219
- Training, validation, and testing set preparation — 223
- Normalization — 224
- Autoencoder as an unsupervised feature learning algorithm — 224
- Evaluating the model — 229
Summary — 233

Chapter 6: Recurrent Neural Networks — 235
Working principles of RNNs — 236
- Implementing basic RNNs in TensorFlow — 239
- RNN and the long-term dependency problem — 243
  - Bi-directional RNNs — 244
RNN and the gradient vanishing-exploding problem — 246
  - LSTM networks — 249
  - GRU cell — 252
Implementing an RNN for spam prediction — 253
- Data description and preprocessing — 253
Developing a predictive model for time series data — 260
- Description of the dataset — 260
- Pre-processing and exploratory analysis — 262
- LSTM predictive model — 264
- Model evaluation — 267
An LSTM predictive model for sentiment analysis — 270
- Network design — 270
- LSTM model training — 271
- Visualizing through TensorBoard — 289
- LSTM model evaluation — 291
Human activity recognition using LSTM model — 294
- Dataset description — 294
- Workflow of the LSTM model for HAR — 296

Implementing an LSTM model for HAR	297
Summary	**307**
Chapter 7: Heterogeneous and Distributed Computing	**309**
GPGPU computing	**310**
The GPGPU history	310
The CUDA architecture	311
The GPU programming model	312
The TensorFlow GPU setup	**313**
Update TensorFlow	313
GPU representation	314
Using a GPU	314
GPU memory management	316
Assigning a single GPU on a multi-GPU system	316
The source code for GPU with soft placement	317
Using multiple GPUs	318
Distributed computing	**320**
Model parallelism	320
Data parallelism	321
The distributed TensorFlow setup	**323**
Summary	**325**
Chapter 8: Advanced TensorFlow Programming	**327**
tf.estimator	**327**
Estimators	328
Graph actions	328
Parsing resources	328
Flower predictions	329
TFLearn	**333**
Installation	334
Titanic survival predictor	334
PrettyTensor	**337**
Chaining layers	337
Normal mode	337
Sequential mode	338
Branch and join	338
Digit classifier	338
Keras	**342**
Keras programming models	343
Sequential model	343
Functional API	348
Summary	**353**

Chapter 9: Recommendation Systems Using Factorization Machines — 355
Recommendation systems — 356
Collaborative filtering approaches — 356
Content-based filtering approaches — 358
Hybrid recommender systems — 358
Model-based collaborative filtering — 358
Movie recommendation using collaborative filtering — 359
The utility matrix — 359
Description of the dataset — 362
Ratings data — 362
Movies data — 362
Users data — 363
Exploratory analysis of the MovieLens dataset — 364
Implementing a movie RE — 370
Training the model with the available ratings — 371
Inferencing the saved model — 380
Generating the user-item table — 380
Clustering similar movies — 382
Movie rating prediction by users — 386
Finding top k movies — 387
Predicting top k similar movies — 388
Computing user-user similarity — 389
Evaluating the recommender system — 390
Factorization machines for recommendation systems — 393
Factorization machines — 394
Cold-start problem and collaborative-filtering approaches — 395
Problem definition and formulation — 397
Dataset description — 398
Workflow of the implementation — 399
Preprocessing — 401
Training the FM model — 408
Improved factorization machines — 413
Neural factorization machines — 414
Dataset description — 414
Using NFM for the movie recommendation — 415
Summary — 420

Chapter 10: Reinforcement Learning — 421
The RL problem — 422
OpenAI Gym — 423
OpenAI environments — 424
The env class — 424
Installing and running OpenAI Gym — 425

The Q-Learning algorithm	**426**
The FrozenLake environment	427
Deep Q-learning	**431**
Deep Q neural networks	431
The Cart-Pole problem	433
Deep Q-Network for the Cart-Pole problem	435
The Experience Replay method	436
Exploitation and exploration	437
The Deep Q-Learning training algorithm	438
Summary	**445**
Other Books You May Enjoy	**447**
Leave a review – let other readers know what you think	449
Index	**451**

Preface

Every week, we follow news of applications and the shocking results obtained from them, thanks to the artificial intelligence algorithms applied in different fields. What we are witnessing is one of the biggest accelerations in the entire history of this sector, and the main suspect behind these important developments is called **deep learning**.

Deep learning comprises a vast set of algorithms that are based on the concept of neural networks and expand to contain a huge number of nodes that are disseminated at several levels of depth.

Though the concept of neural networks, the so-called Artificial Neural Network (ANN), dates back to the late 1940s, initially, they were difficult to be used because of the need for huge computational power resources and the lack of data required to train the algorithms. Presently, the ability to use graphics processors (GPUs) in parallel to perform intensive calculation operations has completely opened the way to the use of deep learning.

In this context, we propose the second edition of this book, with expanded and revised contents that introduce the core concepts of deep learning, using the last version of **TensorFlow**.

TensorFlow is Google's open-source framework for the mathematical, Machine Learning, and Deep Learning capabilities, released in 2011. Subsequently, TensorFlow has been widely adopted in academia, research, and industry. The most stable version of TensorFlow at the time of writing was version 1.6, which was released with a unified API and is thus a significant and stable version in the TensorFlow roadmap. This book also discusses and is compliant with the pre-release version, 1.7, which was available during the production stages of this book.

TensorFlow provides the flexibility needed to implement and research cutting-edge architectures, while allowing users to focus on the structure of their models as opposed to mathematical details.

Preface

You will learn deep learning programming techniques with hands-on model building, data collection, transformation, and much more!

Enjoy reading!

Who this book is for

This book is dedicated to developers, data analysts, and deep learning enthusiasts who do not have much background with complex numerical computations, but want to know what deep learning is. The book majorly appeals to beginners who are looking for a quick guide to gain some hands-on experience with deep learning.

What this book covers

Chapter 1, Getting Started with Deep Learning, covers the concepts that will be found in all the subsequent chapters. The basics of machine learning and deep learning are also discussed. We will also look at Deep learning architectures that are distinguished from the more commonplace single-hidden-layer neural networks by their depth, that is, the number of node layers through which data passes in a multistep process of pattern recognition. We will also analyze these architectures with a chart summarizing all the neural networks from where most of the deep learning algorithm evolved. The chapter ends with an analysis of the major deep learning frameworks.

Chapter 2, A First Look at TensorFlow, gives a detailed description of the main TensorFlow features based on a real-life problem, followed by a detailed discussion on TensorFlow installation and configurations. We then look at a computation graph, data, and programming model before getting started with TensorFlow. Toward the end of the chapter, we will look at an example of implementing the linear regression model for predictive analytics.

Chapter 3, Feed-Forward Neural Networks with TensorFlow, demonstrates the theoretical background of different Feed-Forward Neural Networks' (FFNNs) architectures such as Deep Belief Networks (DBNs) and Multilayer Perceptron (MLP). We will then see how to train and analyze the performance metrics that are needed to evaluate the models; also, how to tune the hyperparameters for FFNNs for better and optimized performance. We will also look at two examples using MLP and DBN on how to build very robust and accurate predictive models for predictive analytics on a bank marketing dataset.

Chapter 4, *Convolutional Neural Networks*, introduces the networks of CNNs that are the basic blocks of a Deep Learning-based image classifier. We will consider the most important CNN architectures, such as **Lenet, AlexNet, Vgg,** and **Inception** with hands-on examples, specifically for AlexNet and Vgg. We will then examine the *transfer learning* and *style learning* techniques. We will end the chapter by developing a CNN to train a network on a series of facial images to classify their *emotional stretch*.

Chapter 5, *Optimizing TensorFlow Autoencoders*, provides sound theoretical background on optimizing autoencoders for data denoising and dimensionality reduction. We will then look at how to implement an autoencoder, gradually moving over to more robust autoencoder implementation, such as denoising autoencoders and convolutional autoencoders. Finally, we will look at a real-life example of fraud analytics using an autoencoder.

Chapter 6, *Recurrent Neural Networks*, provides some theoretical background of RNNs. We will also look at a few examples for implementing predictive models for classification of images, sentiment analysis of movies, and products spam prediction for NLP. Finally, we'll see how to develop predictive models for time series data.

Chapter 7, *Heterogeneous and Distributed Computing*, shows the fundamental topic to execute TensorFlow models on GPU cards and distributed systems. We will also look at basic concepts with application examples.

Chapter 8, *Advanced TensorFlow Programming*, gives an overview of the following TensorFlow-based libraries: `tf.contrib.learn`, `Pretty Tensor`, `TFLearn`, and `Keras`. For each library, we will describe the main features with applications.

Chapter 9, *Recommendation Systems using Factorization Machines*, provides several examples on how to develop recommendation system for predictive analytics followed by some theoretical background of recommendation systems. We will then look at an example of developing a movie recommendation engine using collaborative filtering and K-means. Considering the limitations of classical approaches, we'll see how to use Neural Factorization Machines for developing more accurate and robust recommendation systems.

Chapter 10, *Reinforcement Learning*, covers the basic concepts of RL. We will experience the Q-learning algorithm, which is one of the most popular reinforcement learning algorithms. Furthermore, we'll introduce the OpenAI gym framework that is a TensorFlow compatible toolkit for developing and comparing reinforcement learning algorithms. We end the chapter with the implementation of a Deep Q-Learning algorithm to resolve the cart-pole problem.

To get the most out of this book

- A rudimentary level of programming in one language is assumed, as is a basic familiarity with computer science techniques and technologies, including a basic awareness of computer hardware and algorithms. Some competence in mathematics is needed to the level of elementary linear algebra and calculus.
- Software: Python 3.5.0, Pip, pandas, numpy, tensorflow, Matplotlib 2.1.1, IPython, Scipy 0.19.0, sklearn, seaborn, tffm, and many more
- Step: Issue the following command on Terminal on Ubuntu:

    `$ sudo pip3 install pandas numpy tensorflow sklearn seaborn tffm`

    Nevertheless, installing guidelines are provided in the chapters.

Download the example code files

You can download the example code files for this book from your account at http://www.packtpub.com. If you purchased this book elsewhere, you can visit http://www.packtpub.com/support and register to have the files emailed directly to you.

You can download the code files by following these steps:

1. Log in or register at http://www.packtpub.com.
2. Select the **SUPPORT** tab.
3. Click on **Code Downloads & Errata**.
4. Enter the name of the book in the **Search** box and follow the on-screen instructions.

Once the file is downloaded, please make sure that you unzip or extract the folder using the latest version of any of the following:

- WinRAR / 7-Zip for Windows
- Zipeg / iZip / UnRarX for macOS
- 7-Zip / PeaZip for Linux

The code bundle for the book is also hosted on GitHub at https://github.com/PacktPublishing/Deep-Learning-with-TensorFlow-Second-Edition. We also have other code bundles from our rich catalog of books and videos available at https://github.com/PacktPublishing/. Check them out!

Download the color images

We also provide a PDF file that has color images of the screenshots/diagrams used in this book. You can download it here: https://www.packtpub.com/sites/default/files/downloads/DeepLearningwithTensorFlowSecondEdition_ColorImages.pdf.

Conventions used

There are a number of text conventions used throughout this book.

`CodeInText`: Indicates code words in text, database table names, folder names, filenames, file extensions, pathnames, dummy URLs, user input, and Twitter handles. For example; " This means that using `tf.enable_eager_execution()` is recommended."

A block of code is set as follows:

```
import tensorflow as tf # Import TensorFlow

x = tf.constant(8) # X op
y = tf.constant(9) # Y op
z = tf.multiply(x, y) # New op Z

sess = tf.Session() # Create TensorFlow session

out_z = sess.run(z) # execute Z op
sess.close() # Close TensorFlow session
print('The multiplication of x and y: %d' % out_z)# print result
```

When we wish to draw your attention to a particular part of a code block, the relevant lines or items are set in bold:

```
import tensorflow as tf # Import TensorFlow

x = tf.constant(8) # X op
y = tf.constant(9) # Y op
z = tf.multiply(x, y) # New op Z

sess = tf.Session() # Create TensorFlow session

out_z = sess.run(z) # execute Z op
sess.close() # Close TensorFlow session
print('The multiplication of x and y: %d' % out_z)# print result
```

Preface

Any command-line input or output is written as follows:

```
>>>
MSE: 27.3749
```

Bold: Indicates a new term, an important word, or words that you see on the screen, for example, in menus or dialog boxes, also appear in the text like this. For example: " Now let's move to `http://localhost:6006` and on click on the **GRAPH** tab."

> Warnings or important notes appear in a box like this.

> Tips and tricks appear like this.

Get in touch

Feedback from our readers is always welcome.

General feedback: Email `feedback@packtpub.com`, and mention the book's title in the subject of your message. If you have questions about any aspect of this book, please email us at `questions@packtpub.com`.

Errata: Although we have taken every care to ensure the accuracy of our content, mistakes do happen. If you have found a mistake in this book we would be grateful if you would report this to us. Please visit, `http://www.packtpub.com/submit-errata`, selecting your book, clicking on the Errata Submission Form link, and entering the details.

Piracy: If you come across any illegal copies of our works in any form on the Internet, we would be grateful if you would provide us with the location address or website name. Please contact us at `copyright@packtpub.com` with a link to the material.

If you are interested in becoming an author: If there is a topic that you have expertise in and you are interested in either writing or contributing to a book, please visit `http://authors.packtpub.com`.

Reviews

Please leave a review. Once you have read and used this book, why not leave a review on the site that you purchased it from? Potential readers can then see and use your unbiased opinion to make purchase decisions, we at Packt can understand what you think about our products, and our authors can see your feedback on their book. Thank you!

For more information about Packt, please visit `packtpub.com`.

1
Getting Started with Deep Learning

This chapter explains some of the basic concepts of **Machine Learning** (**ML**) and **Deep Learning** (**DL**) that will be used in all the subsequent chapters. We will start with a brief introduction to ML. Then we will move to DL, which is a branch of ML based on a set of algorithms that attempt to model high-level abstractions in data.

We will briefly discuss some of the most well-known and widely used neural network architectures, before moving on to coding with TensorFlow in *Chapter 2, A First Look at TensorFlow*. In this chapter, we will look at various features of DL frameworks and libraries, such as the native language of the framework, multi-GPU support, and aspects of usability.

In a nutshell, the following topics will be covered:

- A soft introduction to ML
- Artificial neural networks
- ML versus DL
- DL neural network architectures
- Available DL frameworks

A soft introduction to machine learning

ML is about using a set of statistical and mathematical algorithms to perform tasks such as concept learning, predictive modeling, clustering, and mining useful patterns. The ultimate goal is to improve the learning in such a way that it becomes automatic, so that no more human interactions are needed, or at least to reduce the level of human interaction as much as possible.

We now refer to a famous definition of ML by *Tom M. Mitchell* (*Machine Learning, Tom Mitchell, McGraw Hill*), where he explained what learning really means from a computer science perspective:

> "*A computer program is said to learn from experience E with respect to some class of tasks T and performance measure P, if its performance at tasks in T, as measured by P, improves with experience E.*"

Based on this definition, we can conclude that a computer program or machine can do the following:

- Learn from data and histories called training data
- Improve with experience
- Interactively enhance a model that can be used to predict outcomes of questions

Almost every machine-learning algorithm we use can be treated as an optimization problem. This is about finding parameters that minimize some objective function, such as a weighted sum of two terms such as a cost function and regularization (log-likelihood and log-prior, respectively, in statistics).

Typically, an objective function has two components: a regularizer, which controls the complexity of the model, and the loss, which measures the error of the model on the training data (we'll look into the details).

On the other hand, the regularization parameter defines the trade-off between the two goals of minimizing the loss of the training error and of minimizing the model's complexity in an effort to avoid overfitting. Now if both of these components are convex, then their sum is also convex; else it is nonconvex.

> In machine learning, overfitting is when the predictor model fits perfectly on the training examples, but does badly on the test examples. This often happens when the model is too complex and trivially fits the data (too many parameters), or when there is not enough data to accurately estimate the parameters. When the ratio of model complexity to training set size is too high, overfitting will typically occur.

More elaborately, while using an ML algorithm, our goal is to obtain the hyperparameters of a function that returns the minimum error when making predictions. The error loss function has a typically U-shaped curve, when visualized on a two-dimensional plane, and there exists a point, which gives the minimum error.

Therefore, using a convex optimization technique, we can minimize the function until it converges toward the minimum error (that is, it tries to reach the middle region of the curve), which represents the minimum error. Now that a problem is convex, it is usually easier to analyze the asymptotic behavior of the algorithm that shows how fast it converges as the model observes more and more training data.

The challenge of ML is to allow a computer to learn how to automatically recognize complex patterns and make decisions as intelligently as possible. The entire learning process requires a *dataset*, as follows:

- **Training set**: This is the knowledge base used to fit the parameters of the machine-learning algorithm. During this phase, we would use the training set to find the optimal weights, with the *back-prop rule*, and all the parameters to set before the learning process begins (**hyperparameters**).
- **Validation set**: This is a set of examples used *to tune* the parameters of an ML model. For example, we would use the validation set to find the optimal number of *hidden units*, or determine a *stopping point* for the back-propagation algorithm. Some ML practitioners refer to it as **development set** or **dev set**.
- **Test set**: This is used for evaluating the performance of the model on unseen data, which is called **model inferencing**. After assessing the final model on the test set, we don't have to tune the model any further.

Learning theory uses mathematical tools that derive from probability theory and information theory. Three learning paradigms will be briefly discussed:

- Supervised learning
- Unsupervised learning
- Reinforcement learning

The following diagram summarizes the three types of learning, along with the problems they address:

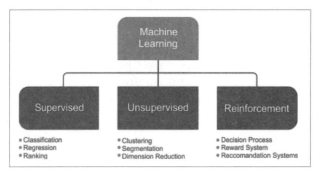

Figure 1: Types of learning and related problems.

Supervised learning

Supervised learning is the simplest and most well-known automatic learning task. It is based on a number of pre-defined examples, in which the category where each of the inputs should belong is already known. In this case, the crucial issue is the problem of generalization. After the analysis of a typical small sample of examples, the system should produce a model that should work well for all possible inputs.

The following figure shows a typical workflow of supervised learning. An actor (for example, an ML practitioner, data scientist, data engineer, or ML engineer) performs **ETL** (**Extraction, Transformation, and Load**) and necessary feature engineering (including feature extraction, selection) to get the appropriate data, with features and labels.

Then he does the following:

- Splits the data into the training, development, and test set
- Uses the training set to train an ML model
- Uses the validation set for validating the training against the overfitting problem, and regularization
- Evaluates the model's performance on the test set (that is, unseen data)
- If the performance is not satisfactory, he performs additional tuning to get the best model, based on hyperparameter optimization
- Finally, he deploys the best model into a production-ready environment

In the overall lifecycle, there might be many actors involved (for example, data engineer, data scientist, or ML engineer) to perform each step independently or collaboratively:

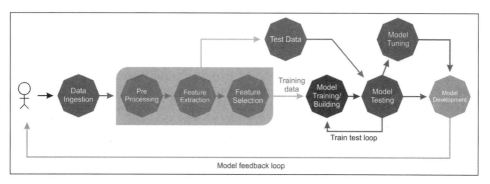

Figure 2: Supervised learning in action.

In supervised ML, the set consists of labeled data, that is, objects and their associated values for regression. This set of labeled examples, therefore, constitutes the training set. Most supervised learning algorithms share one characteristic: the training is performed by the minimization of a particular loss or cost function, representing the output error provided by the system, with respect to the desired output.

The supervised learning context includes **classification** and **regression** tasks: classification is used to predict which class a data point is a part of (*discrete value*) while regression is used to predict *continuous values*:

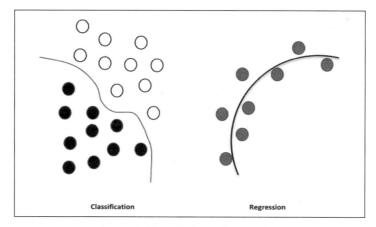

Figure 3: Classification and regression

In other words, the classification task predicts the label of the class attribute, while the regression task makes a numeric prediction of the class attribute.

Unbalanced data

In the context of supervised learning, *unbalanced data* refers to classification problems where we have unequal instances for different classes. For example, if we have a classification task for only two classes, *balanced data* would mean 50% preclassified examples for each of the classes.

If the input dataset is a *little unbalanced* (for example, 60% for one class and 40% for the other class) the learning process will be required to *randomly split* the input dataset into three sets, with 50% for the training set, 20% for the validation set, and the remaining 30% for the testing set.

Unsupervised learning

In *unsupervised learning*, an input set is supplied to the system during the training phase. In contrast with supervised learning, the input objects are *not labeled* with their class. This type of learning is important because, in the human brain, it is probably far more common than supervised learning.

For the classification, we assume that we are given a training dataset of correctly labeled data. Unfortunately, we do not always have that luxury when we collect data in the real world. The only object in the domain of learning models, in this case, is the observed data input, which is often assumed to be independent samples of an *unknown underlying probability distribution*.

For example, suppose that you have a large collection of non-pirated and totally legal MP3s in a crowded and massive folder on your hard drive. How could you possibly group together songs without direct access to their metadata? One possible approach could be a mixture of various ML techniques, but clustering is often at the heart of the solution.

Now, what if you could build a clustering predictive model that could automatically group together similar songs, and organize them into your favorite categories such as "country", "rap" and "rock"? The MP3 would be added to the respective playlist in an unsupervised way. In short, unsupervised learning algorithms are commonly used in *clustering problems*:

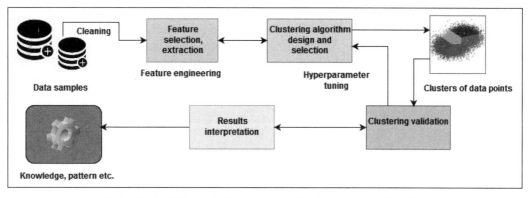

Figure 4: Clustering techniques: an example of unsupervised learning

See the preceding diagram to get an idea of a clustering technique being applied to solve this kind of problem. Although the data points are not labeled, we can still do the necessary feature engineering, and group a set of objects in such a way that objects in the same group (called a *cluster*) are more similar (in some sense) to each other, than to those in other groups (*clusters*).

This is not easy for a human, because a standard approach is to define a *similarity measure* between two objects and then look for any cluster of objects that are more similar to each other than they are to the objects in the other clusters. Once we do the clustering, the validation of data points (that is, MP3 files) is completed and we know the pattern of the data (that is, what type of MP3 files fall in to which group).

Reinforcement learning

Reinforcement learning is an artificial intelligence approach that focuses on the learning of the system through its interactions with the environment. With reinforcement learning, the system adapts its parameters based on feedback received from the environment, which then provides feedback on the decisions made. The following diagram shows a person making decisions in order to arrive at their destination. Suppose that, on your drive from home to work, you always choose the same route. However, one day your curiosity takes over and you decide to try a different route, in the hope of finding a shorter commute. This dilemma of trying out new routes, or sticking to the best-known route, is an example of **exploration versus exploitation**:

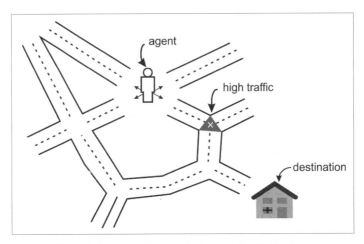

Figure 5: An agent always tries to reach the destination.

Another example is a system that models a chess player, that uses the result of its preceding moves to improve its performance. This is a system that learns with reinforcement.

Current research on reinforcement learning is highly interdisciplinary, including researchers specializing in genetic algorithms, neural networks, psychology, and control engineering.

What is deep learning?

Simple ML methods that were used in the *normal size* data analysis are not effective anymore, and should be substituted for more robust ML methods. Although classical ML techniques allow researchers to identify groups, or clusters, of related variables, the accuracy and effectiveness of these methods diminishes with large and high-dimensional datasets.

Therefore, here comes DL, which is one of the most important developments in artificial intelligence in the last few years. DL is a branch of ML based on a set of algorithms that attempt to model high-level abstractions in data.

The development of DL occurred in parallel with the study of artificial intelligence, and especially with the study of neural networks. It was mainly in the 1980s that this area grew, thanks largely to *Geoff Hinton* and the ML specialists who collaborated with him. At that time, computer technology was not sufficiently advanced to allow a real improvement in this direction, so we had to wait for a greater availability of data and vastly improved *computing power* to see significant developments.

In short, DL algorithms are a set of **Artificial Neural Networks** (**ANNs**), which we will explore later, that can make better representations of large-scale datasets, in order to build models that learn these representations extensively. In this regard, *Ian Goodfellow* and others defined DL as follows:

> "Deep learning is a particular kind of machine learning that achieves great power and flexibility by learning to represent the world as a nested hierarchy of concepts, with each concept defined in relation to simpler concepts, and more abstract representations computed in terms of less abstract ones".

Let's give an example. Suppose we want to develop a predictive analytics model, such as an animal recognizer, where our system has to resolve two problems:

1. *Classify* if an image represents a cat or a dog
2. *Cluster* dog and cat images

If we solve the first problem using a typical ML method, we must define the *facial features* (ears, eyes, whiskers, and so on), and write a method to identify which features (typically *non-linear*) are more important when classifying a particular animal.

However, at the same time, we cannot address the second problem, because classical ML algorithms for clustering images (such as **K-means**) cannot handle non-linear features.

DL algorithms will take these two problems one step further and the most important features will be extracted automatically, after determining *which features are the most important for classification or clustering*. In contrast, using a classic ML algorithm, we would have *to manually provide* the features.

In summary, the DL workflow would be as follows:

- A DL algorithm would first identify the edges that are most relevant when clustering cats or dogs
- It would then build on this hierarchically to find the various combinations of shapes and edges
- After consecutive hierarchical identification of complex concepts and features, it decides which of these features can be used to classify the animal, then takes out the label column and performs unsupervised training using an autoencoder, before doing the clustering.

Up to this point, we have seen that DL systems are able to recognize what an image represents. A computer does not see an image as we see it because it only knows the position of each pixel and its color. Using DL techniques, the image is divided into various *layers of analysis*. At a lower level, the software analyzes, for example, a grid of a few pixels, with the task of detecting a type of color or various nuances. If it finds something, it informs the next level, which at this point verifies whether that given color belongs to a larger form, such as a line.

The process continues to the *upper levels* until you understand what is shown in the image. Software capable of doing these things is now widespread and is found in systems for recognizing faces or searching for an image on **Google**, for example. In many cases, these are hybrid systems, that work with more traditional IT solutions, that are mixed with generation artificial intelligence.

The following diagram shows what we have discussed in the case of an *image classification system*. Each block gradually extracts the features of the input image and goes on to process data from the previous blocks, that have already been processed, extracting increasingly abstract features of the image, and thus building the hierarchical representation of data that comes with a DL-based system.

More precisely, it builds the layers as follows:

- **Layer 1**: The system starts identifying the dark and light pixels
- **Layer 2**: The system identifies edges and shapes
- **Layer 3**: The system learns more complex shapes and objects
- **Layer 4**: The system learns which objects define a human face

This is shown in the following diagram:

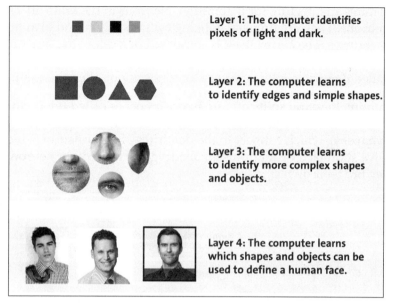

Figure 6: A DL system at work on a facial classification problem.

In the previous section, we saw that using a linear ML algorithm, we typically handle only a few hyperparameters.

However, when neural networks come in the party, things become too complex. In each layer, there are so many hyperparameters, and the cost function always becomes nonconvex.

Another reason is that the activations functions used in the hidden layers are nonlinear, so the cost is nonconvex. We'll discuss this phenomenon in more detail in the later chapters.

Artificial neural networks

ANNs take advantage of the concept of DL. They are an abstract representation of the human nervous system, which contains a collection of neurons that communicate with each other through connections called **axons**.

Warren McCulloch and *Walter Pitts* proposed the first artificial neuron model in 1943 in terms of a computational model of nervous activity. This model was followed by another proposed by *John von Neumann, Marvin Minsky, Frank Rosenblatt* (the so-called perceptron), and many others.

The biological neurons

Look at the brain's architecture for inspiration. Neurons in the brain are called **biological neurons**. They are unusual-looking cells, mostly found in animal brains, consisting of cortexes. The cortex itself is composed of a cell body, containing the nucleus and most of the cell's complex components. There are many branching extensions called **dendrites**, plus one very long extension called the **axon**.

Near its extremity, the axon splits off into many branches called **telodendria** and at the top of these branches are minuscule structures called **synaptic terminals** (or simply **synapses**), which connect to the dendrites of other neurons. Biological neurons receive short electrical impulses called signals from other neurons, and in response, they fire their own signals:

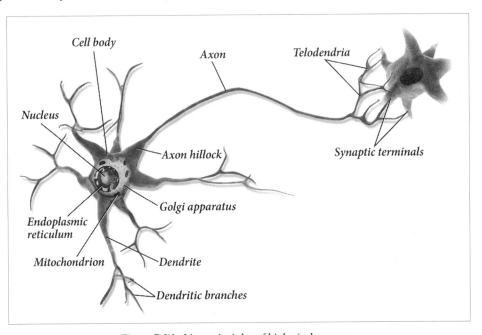

Figure 7: Working principles of biological neurons.

In biology, a neuron is composed of the following:

- A cell body or *soma*
- One or more *dendrites*, whose responsibility it is to receive signals from other neurons
- An *axon*, which in turn conveys the signals generated by the same neuron to the other connected neurons

The neuron's activity alternates between sending a signal (*active state*) and rest/receiving signals from other neurons (*inactive state*). The transition from one phase to another is caused by the external stimuli, represented by signals that are picked up by the dendrites. Each signal has an excitatory or inhibitory effect, conceptually represented by a weight associated with the stimulus.

A neuron in *idle state* accumulates all the signals it has received until it reaches a certain activation threshold.

The artificial neuron

Based on the concept of biological neurons, the term and the idea of artificial neurons arose, and they have been used to build intelligent machines for DL-based predictive analytics. This is the key idea that inspired ANNs. Similarly to biological neurons, the *artificial neuron* consists of the following:

- One or more incoming connections, with the task of collecting numerical signals from other neurons: each connection is assigned a weight that will be used to consider each signal sent
- One or more output connections that carry the signal to the other neurons
- An activation function, which determines the numerical value of the output signal, based on the signals received from the input connections with other neurons, and suitably collected from the weights associated with each received signal, and the activation threshold of the neuron itself:

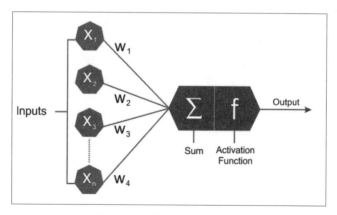

Figure 8: Artificial neuron model.

Getting Started with Deep Learning

The output, that is, the signal that the neuron transmits, is calculated by applying the activation function, also called the transfer function, to the weighted sum of the inputs. These functions have a dynamic range between -1 and 1, or between 0 and 1. Many activation functions differ in terms of complexity and output. Here, we briefly present the three simplest forms:

- **Step function**: Once we fix the threshold value x (for example, $x = 10$), the function will return zero, or one if the mathematical sum of the inputs is at, above, or below the threshold value.
- **Linear combination**: Instead of managing a threshold value, the weighted sum of the input values is subtracted from a default value. We will have a binary outcome that will be expressed by a positive (+b) or negative (-b) output of the subtraction.
- **Sigmoid**: This produces a sigmoid curve, which is a curve with an S trend. Often, the sigmoid function refers to a special case of the logistic function.

From the simplest forms used in the prototyping of the first artificial neurons, we move to more complex forms that allow greater characterization of the functioning of the neuron:

- Hyperbolic tangent function
- Radial basis function
- Conic section function
- Softmax function:

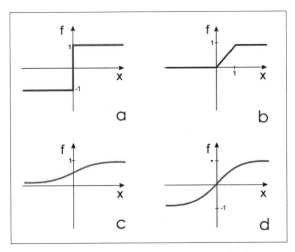

Figure 9: The most commonly used artificial neuron model transfer functions. a. step function. b. linear function c. computed sigmoid function with values between 0 and 1. d. sigmoid function with computed values between -1 and 1.

Choosing proper activation functions (also weights initialization) is key to making a network perform at its best and to obtain good training. These topics are under a lot of research, and studies indicate marginal differences in terms of output quality if the training phase is carried out properly.

 There is no rule of thumb in the field of neural networks. It all depends on your data and in what form you want the data to be transformed, after passing through the activation function. If you want to choose a particular activation function, you need to study the graph of the function to see how the result changes with respect to the values given to it.

How does an ANN learn?

The learning process of a neural network is configured as an *iterative process* of the *optimization* of the *weights* and is therefore of the *supervised* type. The weights are modified because of the network's performance on a set of examples belonging to the training set, that is, the set where you know the classes that the examples belong to.

The aim is to *minimize the loss function*, which indicates the degree to which the behavior of the network deviates from the desired behavior. The performance of the network is then verified on a testing set consisting of objects (for example, images in an image classification problem) other than those in the training set.

ANNs and the backpropagation algorithm

A commonly used supervised learning algorithm is the *backpropagation* algorithm. The basic steps of the training procedure are as follows:

1. Initialize the net with random weights
2. For all training cases, follow these steps:
    - **Forward pass**: Calculates the network's error, that is, the difference between the desired output and the actual output
    - **Backward pass**: For all layers, starting with the output layer back to input layer:

      i: Shows the network layer's output with the correct input (*error function*).

      ii: *Adapts* the weights in the current layer to minimize the error function. This is backpropagation's *optimization* step.

Getting Started with Deep Learning

The training process ends when the error on the validation set begins to increase because this could mark the beginning of a phase overfitting, that is, the phase in which the network tends to interpolate the training data at the expense of generalizability.

Weight optimization

The availability of efficient algorithms to optimize weights, therefore, constitutes an essential tool for the construction of neural networks. The problem can be solved with an iterative numerical technique called **Gradient Descent** (**GD**). This **technique** works according to the following algorithm:

1. Randomly choose initial values for the parameters of the model
2. Compute the gradient **G** of the error function with respect to each parameter of the model
3. Change the model's parameters so that they move in the direction of decreasing the error, that is, in the direction of **-G**
4. Repeat steps 2 and 3 until the value of **G** approaches zero

The *gradient* (**G**) of the error function **E** provides the direction in which the error function with the current values has the steeper slope; so to decrease **E**, we have to make some small steps in the opposite direction, **-G**.

By repeating this operation several times in an iterative manner, we move *down* towards the minimum of **E**, to reach a point where **G = 0**, in such a way that no further progress is possible:

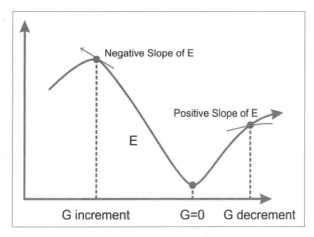

Figure 10: Searching for the minimum for the error function E.
We move in the direction in which the gradient G of the function E is minimal.

[16]

Stochastic gradient descent

In GD optimization, we compute the cost gradient based on the complete training set, so we sometimes also call it batch GD. In the case of very large datasets, using GD can be quite costly, since we are only taking a single step for one pass over the training set. The larger the training set, the more slowly our algorithm updates the weights, and the longer it may take until it converges at the global cost minimum.

The fastest method of gradient descent is **Stochastic Gradient Descent (SGD)**, and for this reason, it is widely used in deep neural networks. In SGD, we use only one training sample from the training set to do the update for a parameter in a particular iteration.

Here, the term stochastic comes from the fact that the gradient based on a single training sample is a stochastic approximation of the true cost gradient. Due to its stochastic nature, the path towards the global cost minimum is not direct, as in GD, but may zigzag if we are visualizing the cost surface in a 2D space:

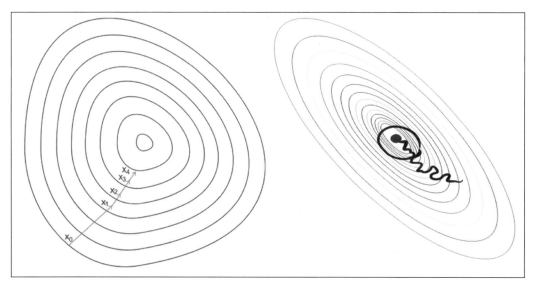

Figure 11: GD versus SGD: the gradient descent (left figure) ensures that each update in the weights is done in the right direction: the direction that minimizes the cost function. With the growth in the dataset's size, and more complex computations in each step, SGD (right figure) is preferred in these cases. Here, updates to the weights are done as each sample is processed and, as such, subsequent calculations already use improved weights. Nonetheless, this very reason leads to some misdirection in minimizing the error function.

Neural network architectures

The way that we connect the nodes and the number of layers present (that is, the levels of nodes between input and output, and the number of neurons per layer), defines the architecture of a neural network.

There are various types of architectures in neural networks. We can categorize DL architectures into four groups: **Deep Neural Networks (DNNs)**, **Convolutional Neural Networks (CNNs)**, **Recurrent Neural Networks (RNNs)**, and **Emergent Architectures (EAs)**. The following sections of this chapter will offer a brief introduction to these architectures. A more detailed analysis, with examples of applications, will be the subject of the following chapters of this book.

Deep Neural Networks (DNNs)

DNNs are ANNs which are strongly oriented to DL. Where normal procedures of analysis are inapplicable, due to the complexity of the data to be processed, such networks are therefore an excellent modeling tool. DNNs are neural networks that are very similar to those we have discussed, but they must implement a more complex model (a greater number of neurons, hidden layers, and connections), although they follow the learning principles that apply to all ML problems (such as supervised learning). The computation in each layer transforms the representations in the layer below into *slightly more abstract representations*.

We will use the term DNN to refer specifically to **Multilayer Perceptron (MLP)**, **Stacked Auto-Encoder (SAE)**, and **Deep Belief Networks (DBNs)**. SAEs and DBNs use **AutoEncoders (AEs)** and RBMs as building blocks of the architectures. The main difference between them and MLP is that training is executed in two phases: *unsupervised pre-training* and *supervised fine-tuning*:

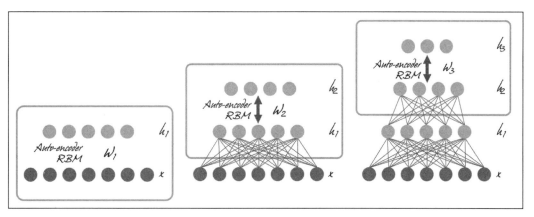

Figure 12: SAE and DBN using AE and RBM respectively.

In unsupervised pre-training, shown in the preceding diagram, the layers are stacked sequentially and trained in a layer-wise manner, like an AE or RBM using unlabeled data. Afterwards, in supervised fine-tuning, an output classifier layer is stacked, and the complete neural network is optimized, by retraining with labeled data.

In this chapter, we will not discuss SAEs (see more details in *Chapter 5, Optimizing TensorFlow Autoencoders*), but will stick to MLPs and DBNs and use these two DNN architectures. We will see how to develop predictive models to deal with high-dimensional datasets.

Multilayer perceptron

In multilayer networks, you can identify the artificial neurons of the layers, so that each neuron is connected to all those in the next layer, ensuring that:

- There are no connections between neurons belonging to the same layer
- There are no connections between neurons belonging to non-adjacent layers
- The number of layers and neurons per layer depends on the problem to be solved

The *input* and *output layers* define inputs and outputs, and there are *hidden layers*, whose complexity realizes different behaviors of the network. Finally, the connections between neurons are represented by as many matrices as the pairs of adjacent layers.

Each array contains the weights of the connections between the pairs of nodes of two adjacent layers. The feedforward networks are networks with *no loops* within the layers.

We will describe feedforward networks in more detail in *Chapter 3, Feed-Forward Neural Networks with TensorFlow*:

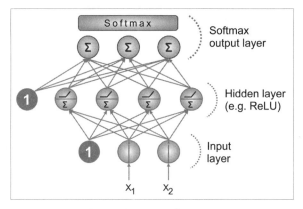

Figure 13: MLP architecture

Deep Belief Networks (DBNs)

To overcome the overfitting problem in MLP, we set up a DBN, do unsupervised pre-training to get a decent set of feature representations for the inputs, then fine-tune the training set to get actual predictions from the network. While the weights of an MLP are initialized randomly, a DBN uses a greedy layer-by-layer pre-training algorithm to initialize the network weights through probabilistic generative models. The models are composed of a visible layer and multiple layers of stochastic and latent variables, which are called hidden units or feature detectors.

DBNs are *Deep Generative Models*, which are neural network models that can replicate the data distribution that you provide. This allows you to generate "fake-but-realistic" data points from real data points.

DBNs are composed of a visible layer and multiple layers of stochastic, latent variables, which are called *hidden units* or *feature detectors*. The top two layers have undirected, symmetric connections between them and form an associative memory, whereas lower layers receive top-down, directed connections from the preceding layer. The building blocks of DBNs are **Restricted Boltzmann Machines** (**RBMs**). As you can see in the following figure, several RBMs are *stacked* one after another to form DBNs:

Chapter 1

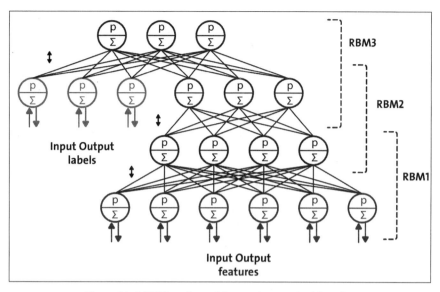

Figure 14: A DBN configured for semi-supervised learning

A single **RBM** consists of two layers. The first layer is composed of visible neurons, and the second layer consists of hidden neurons. The following figure shows the structure of a simple RBM. Visible units accept inputs, and hidden units are nonlinear feature detectors. Each visible neuron is connected to all the hidden neurons, but there is no internal connection among neurons in the same layer.

An RBM consists of a visible layer node and a hidden layer node, but without visible-visible and hidden-hidden connections, hence the term *restricted*. They allow more efficient network training that can be supervised or unsupervised. This type of neural network is able to represent a large number of features of the inputs, then hidden nodes can represent up to 2n features. The network can be trained to respond to a single question (for example, yes or no to the question: *Is it a cat?*) until it can respond (again in binary terms) to a total of 2n questions (*Is it a cat?*, *It is Siamese?*, *Is it white?*).

The architecture of the RBM is as follows, with neurons arranged according to a *symmetrical bipartite graph*:

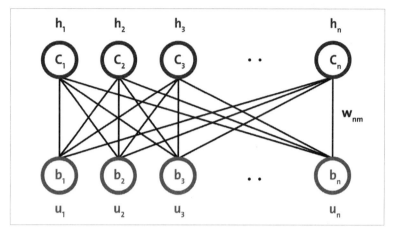

Figure 15: RBM architecture.

A single hidden layer RBM cannot extract all the features from the input data, due to its inability to model the relationship between variables. Hence, multiple layers of RBMs are used one after another to extract nonlinear features. In DBNs, an RBM is trained first with input data, and the hidden layer represents the features learned using a greedy learning approach. These learned features of the first RBM, that is, a hidden layer of the first RBM, are used as the input to the second RBM, as another layer in the DBN.

Similarly, the learned features of the second layer are used as input for another layer. This way, DBNs can extract deep and nonlinear features from input data. The hidden layer of the last RBM represents the learned features of the whole network.

Convolutional Neural Networks (CNNs)

CNNs have been specifically designed for image recognition. Each image used in learning is divided into compact topological portions, each of which will be processed by filters to search for particular patterns. Formally, each image is represented as a three-dimensional matrix of pixels (width, height, and color), and every sub-portion can be placed on convolution with the filter set. In other words, scrolling each filter along the image computes the inner product of the same filter and input.

This procedure produces a set of feature maps (activation maps) for the various filters. Superimposing the various feature maps onto the same portion of the image, we get an output volume. This type of layer is called the convolutional layer. The following diagram is a schematic of the architecture of a CNN:

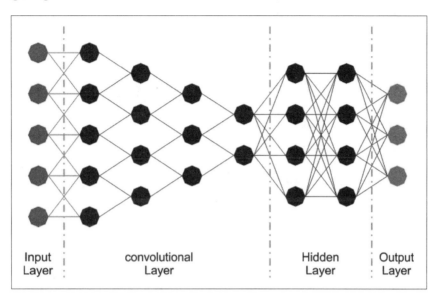

Figure 16: CNN architecture.

Although regular DNNs work fine for small images (for example, MNIST and CIFAR-10), they break down with larger images because of the huge number of parameters required. For example, a 100×100 image has 10,000 pixels, and if the first layer has just 1,000 neurons (which already severely restricts the amount of information transmitted to the next layer), this means 10 million connections. In addition, that is just for the first layer.

CNNs solve this problem using partially connected layers. Because consecutive layers are only partially connected and because it heavily reuses its weights, a CNN has far fewer parameters than a fully connected DNN, which makes it much faster to train. This reduces the risk of overfitting and requires much less training data. Moreover, when a CNN has learned a kernel that can detect a particular feature, it can detect that feature anywhere on the image. In contrast, when a DNN learns a feature in one location, it can detect it only in that particular location. Since images typically have very repetitive features, CNNs are able to generalize much better than DNNs on image processing tasks such as classification and use fewer training examples.

Importantly, the DNN has no prior knowledge of how the pixels are organized; it does not know that nearby pixels are close. A CNN's architecture embeds this prior knowledge. Lower layers typically identify features in small areas of the images, while higher layers combine the lower-level features into larger features. This works well with most natural images, giving CNNs a decisive head-start over DNNs:

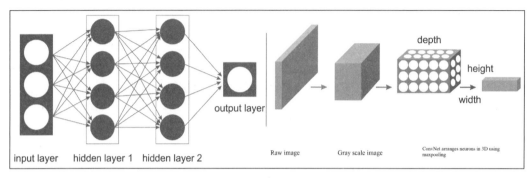

Figure 17: A regular DNN versus a CNN.

For example, in the preceding diagram, on the left, you can see a regular three-layer neural network. On the right, a CNN arranges its neurons in three dimensions (width, height, and depth), as visualized in one of the layers. Every layer of a CNN transforms the 3D input volume to a 3D output volume of neuron activations. The red input layer holds the image, so its width and height would be the dimensions of the image, and the depth would be three (red, green and blue channels).

Therefore, all the multilayer neural networks we looked at had layers composed of a long line of neurons, and we had to flatten input images or data to 1D before feeding them to the neural network. However, what happens when you try to feed them a 2D image directly? The answer is that in a CNN, each layer is represented in 2D, which makes it easier to match neurons with their corresponding inputs. We will see examples of this in upcoming sections.

AutoEncoders

An AE is a network with three or more layers, where the input layer and the output have the *same number* of neurons, and those intermediate (hidden layers) have a *lower number* of neurons. The network is trained to simply reproduce in the output, for each piece of input data, the same pattern of activity in the input.

AEs are ANNs capable of learning efficient representations of the input data without any supervision (that is, the training set is unlabeled). They typically have a much lower dimensionality than the input data, making AEs useful for dimensionality reduction. More importantly, AEs act as powerful feature detectors, and they can be used for unsupervised pre-training of DNNs.

The remarkable aspect of the problem is that, due to the lower number of neurons in the hidden layer, if the network can learn from examples and generalize to an acceptable extent, it performs *data compression*; the status of the hidden neurons provides, for each example, a *compressed version* of the *input* and *output common states*. Useful applications of AEs are *data denoising* and *dimensionality reduction* for data visualization.

The following diagram shows how an AE typically works; it reconstructs the received input through two phases: an *encoding* phase, which corresponds to a *dimensional reduction* for the original input, and a *decoding* phase, which is capable of *reconstructing* the original input from the encoded (*compressed*) representation:

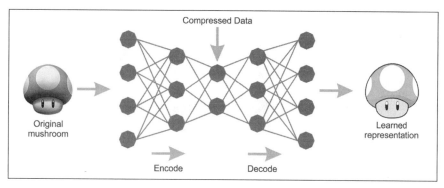

Figure 18: Encoding and decoding phases of an autoencoder.

As an unsupervised neural network, the main characteristic of an autoencoder is its symmetrical structure. An autoencoder has two components: an *encoder* that converts the input to an internal representation, followed by a *decoder* that converts the internal representation to the output.

In other words, an autoencoder can be seen as a combination of an encoder, where we encode some input into a code, and a decoder, where we decode/reconstruct the code back to its original input as the output. Thus, an MLP typically has the same architecture as an autoencoder, except that the number of neurons in the output layer must be equal to the number of inputs.

As mentioned previously, there is more than one way to train an autoencoder. The first way is to train the whole layer at once, similar to MLP. However, instead of using some labeled output when calculating the cost function, as in supervised learning, we use the input itself. Therefore, the cost function shows the difference between the actual input and the reconstructed input.

Recurrent Neural Networks (RNNs)

The fundamental feature of an **RNN** is that the network contains at least one feedback connection, so the activations can flow around in a loop. It enables the networks to do temporal processing and learn sequences, for example performing sequence recognition/reproduction or temporal association/prediction.

RNN architectures can have many different forms. One common type consists of a standard **MLP** plus added loops. These can exploit the powerful non-linear mapping capabilities of the **MLP**, and have some form of memory. Others have more uniform structures, potentially with every neuron connected to all the others, and may have *stochastic activation functions*:

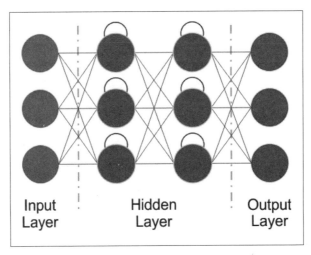

Figure 19: RNN architecture.

For simple architectures and deterministic activation functions, learning can be achieved using similar GD procedures to those leading to the backpropagation algorithm for feedforward networks.

The preceding image looks at a few of the most important types and features of RNNs. RNNs are designed to utilize sequential information of input data with cyclic connections among building blocks such as perceptrons, **Long Short-term memory units (LSTMs)**, or **Gated Recurrent units (GRUs)**. The latter two are used to remove the drawbacks of regular RNNs, such as the gradient vanishing/exploding problem and long-short term dependency. We will look at these architectures in later chapters.

Emergent architectures

Many other emergent DL architectures have been suggested, such as **Deep SpatioTemporal Neural Networks (DST-NNs)**, **Multi-Dimensional Recurrent Neural Networks (MD-RNNs)**, and **Convolutional AutoEncoders (CAEs)**.

Nevertheless, people are talking about and using other emerging networks, such as CapsNets (an improved version of a CNN, designed to remove the drawbacks of regular CNNs), *Factorization Machines* for personalization, and *Deep Reinforcement Learning*.

Deep learning frameworks

In this section, we present some of the most popular DL frameworks. In short, almost all of the libraries provide the possibility of using the graphics processor to speed up the learning process, are released under an open license, and are the result of university research groups.

TensorFlow is mathematical software, and an open source software library, written in Python and C++ for machine intelligence. The Google Brain Team developed it in 2011, and it can be used to help us analyze data, to predict an effective business outcome. Once you have constructed your neural network model, after the necessary feature engineering, you can simply perform the training interactively using plotting or TensorBoard.

The main features offered by the latest release of TensorFlow are faster computing, flexibility, portability, easy debugging, a unified API, transparent use of GPU computing, easy use and extensibility. Other benefits include the fact that it is widely used, supported, and is production-ready at scale.

Keras is a deep-learning library that sits atop TensorFlow and Theano, providing an intuitive API, which is inspired by Torch (perhaps the best Python API in existence). Deeplearning4j relies on Keras as its Python API and imports models from Keras, and through Keras, from Theano and TensorFlow.

François Chollet, a software engineer at Google, created Keras. It runs seamlessly on CPU and GPU. This allows for easy and fast prototyping through user friendliness, modularity, and extensibility. Keras is probably one of the fastest growing frameworks, because it is too easy to construct NN layers. Therefore, Keras is likely to become the standard Python API for NNs.

Theano is probably the most widespread library. Theano is written in Python, which is one of the most widely used languages in the field of ML (**Python** is also used in **TensorFlow**). Moreover, Theano allows the use of GPU, which is 24x faster than a single CPU. Theano lets you efficiently define, optimize, and evaluate complex mathematical expressions, such as multidimensional arrays. Unfortunately, Yoshua Bengio announced on 28th September 2017, that development on Theano would cease. That means Theano is effectively dead.

Neon is a Python-based deep learning framework developed by Nirvana. Neon has a syntax similar to Theano's high-level framework (for example, Keras). Currently, Neon is considered the fastest tool for GPU-based implementation, especially for CNN. Although it's CPU-based implementation is relatively worse than most other libraries.

Torch is a vast ecosystem for ML that offers a large number of algorithms and functions, including for DL and for processing various types of multimedia data, with a particular focus on parallel computing. It provides an excellent interface for the **C** language and has a large community of users. Torch is a library that extends the scripting language Lua and is intended to provide a flexible environment for designing and training ML systems. Torch is a self-contained and highly portable framework on various platforms (**Windows**, **Mac**, **Linux**, and **Android**) and scripts can run on these platforms without modification. Torch provides many uses for different applications.

Caffe, developed primarily by **Berkeley Vision and Learning Center** (**BVLC**), is a framework designed to stand out because of its expression, speed, and modularity. Its unique architecture encourages application and innovation, by allowing an easier transition from CPU to GPU calculations. The large community of users means that considerable development has occurred recently. It is written in Python, but the installation process can be long, due to the numerous support libraries it has to compile.

MXNet is a DL framework that supports many languages, such as R, Python, C++, and Julia. This is helpful because if you know any of these languages, you will not need to step out of your comfort zone at all to train your DL models. Its backend is written in C++ and CUDA and it is able to manage its own memory in a similar way to Theano.

MXNet is also popular because it scales very well and can work with multiple GPUs and computers, which makes it very useful for enterprise. This is why Amazon has made MXNet its reference library for DL. In November 2017, AWS announced the availability of ONNX-MXNet, which is an open source Python package used to import **Open Neural Network Exchange** (**ONNX**) DL models into Apache MXNet.

The **Microsoft Cognitive Toolkit** (**CNTK**) is a unified DL toolkit from Microsoft Research that makes it easy to train and combine popular model types across multiple GPUs and servers. CNTK implements highly efficient CNN and RNN training for speech, image, and text data. It supports cuDNN v5.1 for GPU acceleration. CNTK also supports Python, C++, C#, and command-line interface.

Here is a table summarizing these frameworks:

Framework	Supported programming languages	Training materials community	CNN modeling capability	RNN modeling capability	Usability	Multi-GPU support
Theano	Python, C++	++	Ample CNN tutorials and prebuilt models	Ample RNN tutorials and prebuilt models	Modular architecture	No
Neon	Python,	+	Fastest tools for CNN	Minimal resources	Modular architecture	No
Torch	Lua, Python	+	Minimal resources	Ample RNN tutorials and prebuilt models	Modular architecture	Yes
Caffe	C++	++	Ample CNN tutorials and prebuilt models	Minimal resources	Creating layers takes time	Yes
MXNet	R, Python, Julia, Scala	++	Ample CNN tutorials and prebuilt models	Minimal resources	Modular architecture	Yes
CNTK	C++	+	Ample CNN tutorials and prebuilt models	Ample RNN tutorials and prebuilt models	Modular architecture	Yes
TensorFlow	Python, C++	+++	Ample RNN tutorials and prebuilt models	Ample RNN tutorials and prebuilt models	Modular architecture	Yes
DeepLearning4j	Java, Scala	+++	Ample RNN tutorials and prebuilt models	Ample RNN tutorials and prebuilt models	Modular architecture	Yes
Keras	Python	+++	Ample RNN tutorials and prebuilt models	Ample RNN tutorials and prebuilt models	Modular architecture	Yes

Apart from the preceding libraries, there are some recent initiatives for DL on the cloud. The idea is to bring DL capability to big data, with billions of data points and high dimensional data. For example, **Amazon Web Services** (**AWS**), Microsoft Azure, Google Cloud Platform and **NVIDIA GPU Cloud** (**NGC**) all offer machine and deep learning services (http://searchbusinessanalytics.techtarget.com/feature/Machine-learning-platforms-comparison-Amazon-Azure-Google-IBM) that are native to their public clouds.

Getting Started with Deep Learning

In October 2017, AWS released Deep Learning **AMIs** (**Amazon Machine Images**) for Amazon **Elastic Compute Cloud** (**EC2**) P3 Instances. These AMIs come pre-installed with deep learning frameworks, such as TensorFlow, Gluon and Apache MXNet, that are optimized for the NVIDIA Volta V100 GPUs within Amazon EC2 P3 instances. The deep learning service currently offers three types of AMIs: Conda AMI, Base AMI and AMI with Source Code.

The Microsoft Cognitive Toolkit is Azure's open source, deep learning service. Similar to AWS' offering, it focuses on tools that can help developers build and deploy deep learning applications. The toolkit is installed in Python 2.7, in the root environment. Azure also provides a model gallery (https://www.microsoft.com/en-us/cognitive-toolkit/features/model-gallery/) that includes resources, such as code samples, to help enterprises get started with the service.

On the other hand, NGC empowers AI scientists and researchers with GPU-accelerated containers (see https://www.nvidia.com/en-us/data-center/gpu-cloud-computing/). The NGC features containerized deep learning frameworks such as TensorFlow, PyTorch, and MXNet that are tuned, tested, and certified by NVIDIA to run on the latest NVIDIA GPUs on participating cloud service providers. Nevertheless, there are also third-party services available through their respective marketplaces.

Summary

In this chapter, we introduced some of the fundamental themes of DL. DL consists of a set of methods that allow an ML system to obtain a *hierarchical representation* of data on *multiple levels*. This is achieved by combining simple units, each of which transforms the representation at its own level, starting from the input level, in a representation at a higher and abstraction level.

Recently, these techniques have provided results that have never been seen before in many applications, such as *image recognition* and *speech recognition*. One of the main reasons for the spread of these techniques has been the development of GPU architectures that considerably reduce the training time of DNNs.

There are different **DNN** architectures, each of which has been developed for a specific problem. We will talk more about these architectures in later chapters and show examples of applications created with the **TensorFlow** framework. This chapter ended with a brief overview of the most important DL frameworks.

In the next chapter, we begin our journey into DL, introducing the TensorFlow software library. We will describe the main features of TensorFlow and see how to install it and set up our first working remarketing dataset.

2
A First Look at TensorFlow

TensorFlow is a mathematical software and an open source framework for deep learning developed by the Google Brain Team in 2011. Nevertheless, it can be used to help us analyze data in order to predict an effective business outcome.

Although the initial target of TensorFlow was to conduct research in ML and in **Deep Neural Networks** (**DNNs**), the system is general enough to be applicable to a wide variety of classical machine learning algorithm such as **Support Vector Machine** (**SVM**), logistic regression, decision trees, and random forest.

Keeping in mind your needs and based on all the latest exciting features of the most stable version 1.6 (v1.7 was the pre-release during the production stage of this book), in this chapter, we will describe the main capabilities and core concepts of TensorFlow that will be used in all the subsequent chapters.

The following topics will be covered in this chapter:

- A general overview of TensorFlow
- What is new in TensorFlow v1.6
- The TensorFlow computational graph
- The TensorFlow code structure
- The TensorFlow data model
- Visualizing computations through TensorBoard
- Linear regression and beyond

A general overview of TensorFlow

TensorFlow is an open source framework from Google for scientific and numerical computation using data flow graphs that stand for TensorFlow's execution model. The data flow graphs used in TensorFlow help ML experts to perform more advanced and intensive training on their data to develop **DL** and predictive analytics models.

As the name implies, TensorFlow includes operations that are performed by neural networks on multidimensional data arrays, that is, flow of tensors. Nodes in a flow graph correspond to mathematical operations, that is, addition, multiplication, matrix factorization, and so on; whereas, edges correspond to tensors that ensure communication between edges and nodes – that is, data flow and control flow. This way, TensorFlow provides some widely used and robustly implemented linear models and DL algorithms.

You can perform numerical computations on a CPU. However, with TensorFlow, it is also possible to distribute the training among multiple devices on the same system, especially if you have more than one GPU on your system that can share the computational load.

Deploying a predictive or general-purpose model using TensorFlow is straightforward. Once you have constructed your neural network model after the required feature engineering, you can simply perform the training interactively and use the TensorBoard to visualize your TensorFlow graph, plot quantitative metrics about the execution of your graph, and show additional data like images that pass through it.

If TensorFlow can access GPU devices, it will automatically distribute computations to multiple devices via a greedy process. Therefore, no special configuration is needed to utilize the cores of the CPU. Nevertheless, TensorFlow also allows the program to specify which operations will be on which device via name scope placement. Finally, after evaluating the model, you deploy it by feeding some test data to it. The main features offered by the latest release of TensorFlow are as follows:

- **Faster computing**: The latest release of TensorFlow is incredibly fast. For example, the Inception-v3 model runs 7.3 times faster on 8 GPUs, and distributed Inception-v3 runs 58 times faster on 64 GPUs.
- **Flexibility**: TensorFlow is not just a DL library. It comes with almost everything you need for powerful mathematical operations, thanks to its functions for solving the most difficult problems.
- **Portability**: TensorFlow runs on Windows (only CPU support, though), Linux, and Mac machines, and on mobile computing platforms (that is, Android).

- **Easy debugging**: TensorFlow provides the TensorBoard tool, which is useful for analyzing the models you develop.
- **Unified API**: TensorFlow offers you a very flexible architecture that enables you to deploy computation to one or more CPUs or GPUs on a desktop, server, or mobile device with a single API.
- **Transparent use of GPU computing**: TensorFlow now automates the management and optimization of the memory and the data used. You can now use your machine for large-scale and data-intensive GPU computing with NVIDIA's cuDNN and CUDA toolkits.
- **Easy use**: TensorFlow is for everyone. It is not only suitable for students, researchers, DL practitioners, but also for professionals who work in the industries.
- **Production-ready at scale**: Recently, TensorFlow has evolved into a neural network for machine translation at production scale. TensorFlow 1.6 promises Python API stability, making it easier to choose new features without worrying too much about breaking your existing code.
- **Extensibility**: TensorFlow is a relatively new technology, and it's still in active development. The source codes are available on GitHub (https://github.com/tensorflow/tensorflow).
- **Support**: There is a large community of developers and users working together to make TensorFlow a better product, both by providing feedback and by actively contributing to the source code.
- **Wide adoption**: Numerous tech giants use TensorFlow to increase their business intelligence, such as ARM, Google, Intel, eBay, Qualcomm, SAM, Dropbox, DeepMind, Airbnb, and Twitter.

Now, before we start coding with TensorFlow, let's see what the new features in TensorFlow's latest release.

What's new in TensorFlow v1.6?

In 2015, Google made TensorFlow open source, including all of its reference implementation. All of the source code was made available on GitHub under the Apache 2.0 license. Since then, TensorFlow has been widely adopted in academia and industrial research, and the most stable version, 1.6, has recently been released with a unified API.

It is important to note that the APIs in TensorFlow 1.6 (and higher) are not all backward compatible for pre v1.5 code. This means that some programs that worked on pre v1.5 will not necessarily work on TensorFlow 1.6.

Now let us see the new and exciting features that TensorFlow v1.6 has.

Nvidia GPU support optimized

From TensorFlow v1.5, prebuilt binaries are now built against CUDA 9.0 and cuDNN 7. However, from v1.6's release, TensorFlow prebuilt binaries use AVX instructions, which may break TensorFlow on older CPUs. Nevertheless, since v1.5, an added support for CUDA on NVIDIA Tegra devices has been available.

Introducing TensorFlow Lite

TensorFlow Lite is TensorFlow's lightweight solution for mobile and embedded devices. It enables low-latency inference of on-device machine learning models with a small binary size and fast performance supporting hardware acceleration.

TensorFlow Lite uses many techniques for achieving low latency like optimizing the kernels for specific mobile apps, pre-fused activations, quantized kernels that allow smaller and faster (fixed-point math) models, and in the future, leverage-specialized machine learning hardware to get the best possible performance for a particular model on a particular device.

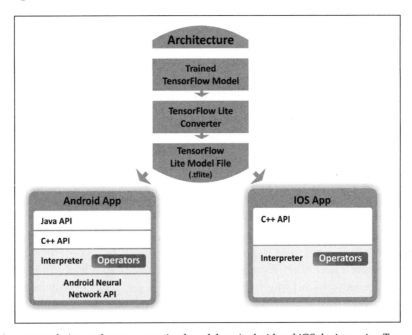

Figure 1: A conceptual view on how to use trained model on Android and iOS devices using TensorFlow Lite

Machine learning is changing the computing paradigm, and we see an emerging trend of new use cases on mobile and embedded devices. Consumer expectations are also trending toward natural, human-like interactions with their devices, driven by the camera and voice interaction models.

Therefore, the user's expectations are no longer limited to the computer, and but the computational power of mobile devices has also increased exponentially due to hardware acceleration, and frameworks such as the Android Neural Networks API and C++ API for iOS. As shown in the preceding figure, a pre-trained model can be converted into a lighter version to be running as an Android or iOS app.

Therefore, widely available smart appliances create new possibilities for on-device intelligence. These allow us to use our smartphones to perform real-time computer vision and **Natural Language Processing** (**NLP**).

Eager execution

Eager execution is an interface for TensorFlow that provides an imperative programming style. When you enable eager execution, TensorFlow operations (defined in a program) execute immediately.

It is to be noted that from TensorFlow v1.7, eager execution will be moved out of contrib. This means that using `tf.enable_eager_execution()` is recommended. We will see an example on this in a later section.

Optimized Accelerated Linear Algebra (XLA)

Pre v1.5 XLA was unstable and had a very limited number of features. However, v1.6 has more support for XLA. This includes the following:

- Complex64 support to XLA compiler has been added
- Now the **Fast Fourier Transformation** (**FFT**) support has been added for both CPU and GPU
- The bfloat support is now added to the XLA infrastructure
- The ClusterSpec propagation work with XLA devices has been enabled
- Android TF can now be built with CUDA acceleration on compatible Tegra devices
- Support for adding the deterministic executor to generate an XLA graph has been enabled

Numerous bugs reported by the open source community have been fixed and a significant amount of API-level changes have been integrated with this version.

However, since we have not explored anything with TensorFlow yet, we will see how to leverage these features for developing real-life deep learning applications later on. Before that, let's see how to prepare your programming environment.

Installing and configuring TensorFlow

You can install and use TensorFlow on a number of platforms such as Linux, macOS, and Windows. Moreover, you can also build and install TensorFlow from the latest GitHub source of TensorFlow. Furthermore, if you have a Windows machine, you can install TensorFlow via native pip or Anacondas. TensorFlow supports Python 3.5.x and 3.6.x on Windows.

In addition, Python 3 comes with the pip3 package manager, which is the program you will use to install TensorFlow. Therefore, you do not need to install pip if you are using this Python version. From our experience, even if you have NVIDIA GPU hardware integrated on your machine, it would be worth installing and trying the CPU-only version first and if you don't experience good performance, you should switch to GPU support then.

The GPU-enabled version of TensorFlow has several requirements such as 64-bit Linux, Python 2.7 (or 3.3+ for Python 3), NVIDIA CUDA® 7.5 or higher (CUDA 8.0 required for Pascal GPUs), and NVIDIA cuDNN (this is GPU accelerated deep learning) v5.1 (or higher is recommended). See more at `https://developer.nvidia.com/cudnn`.

More specifically, the current development of TensorFlow supports only GPU computing using NVIDIA toolkits and software. Therefore, the following software must have to be installed on your machine to get the GPU support on your predictive analytics applications:

- NVIDIA driver
- CUDA with *compute capability* >= 3.0
- CudNN

The NVIDIA CUDA toolkit includes (see more at `https://developer.nvidia.com/cuda-zone`):

- GPU-accelerated libraries such as cuFFT for **FFT**
- cuBLAS for **Basic Linear Algebra Subroutines (BLAS)**
- cuSPARSE for sparse matrix routines
- cuSOLVER for dense and sparse direct solvers

- cuRAND for random number generation, NPP for image, and video processing primitives
- **nvGRAPH** for **NVIDIA Graph Analytics Library**
- Thrust for templated parallel algorithms and data structures and a dedicated CUDA math library

However, we will not cover the installation and configuration of TensorFlow since the documentation provided on TensorFlow is very rich to be followed and acted accordingly. Another reason is that the version will be changed periodically. Therefore, keeping yourself updated with the TensorFlow website `https://www.tensorflow.org/install/` will be a better idea.

If you have already installed and configured your programming environment, let us dive into TensorFlow computation graph.

TensorFlow computational graph

When thinking of executing a TensorFlow program, we should be familiar with the concepts of graph creation and session execution. Basically, the first one is for building the model, and the second one is for feeding the data in and getting the results.

Interestingly, TensorFlow does everything on the C++ engine, which means not even a little multiplication or addition is executed in Python. Python is just a wrapper. Fundamentally, the TensorFlow C++ engine consists of the following two things:

- Efficient implementations of operations, such as convolution, max pool, and sigmoid for a CNN for example
- Derivatives of the forwarding mode operation

The TensorFlow lib is an extraordinary lib in terms of coding and it is not like conventional Python code (for example, you can write statements and they get executed). TensorFlow code consists of different operations. Even variable initialization is special in TensorFlow. When you are performing a complex operation with TensorFlow, such as training a linear regression, TensorFlow internally represents its computation using a data flow graph. The graph is called a computational graph, which is a directed graph consisting of the following:

- A set of nodes, each one representing an operation
- A set of directed arcs, each one representing the data on which the operations are performed

A First Look at TensorFlow

TensorFlow has two types of edges:

- **Normal**: They carry the data structures between the nodes. The output of one operation, that is, from one node, becomes the input for another operation. The edge connecting two nodes carries the values.
- **Special**: This edge doesn't carry values, but only represents a *control dependency* between two nodes, say X and Y. It means that node Y will be executed only if the operation in X has already been executed, but before the relationship between operations on the data.

The TensorFlow implementation defines control dependencies to enforce the order of otherwise independent operations as a way of controlling the peak memory usage.

A computational graph is basically like a data flow graph. Figure 2 shows a computational graph for a simple computation such as $z = d \times c = (a+b) \times c$:

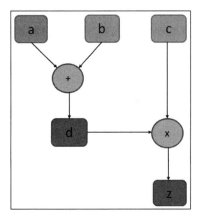

Figure 2: A very simple execution graph that computes a simple equation

In the preceding figure, the circles in the graph indicate the operations, while the rectangles indicate the computational graph. As stated earlier, a TensorFlow graph contains the following:

- tf.Operation objects: These are the nodes in the graph. These are usually simply referred to as ops. An op is simply **TITO (tensor-in-tensor-out)**. One or more tensors input and one or more tensors output.
- tf.Tensor objects: These are the edges of the graph. These are usually simply referred to as tensors.

Tensor objects flow between various ops in the graph. In the preceding figure, **d** is also an op. It can be a "constant" op whose output is a tensor that contains the actual value assigned to **d**.

It is also possible to perform a *deferred execution* using TensorFlow. In a nutshell, once you have composed a highly compositional expression during the building phase of the computational graph, you can still evaluate it in the running session phase. Technically speaking, TensorFlow schedules the job and executes on time in an efficient manner.

For example, parallel execution of independent parts of the code using the GPU is shown in the following figure:

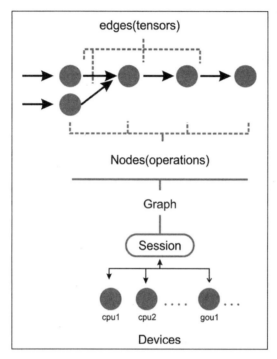

Figure 3: Edges and nodes in a TensorFlow graph to be executed on a session on devices such as CPUs or GPUs

After a computational graph is created, TensorFlow needs to have an active session that is executed by multiple CPUs (and GPUs if available) in a distributed way. In general, you really don't need to specify whether to use a CPU or a GPU explicitly, since TensorFlow can choose which one to use.

By default, a GPU will be picked for as many operations as possible; otherwise, CPU will be used. Nevertheless, generally, it allocates all GPU memory even if does not consume it.

Here are the main components of a TensorFlow graph:

- **Variables**: Used to contain values for the weights and biases between TensorFlow sessions.
- **Tensors**: A set of values that pass between nodes to perform operations (aka. op).
- **Placeholders**: Used to send data between the program and the TensorFlow graph.
- **Session**: When a session is started, TensorFlow automatically calculates gradients for all the operations in the graph and uses them in a chain rule. In fact, a session is invoked when the graph is to be executed.

Don't worry, each of these preceding components will be discussed in later sections. Technically, the program you will be writing can be considered as a client. The client is then used to create the execution graph in C/C++ or Python symbolically, and then your code can ask TensorFlow to execute this graph. The whole concept gets clearer from the following figure:

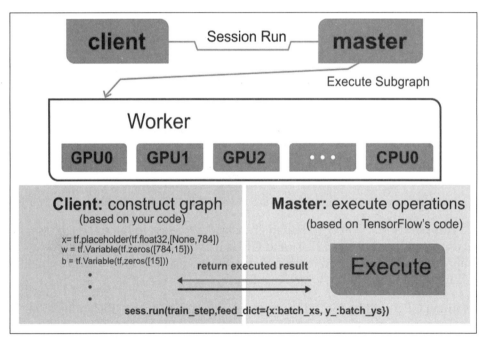

Figure 4: Using a client-master architecture to execute a TensorFlow graph

A computational graph helps to distribute the workload across multiple computing nodes with a CPU or GPU. This way, a neural network can be equated to a composite function where each layer (input, hidden, or output layer) can be represented as a function. To understand the operations performed on the tensors, knowing a good workaround for the TensorFlow programming model is necessary.

TensorFlow code structure

The TensorFlow programming model signifies how to structure your predictive models. A TensorFlow program is generally divided into four phases when you have imported the TensorFlow library:

- Construction of the computational graph that involves some operations on tensors (we will see what a tensor is soon)
- Creation of a session
- Running a session; performed for the operations defined in the graph
- Computation for data collection and analysis

These main phases define the programming model in TensorFlow. Consider the following example, in which we want to multiply two numbers:

```
import tensorflow as tf # Import TensorFlow

x = tf.constant(8) # X op
y = tf.constant(9) # Y op
z = tf.multiply(x, y) # New op Z

sess = tf.Session() # Create TensorFlow session

out_z = sess.run(z) # execute Z op
sess.close() # Close TensorFlow session
print('The multiplication of x and y: %d' % out_z)# print result
```

A First Look at TensorFlow

The preceding code segment can be represented by the following figure:

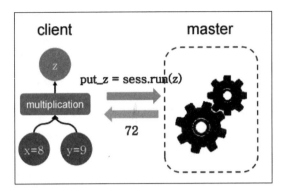

Figure 5: A simple multiplication executed and returned on a client-master architecture

To make the preceding program more efficient, TensorFlow also allows exchanging data in your graph variables through *placeholders* (to be discussed later). Now imagine the following code segment that does the same thing but more efficiently:

```
import tensorflow as tf

# Build a graph and create session passing the graph
with tf.Session() as sess:
    x = tf.placeholder(tf.float32, name="x")
    y = tf.placeholder(tf.float32, name="y")
    z = tf.multiply(x,y)

# Put the values 8,9 on the placeholders x,y and execute the graph
z_output = sess.run(z,feed_dict={x: 8, y:9})
print(z_output)
```

TensorFlow is not necessary for multiplying two numbers. Also, there are many lines of code for this simple operation. The purpose of the example is to clarify how to structure code, from the simplest (as in this instance) to the most complex. Furthermore, the example also contains some basic instructions that we will find in all the other examples given in this book.

This single import in the first line imports TensorFlow for your command; it can be instantiated with `tf`, as stated earlier. The TensorFlow operator will then be expressed by `tf` and by the name of the operator to use. In the next line, we construct the `session` object by means of the `tf.Session()` instruction:

```
with tf.Session() as sess:
```

Chapter 2

 The session object (that is, `sess`) encapsulates the environment for the TensorFlow so that all the operation objects are executed, and Tensor objects are evaluated. We will see them in upcoming sections.

This object contains the computation graph, which, as we said earlier, contains the calculations to be carried out. The following two lines define variables x and y, using a `placeholder`. Through a `placeholder`, you may define both an input (such as the variable x of our example) and an output variable (such as the variable y):

```
x = tf.placeholder(tf.float32, name="x")
y = tf.placeholder(tf.float32, name="y")
```

 Placeholders provide an interface between the elements of the graph and the computational data of the problem. They allow us to create our operations and build our computation graph without needing the data, instead of using a reference to it.

To define a *data* or *tensor* (we will introduce you to the concept of tensor soon) via the `placeholder` function, three arguments are required:

- **Data type** is the type of element in the tensor to be fed.
- **Shape** of the placeholder is the shape of the tensor to be fed (optional). If the shape is not specified, you can feed a tensor of any shape.
- **Name** is very useful for debugging and code analysis purposes, but it is optional.

 For more on tensors, refer to https://www.tensorflow.org/api_docs/python/tf/Tensor.

So, we can introduce the model that we want to compute with two arguments, the placeholder and the constant, that were previously defined. Next, we define the computational model.

The following statement, inside the *session*, builds the data structure of the product of x and y, and the subsequent assignment of the result of the operation to tensor z. Then it goes as follows:

```
z = tf.multiply(x, y)
```

Since the result is already held by the placeholder z, we execute the graph through the `sess.run` statement. Here, we feed two values to patch a tensor into a graph node. It temporarily replaces the output of an operation with a tensor value:

```
z_output = sess.run(z,feed_dict={x: 8, y:9})
```

In the final instruction, we print the result:

```
print(z_output)
```

This prints the output, `72.0`.

Eager execution with TensorFlow

As described earlier, with eager execution for TensorFlow enabled, we can execute TensorFlow operations *immediately* as they are called from Python in an imperative way.

With eager execution enabled, TensorFlow functions execute operations immediately and return concrete values. This is opposed to adding to a graph to be executed later in a `tf.Session` (https://www.tensorflow.org/versions/master/api_docs/python/tf/Session) and creating symbolic references to a node in a computational graph.

TensorFlow serves eager execution features through `tf.enable_eager_execution`, which is aliased with the following:

- `tf.contrib.eager.enable_eager_execution`
- `tf.enable_eager_execution`

The `tf.enable_eager_execution` has the following signature:

```
tf.enable_eager_execution(
        config=None,
        device_policy=None
)
```

In the above signature, `config` is a `tf.ConfigProto` used to configure the environment in which operations are executed but this is an optional argument. On the other hand, `device_policy` is also an optional argument used for controlling the policy on how operations requiring inputs on a specific device (for example, GPU0) handle inputs on a different device (for example, GPU1 or CPU).

Now invoking the preceding code will enable the eager execution for the lifetime of your program. For example, the following code performs a simple multiplication operation in TensorFlow:

```
import tensorflow as tf

x = tf.placeholder(tf.float32, shape=[1, 1]) # a placeholder for variable x
y = tf.placeholder(tf.float32, shape=[1, 1]) # a placeholder for variable y
m = tf.matmul(x, y)

with tf.Session() as sess:
    print(sess.run(m, feed_dict={x: [[2.]], y: [[4.]]}))
```

The following is the output of the preceding code:

```
>>>
8.
```

However, with the eager execution, the overall code looks much simpler:

```
import tensorflow as tf

# Eager execution (from TF v1.7 onwards):
tf.eager.enable_eager_execution()
x = [[2.]]
y = [[4.]]
m = tf.matmul(x, y)

print(m)
```

The following is the output of the preceding code:

```
>>>
tf.Tensor([[8.]], shape=(1, 1), dtype=float32)
```

Can you understand what happens when the preceding code block is executed? Well, after eager execution is enabled, operations are executed as they are defined and Tensor objects hold concrete values, which can be accessed as `numpy.ndarray` through the `numpy()` method.

Note that eager execution cannot be enabled after TensorFlow APIs have been used to create or execute graphs. It is typically recommended to invoke this function at program startup and not in a library. Although this sounds fascinating, we will not use this feature in upcoming chapters since this a new feature and not well explored yet.

Data model in TensorFlow

The data model in TensorFlow is represented by tensors. Without using complex mathematical definitions, we can say that a tensor (in TensorFlow) identifies a multidimensional numerical array. We will see more details on tensors in the next subsection.

Tensor

Let's see the formal definition of tensor from Wikipedia (https://en.wikipedia.org/wiki/Tensor):

> "Tensors are geometric objects that describe linear relations between geometric vectors, scalars, and other tensors. Elementary examples of such relations include the dot product, the cross product, and linear maps. Geometric vectors, often used in physics and engineering applications, and scalars themselves are also tensors."

This data structure is characterized by three parameters: rank, shape, and type, as shown in the following figure:

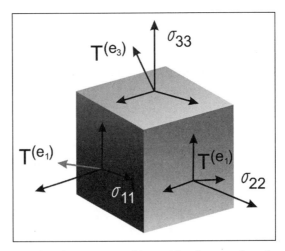

Figure 6: Tensors are nothing but geometric objects with a shape, rank, and type, used to hold a multidimensional array

A tensor can thus be thought of as the generalization of a matrix that specifies an element with an arbitrary number of indices. The syntax for tensors is more or less the same as nested vectors.

> Tensors just define the type of this value and the means by which this value should be calculated during the session. Therefore, they do not represent or hold any value produced by an operation.

Some people love to compare NumPy and TensorFlow. However, in reality, TensorFlow and NumPy are quite similar in the sense that both are N-d array libraries!

Well, it's true that NumPy has n-dimensional array support, but it doesn't offer methods to create tensor functions and automatically compute derivatives (and it has no GPU support). The following figure is a short and one-to-one comparison of NumPy and TensorFlow:

Numpy	TensorFlow
a=np.zeros((2,2));b=np.ones((2,2))	a=tf.zeros((2,2)),b=tf.ones((2,2))
np.sum(b,axis=1)	tf.reduce_sum(a,reduction_indices=[1])
a.shape	a.get_shape()
np.reshape(a,(1,4))	tf.reshape(a,(1,4))
b*5+1	b*5+1
np.dot(a,b)	tf.matmul(a,b)
a[0,0], a[:,0], a[0,:]	a[0,0],a[:,0],a[0,:]

Figure 7: NumPy versus TensorFlow: a one-to-one comparison

Now let's see an alternative way of creating tensors before they could be fed (we will see other feeding mechanisms later on) by the TensorFlow graph:

```
>>> X = [[2.0, 4.0],
        [6.0, 8.0]] # X is a list of lists
>>> Y = np.array([[2.0, 4.0],
                [6.0, 6.0]], dtype=np.float32)#Y is a Numpy array
>>> Z = tf.constant([[2.0, 4.0],
        [6.0, 8.0]]) # Z is a tensor
```

A First Look at TensorFlow

Here, X is a list, Y is an n-dimensional array from the NumPy library, and Z is a TensorFlow tensor object. Now let's see their types:

```
>>> print(type(X))
>>> print(type(Y))
>>> print(type(Z))

#Output
<class 'list'>
<class 'numpy.ndarray'>
<class 'tensorflow.python.framework.ops.Tensor'>
```

Well, their types are printed correctly. However, a more convenient function that we're formally dealing with tensors as opposed to the other types is `tf.convert_to_tensor()` function as follows:

```
t1 = tf.convert_to_tensor(X, dtype=tf.float32)
t2 = tf.convert_to_tensor(Z, dtype=tf.float32)
```

Now let's see their types using the following code:

```
>>> print(type(t1))
>>> print(type(t2))

#Output:
<class 'tensorflow.python.framework.ops.Tensor'>
<class 'tensorflow.python.framework.ops.Tensor'>
```

Fantastic! That's enough discussion about tensors for now. So, we can think about the structure that is characterized by the term rank.

Rank and shape

A unit of dimensionality called rank describes each tensor. It identifies the number of dimensions of the tensor. For this reason, a rank is known as order or n–dimensions of a tensor. A rank zero tensor is a **scalar**, a rank one tensor is a **vector**, and a rank two tensor is a **matrix**.

The following code defines a TensorFlow `scalar`, `vector`, `matrix`, and `cube_matrix`. In the next example, we will show how rank works:

```
import tensorflow as tf
scalar = tf.constant(100)
vector = tf.constant([1,2,3,4,5])
```

```
matrix = tf.constant([[1,2,3],[4,5,6]])

cube_matrix =
tf.constant([[[1],[2],[3]],[[4],[5],[6]],[[7],[8],[9]]])

print(scalar.get_shape())
print(vector.get_shape())
print(matrix.get_shape())
print(cube_matrix.get_shape())
```

The results are printed here:

```
>>>
()
(5,)
(2, 3)
(3, 3, 1)
>>>
```

The shape of a tensor is the number of rows and columns it has. Now we will see how to relate the shape of a tensor to its rank:

```
>>scalar.get_shape()
TensorShape([])

>>vector.get_shape()
TensorShape([Dimension(5)])

>>matrix.get_shape()
TensorShape([Dimension(2), Dimension(3)])

>>cube.get_shape()
TensorShape([Dimension(3), Dimension(3), Dimension(1)])
```

A First Look at TensorFlow

Data type

In addition to rank and shape, tensors have a data type. Here is a list of the data types:

Data type	Python type	Description
DT_FLOAT	tf.float32	32-bit floating point
DT_DOUBLE	tf.float64	64-bit floating point
DT_INT8	tf.int8	8-bit signed integer
DT_INT16	tf.int16	16-bit signed integer
DT_INT32	tf.int32	32-bit signed integer
DT_INT64	tf.int64	64-bit signed integer
DT_UINT8	tf.uint8	8-bit unsigned integer
DT_STRING	tf.string	Variable length byte arrays. Each element of a tensor is a byte array
DT_BOOL	tf.bool	Boolean
DT_COMPLEX64	tf.complex64	Complex number made of two 32-bit floating points: real and imaginary parts
DT_COMPLEX128	tf.complex128	Complex number made of two 64-bit floating points: real and imaginary parts
DT_QINT8	tf.qint8	8-bit signed integer used in quantized Ops
DT_QINT32	tf.qint32	32-bit signed integer used in quantized Ops
DT_QUINT8	tf.quint8	8-bit unsigned integer used in quantized Ops

The preceding table is self-explanatory, so we have not provided a detailed discussion of the data types. The TensorFlow APIs are implemented to manage data *to* and *from* NumPy arrays.

Thus, to build a tensor with a constant value, pass a NumPy array to the tf.constant() operator, and the result will be a tensor with that value:

```
import tensorflow as tf

import numpy as np
array_1d = np.array([1,2,3,4,5,6,7,8,9,10])
tensor_1d = tf.constant(array_1d)

with tf.Session() as sess:
    print(tensor_1d.get_shape())
    print(sess.run(tensor_1d))
```

Running the example, we obtain the following:

```
>>>
 (10,)
[ 1  2  3  4  5  6  7  8  9 10]
```

To build a tensor with variable values, use a NumPy array and pass it to the `tf.Variable` constructor. The result will be a variable tensor with that initial value:

```
import tensorflow as tf
import numpy as np

# Create a sample NumPy array
array_2d = np.array([(1,2,3),(4,5,6),(7,8,9)])

# Now pass the preceding array to tf.Variable()
tensor_2d = tf.Variable(array_2d)

# Execute the preceding op under an active session
with tf.Session() as sess:
    sess.run(tf.global_variables_initializer())
    print(tensor_2d.get_shape())
    print sess.run(tensor_2d)
# Finally, close the TensorFlow session when you're done
sess.close()
```

In the preceding code block, `tf.global_variables_initializer()` is used to initialize all the ops we created before. If you need to create a variable with an initial value dependent on another variable, use the other variable's `initialized_value()`. This ensures that variables are initialized in the right order.

The result is as follows:

```
>>>
 (3, 3)
[[1 2 3]
 [4 5 6]
 [7 8 9]]
```

For ease of use in interactive Python environments, we can use the `InteractiveSession` class, and then use that session for all `Tensor.eval()` and `Operation.run()` calls:

```
import tensorflow as tf # Import TensorFlow
import numpy as np # Import numpy

# Create an interactive TensorFlow session
interactive_session = tf.InteractiveSession()

# Create a 1d NumPy array
array1 = np.array([1,2,3,4,5]) # An array

# Then convert the preceding array into a tensor
tensor = tf.constant(array1) # convert to tensor
print(tensor.eval()) # evaluate the tensor op

interactive_session.close() # close the session
```

 `tf.InteractiveSession()` is just convenient syntactic sugar for keeping a default session open in IPython.

The result is as follows:

```
>>>
[1 2 3 4 5]
```

This can be easier in an interactive setting, such as the shell or an IPython Notebook, as it can be tedious to pass around a session object everywhere.

 The IPython Notebook is now known as the Jupyter Notebook. It is an interactive computational environment in which you can combine code execution, rich text, mathematics, plots, and rich media. For more information, interested readers should refer to `https://ipython.org/notebook.html`.

Another way to define a tensor is using the `tf.convert_to_tensor` statement:

```
import tensorflow as tf
import numpy as np
tensor_3d = np.array([[[0, 1, 2], [3, 4, 5], [6, 7, 8]],
                      [[9, 10, 11], [12, 13, 14], [15, 16, 17]],
                      [[18, 19, 20], [21, 22, 23], [24, 25, 26]]])
tensor_3d = tf.convert_to_tensor(tensor_3d, dtype=tf.float64)

with tf.Session() as sess:
    print(tensor_3d.get_shape())
    print(sess.run(tensor_3d))
# Finally, close the TensorFlow session when you're done
sess.close()
```

The following is the output of the preceding code:

```
>>>
(3, 3, 3)
[[[  0.   1.   2.]
  [  3.   4.   5.]
  [  6.   7.   8.]]
 [[  9.  10.  11.]
  [ 12.  13.  14.]
  [ 15.  16.  17.]]
 [[ 18.  19.  20.]
  [ 21.  22.  23.]
  [ 24.  25.  26.]]]
```

Variables

Variables are TensorFlow objects used to hold and update parameters. A variable must be initialized so that you can save and restore it to analyze your code later on. Variables are created by using either `tf.Variable()` or `tf.get_variable()` statements. Whereas `tf.get_varaiable()` is recommended but `tf.Variable()` is lower-label abstraction.

In the following example, we want to count the numbers from 1 to 10, but let's import TensorFlow first:

```
import tensorflow as tf
```

A First Look at TensorFlow

We created a variable that will be initialized to the scalar value 0:

```
value = tf.get_variable("value", shape=[], dtype=tf.int32,
initializer=None, regularizer=None, trainable=True,
collections=None)
```

The `assign()` and `add()` operators are just nodes of the computation graph, so they do not execute the assignment until the session is run:

```
one = tf.constant(1)
update_value = tf.assign_add(value, one)
initialize_var = tf.global_variables_initializer()
```

We can instantiate the computation graph:

```
with tf.Session() as sess:
    sess.run(initialize_var)
    print(sess.run(value))
    for _ in range(5):
        sess.run(update_value)
        print(sess.run(value))
# Close the session
```

Let's recall that a tensor object is a symbolic handle to the result of an operation, but it does not actually hold the values of the operation's output:

```
>>>
0
1
2
3
4
5
```

Fetches

To fetch the output of an operation, the graph can be executed by calling `run()` on the session object and passing in the tensors. Apart from fetching a single tensor node, you can also fetch multiple tensors.

In the following example, the sum and multiply tensors are fetched together using the `run()` call:

```
import tensorflow as tf
constant_A = tf.constant([100.0])
constant_B = tf.constant([300.0])
constant_C = tf.constant([3.0])

sum_ = tf.add(constant_A,constant_B)
mul_ = tf.multiply(constant_A,constant_C)

with tf.Session() as sess:
    result = sess.run([sum_,mul_])# _ means throw away afterwards
    print(result)
```

The output is as follows:

```
>>>
[array(400.],dtype=float32),array([ 300.],dtype=float32)]
```

It should be noted that all the ops that need to be executed (that is, in order to produce tensor values) are run once (not once per requested tensor).

Feeds and placeholders

There are four methods of getting data into a TensorFlow program (for more information, see https://www.tensorflow.org/api_guides/python/reading_data):

- **The Dataset API**: This enables you to build complex input pipelines from simple and reusable pieces of distributed filesystems and perform complex operations. Using the Dataset API is recommended if you are dealing with large amounts of data in different data formats. The Dataset API introduces two new abstractions to TensorFlow for creating a feedable dataset: `tf.contrib.data.Dataset` (by creating a source or applying transformation operations) and `tf.contrib.data.Iterator`.
- **Feeding**: This allows us to inject data into any tensor in a computation graph.

- **Reading from files**: This allows us to develop an input pipeline using Python's built-in mechanism for reading data from data files at the beginning of the graph.
- **Preloaded data**: For a small dataset, we can use either constants or variables in the TensorFlow graph to hold all the data.

In this section, we will see an example of a feeding mechanism. We will see the other methods in upcoming chapters. TensorFlow provides a feed mechanism that allows us to inject data into any tensor in a computation graph. You can provide the feed data through the `feed_dict` argument to a `run()` or `eval()` invocation that initiates the computation.

> Feeding using `feed_dict` argument is the least efficient way to feed data into a TensorFlow execution graph and should only be used for small experiments needing small dataset. It can also be used for debugging.

We can also replace any tensor with feed data (that is, variables and constants). Best practice is to use a TensorFlow placeholder node using `tf.placeholder()` (https://www.tensorflow.org/api_docs/python/tf/placeholder). A placeholder exists exclusively to serve as the target of feeds. An empty placeholder is not initialized, so it does not contain any data.

Therefore, it will always generate an error if it is executed without a feed, so you won't forget to feed it. The following example shows how to feed data to build a random 2×3 matrix:

```
import tensorflow as tf
import numpy as np

a = 3
b = 2
x = tf.placeholder(tf.float32,shape=(a,b))
y = tf.add(x,x)

data = np.random.rand(a,b)
sess = tf.Session()
print(sess.run(y,feed_dict={x:data}))

sess.close()# close the session
```

The output is as follows:

```
>>>
[[ 1.78602004  1.64606333]
 [ 1.03966308  0.99269408]
 [ 0.98822606  1.50157797]]
>>>
```

Visualizing computations through TensorBoard

TensorFlow includes functions that allow you to debug and optimize programs in a visualization tool called TensorBoard. With TensorBoard, you can graphically observe different types of statistics concerning the parameters and details of any part of the graph.

Moreover, while doing predictive modeling using a complex DNN, the graph can be complex and confusing. To make it easier to understand, debug, and optimize TensorFlow programs, you can use TensorBoard to visualize your TensorFlow graph, plot quantitative metrics about the execution of your graph, and show additional data, such as images that pass through it.

Therefore, **TensorBoard** can be thought of as a framework designed for analyzing and debugging predictive models. TensorBoard uses the so-called summaries to view the parameters of the model: once a TensorFlow code is executed, we can call TensorBoard to view the summaries in a GUI.

How does TensorBoard work?

TensorFlow uses the computation graph to execute an application. In a computation graph, nodes represent an operation and the arcs are the data between operations.

The main idea of TensorBoard is to associate the summary with the nodes (operations) on the graph. When the code is running, summary operations will serialize the data of the node and output the data into a file. Later on, TensorBoard will visualize the summarized operations. For more detailed discussion, readers can refer to `https://github.com/tensorflow/tensorboard`.

A First Look at TensorFlow

Putting it simply, TensorBoard is a suite of web applications for inspecting and understanding your TensorFlow runs and graphs. The workflow when using TensorBoard is as follows:

1. Build your computational graph/code
2. Attach summary ops to the nodes you are interested in examining
3. Start running your graph as you normally would
4. Run the summary ops
5. When the execution is finished, run TensorBoard to visualize the summary outputs

For step 2 (that is, before running TensorBoard), make sure you have generated summary data in a log directory by creating a summary writer:

```
# sess.graph contains the graph definition; that enables the graph visualizer
    file_writer = tf.summary.FileWriter('/path/to/logs',
    sess.graph)
```

Now if you type `$ which tensorboard` in Terminal, it should exist if you installed it with pip:

```
root@ubuntu:~$ which tensorboard
/usr/local/bin/tensorboard
```

You need to give it a log directory. When you are in the directory where you ran your graph, you can launch it from Terminal with something like this:

```
tensorboard --logdir path/to/logs
```

When TensorBoard is fully configured, it can be accessed by issuing the following command:

```
# Make sure there's no space before or after '="
$ tensorboard -logdir=<trace_file_name>
```

Now you simply need to access localhost 6006 from the browser by typing `http://localhost:6006/`. Then it should look like this:

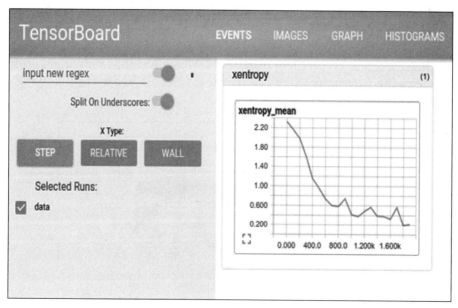

Figure 8: Using TensorBoard on a browser

 TensorBoard can be used in Google Chrome or Firefox. Other browsers might work, but there may be bugs or performance issues.

Is this already too much? Don't worry, in the last section, we'll combine all the ideas previously explained to build a single input neuron model and to analyze it with TensorBoard.

Linear regression and beyond

In this section, we will take a closer look at the main concepts of TensorFlow and TensorBoard and try to do some basic operations to get you started. The model we want to implement simulates linear regression.

In statistics and ML, linear regression is a technique that's frequently used to measure the relationship between variables. This is a quite simple but effective algorithm that can be used in predictive modeling as well.

Linear regression models the relationship between a dependent variable, y_i, an interdependent variable, x_i, and a random term, b. This can be seen as follows:

$$y = W * x + b$$

A typical linear regression problem using TensorFlow has the following workflow, which updates the parameters to minimize the given cost (see in the following figure) function:

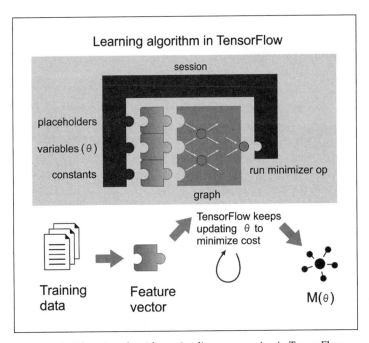

Figure 9: A learning algorithm using linear regression in TensorFlow

Now, let's try to follow the preceding figure and reproduce it for the linear regression by conceptualizing the preceding equation. For this, we're going to write a simple Python program for creating data in a 2D space. Then we will use TensorFlow to look for the line that best fits the data points (as shown in the following figure):

```
# Import libraries (Numpy, matplotlib)

import numpy as np
import matplotlib.pyplot as plot

# Create 1000 points following a function y=0.1 * x + 0.4z
```

(i.e. # y = W * x + b) with some normal random distribution:

```
num_points = 1000
vectors_set = []

# Create a few random data points
for i in range(num_points):
    W = 0.1 # W
    b = 0.4 # b
    x1 = np.random.normal(0.0, 1.0)#in: mean, standard deviation
    nd = np.random.normal(0.0, 0.05)#in:mean,standard deviation
    y1 = W * x1 + b

 # Add some impurity with normal distribution -i.e. nd
    y1 = y1 + nd

 # Append them and create a combined vector set:
    vectors_set.append([x1, y1])

# Separate the data point across axises:
x_data = [v[0] for v in vectors_set]
y_data = [v[1] for v in vectors_set]

# Plot and show the data points in a 2D space
plot.plot(x_data, y_data, 'ro', label='Original data')
plot.legend()
plot.show()
```

If your compiler does not complain, you should get the following graph:

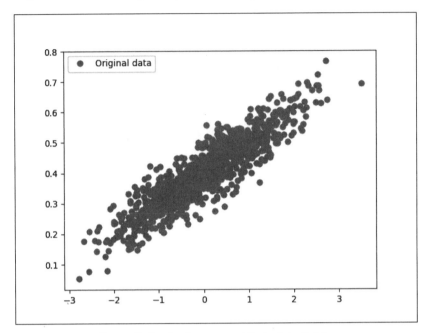

Figure 10: Randomly generated (but original) data

Well, so far we have just created a few data points without an associated model that could be executed through TensorFlow. So, the next step is to create a linear regression model that can obtain the output values y that is estimated from the input data points, that is, x_data. In this context, we have only two associated parameters, W and b.

Now the objective is to create a graph that allows us to find the values for these two parameters based on the input data, x_data, by adjusting them to y_data. So, the target function in our case would be as follows:

$$y\_data = W * x\_data + b$$

If you recall, we defined W = 0.1 and b = 0.4 while creating the data points in the 2D space. TensorFlow has to optimize these two values so that W tends to 0.1 and b to 0.4.

A standard way to solve such optimization problems is to iterate through each value of the data points and adjust the values of W and b in order to get a more precise answer for each iteration. To see if the values really are improving, we need to define a cost function that measures how good a certain line is.

In our case, the cost function is the mean squared error, which helps us find the average of the errors based on the distance function between the real data points and the estimated ones on each iteration. We start by importing the TensorFlow library:

```
import tensorflow as tf
W = tf.Variable(tf.zeros([1]))
b = tf.Variable(tf.zeros([1]))
y = W * x_data + b
```

In the preceding code segment, we are generating a random point using a different strategy and storing it in the variable *W*. Now, let's define a loss function, $loss = mean\left[(y - y\,data)2\right]$, and this returns a scalar value with the mean of all distances between our data and the model prediction. In terms of TensorFlow convention, the loss function can be expressed as follows:

```
loss = tf.reduce_mean(tf.square(y - y_data))
```

The preceding line actually computes **mean square error** (**MSE**). Without going into further detail, we can use some widely used optimization algorithms, such as GD. At a minimal level, GD is an algorithm that works on a set of given parameters that we already have.

It starts with an initial set of parameter values and iteratively moves toward a set of values that minimize the function by taking another parameter called the learning rate. This iterative minimization is achieved by taking steps in the negative direction of the gradient function:

```
optimizer = tf.train.GradientDescentOptimizer(0.6)
train = optimizer.minimize(loss)
```

Before running this optimization function, we need to initialize all the variables that we have so far. Let's do it using a conventional TensorFlow technique, as follows:

```
init = tf.global_variables_initializer()
sess = tf.Session()
sess.run(init)
```

Since we have created a TensorFlow session, we are ready for the iterative process that helps us find the optimal values of W and b:

```
for i in range(6):
  sess.run(train)
  print(i, sess.run(W), sess.run(b), sess.run(loss))
```

A First Look at TensorFlow

You should observe the following output:

```
>>>
0 [ 0.18418592] [ 0.47198644] 0.0152888
1 [ 0.08373772] [ 0.38146532] 0.00311204
2 [ 0.10470386] [ 0.39876288] 0.00262051
3 [ 0.10031486] [ 0.39547175] 0.00260051
4 [ 0.10123629] [ 0.39609471] 0.00259969
5 [ 0.1010423]  [ 0.39597753] 0.00259966
6 [ 0.10108326] [ 0.3959994]  0.00259966
7 [ 0.10107458] [ 0.39599535] 0.00259966
```

You can see the algorithm starts with the initial values of W = 0.18418592 and b = 0.47198644, and the loss is pretty high. Then, the algorithm iteratively adjusts the values by minimizing the cost function. In the eighth iteration, all the values tend to our desired values.

Now, what if we could plot them? Let's do it by adding a plotting line under the `for` loop, as follows:

```
for i in range(6):
    sess.run(train)
    print(i, sess.run(W), sess.run(b), sess.run(loss))
    plot.plot(x_data, y_data, 'ro', label='Original data')
    plot.plot(x_data, sess.run(W)*x_data + sess.run(b))
    plot.xlabel('X')
    plot.xlim(-2, 2)
    plot.ylim(0.1, 0.6)
    plot.ylabel('Y')
    plot.legend()
    plot.show()
```

The preceding code block, should produce the following figure (merged together, though):

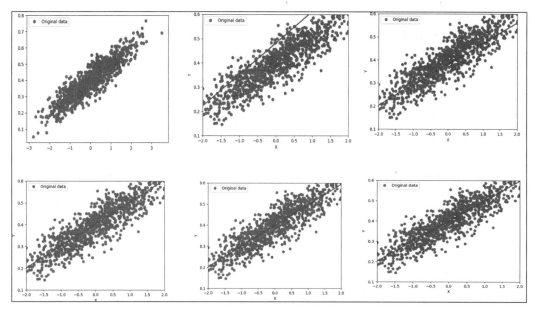

Figure 11: Linear regression optimizing the loss function after the sixth iteration

Now let's go up to the 16th iteration:

```
>>>
0   [ 0.23306453]   [ 0.47967502]   0.0259004
1   [ 0.08183448]   [ 0.38200468]   0.00311023
2   [ 0.10253634]   [ 0.40177572]   0.00254209
3   [ 0.09969243]   [ 0.39778906]   0.0025257
4   [ 0.10008509]   [ 0.39859086]   0.00252516
5   [ 0.10003048]   [ 0.39842987]   0.00252514
6   [ 0.10003816]   [ 0.39846218]   0.00252514
7   [ 0.10003706]   [ 0.39845571]   0.00252514
8   [ 0.10003722]   [ 0.39845699]   0.00252514
9   [ 0.10003719]   [ 0.39845672]   0.00252514
10  [ 0.1000372]    [ 0.39845678]   0.00252514
11  [ 0.1000372]    [ 0.39845678]   0.00252514
12  [ 0.1000372]    [ 0.39845678]   0.00252514
13  [ 0.1000372]    [ 0.39845678]   0.00252514
14  [ 0.1000372]    [ 0.39845678]   0.00252514
15  [ 0.1000372]    [ 0.39845678]   0.00252514
```

Much better, and we're closer to the optimized values, right? Now, what if we improve our visual analytics further through TensorFlow to help visualize what is happening in these graphs. TensorBoard provides a web page for debugging your graph and inspecting the variables, node, edges, and their corresponding connections.

Also, we need to annotate the preceding graphs with the variables, such as the loss function, W, b, y_data, x_data, and so on. Then you need to generate all the summaries by invoking the tf.summary.merge_all() function.

Now, we need to make the following changes to the preceding code. However, it is good practice to group related nodes on the graph using the tf.name_scope() function. Thus, we can use tf.name_scope() to organize things on the TensorBoard graph view, but let's give it a better name:

```
with tf.name_scope("LinearRegression") as scope:
    W = tf.Variable(tf.zeros([1]))
    b = tf.Variable(tf.zeros([1]))
    y = W * x_data + b
```

Then, let's annotate the loss function in a similar way, but with a suitable name, such as LossFunction:

```
with tf.name_scope("LossFunction") as scope:
    loss = tf.reduce_mean(tf.square(y - y_data))
```

Let's annotate the loss, weights, and bias that are needed for TensorBoard:

```
loss_summary = tf.summary.scalar("loss", loss)
w_ = tf.summary.histogram("W", W)
b_ = tf.summary.histogram("b", b)
```

Once you have annotated the graph, it's time to configure the summary by merging them:

```
merged_op = tf.summary.merge_all()
```

Before running the training (after the initialization), write the summary using the tf.summary.FileWriter() API as follows:

```
writer_tensorboard = tf.summary.FileWriter('logs/', tf.get_default_graph())
```

Chapter 2

Then start TensorBoard as follows:

```
$ tensorboard -logdir=<trace_dir_name>
```

In our case, it could be something like the following:

```
$ tensorboard --logdir=/home/root/LR/
```

Now let's move to `http://localhost:6006` and click on the **GRAPH** tab. You should see the following graph:

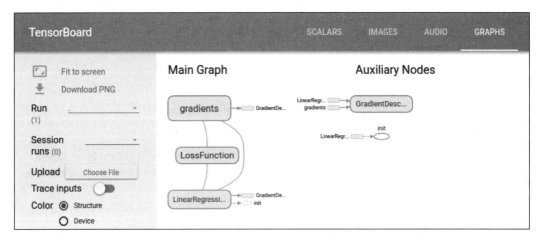

Figure 12: The main graph and auxiliary nodes on TensorBoard

Note that Ubuntu may ask you to install the `python-tk` package. You can do it by executing the following command on Ubuntu:
```
$ sudo apt-get install python-tk
# For Python 3.x, use the following
$ sudo apt-get install python3-tk
```

Linear regression revisited for a real dataset

In the previous section, we saw an example of linear regression. We saw how to use TensorFlow with a randomly generated dataset, that is, fake data. We have seen that regression is a type of supervised machine learning for predicting continuous (rather than discrete) output.

A First Look at TensorFlow

However, running a linear regression on fake data is just like buying a new car but never driving it. This awesome machinery begs to be used in the real world! Fortunately, many datasets are available online to test your new-found knowledge of regression.

One of them is the Boston housing dataset, which can be downloaded from the UCI Machine Learning Repository at `https://archive.ics.uci.edu/ml/datasets/Housing`. It is also available as a preprocessed dataset with scikit-learn.

So, let's get started by importing all the required libraries, including TensorFlow, NumPy, Matplotlib, and scikit-learn:

```
import matplotlib.pyplot as plt
import tensorflow as tf
import numpy as np
from numpy import genfromtxt
from sklearn.datasets import load_boston
from sklearn.model_selection import train_test_split
```

Next, we need to prepare the training set consisting of features and labels from the Boston housing dataset. The `read_boston_data ()` method reads from scikit-learn and returns the features and labels separately:

```
def read_boston_data():
    boston = load_boston()
    features = np.array(boston.data)
    labels = np.array(boston.target)
    return features, labels
```

Now that we have the features and labels, we need to normalize the features as well, using the `normalizer()` method. Here is the signature of the method:

```
def normalizer(dataset):
    mu = np.mean(dataset,axis=0)
    sigma = np.std(dataset,axis=0)
    return(dataset - mu)/sigma
```

`bias_vector()` is used to append the bias term (that is all 1s) to the normalized features that we prepared in the preceding step. It corresponds to the **b** term in the equation of straight line in the previous example:

```
def bias_vector(features,labels):
    n_training_samples = features.shape[0]
    n_dim = features.shape[1]
```

```
    f = 
np.reshape(np.c_[np.ones(n_training_samples),features],
[n_training_samples,n_dim + 1])
    l = np.reshape(labels,[n_training_samples,1])
    return f, l
```

We will now invoke these methods and split the dataset into training and testing, 75% for training and rest for testing:

```
features,labels = read_boston_data()
normalized_features = normalizer(features)
data, label = bias_vector(normalized_features,labels)
n_dim = data.shape[1]
# Train-test split
train_x, test_x, train_y, test_y =
train_test_split(data,label,test_size = 0.25,random_state =
100)
```

Now let's use TensorFlow's data structures (such as placeholders, labels, and weights):

```
learning_rate = 0.01
training_epochs = 100000
log_loss = np.empty(shape=[1],dtype=float)
X = tf.placeholder(tf.float32,[None,n_dim]) #takes any number
of rows but n_dim columns
Y = tf.placeholder(tf.float32,[None,1]) # #takes any number
of rows but only 1 continuous column
W = tf.Variable(tf.ones([n_dim,1])) # W weight vector
```

Well done! We have prepared the data structure required to construct the TensorFlow graph. Now it's time to construct the linear regression, which is pretty straightforward:

```
y_ = tf.matmul(X, W)
cost_op = tf.reduce_mean(tf.square(y_ - Y))
training_step =
tf.train.GradientDescentOptimizer(learning_rate).minimize
(cost_op)
```

In the preceding code segment, the first line multiplies the features matrix by the weights matrix that can be used for prediction. The second line computes the loss, which is the squared error of the regression line. Finally, the third line performs one-step of GD optimization to minimize the square error.

> Which optimizer to use: the main objective of using optimizer is to minimize the evaluated cost; therefore, we must define an optimizer. Using the most common optimizer like SGD, the learning rates must scale with *1/T* to get convergence, where *T* is the number of iteration.
>
> Adam or RMSProp tries to overcome this limitation automatically by adjusting the step size so that the step is on the same scale as the gradients. In addition, in the previous example, we have used Adam optimizer, which performs well in most of the cases.
>
> Nevertheless, if you are training a neural network computing the gradients is mandatory, using the RMSPropOptimizer function, which implements the RMSProp algorithm is a better idea, since it would be the faster way of learning in a mini-batch setting. Researchers also recommend using the Momentum optimizer while training a deep CNN or DNN.
>
> Technically, RMSPropOptimizer is an advanced form of gradient descent that divides the learning rate by an exponentially decaying average of squared gradients. The suggested setting value of the decay parameter is 0.9, while a good default value for the learning rate is 0.001.
>
> For example in TensorFlow, tf.train.RMSPropOptimizer() helps us use this with ease:
>
> ```
> optimizer = tf.train.RMSPropOptimizer(0.001,
> 0.9).minimize(cost_op)
> ```

Now, before we start training the model, we need to initialize all the variables using the initialize_all_variables() method, as follows:

```
init = tf.initialize_all_variables()
```

Fantastic! Now that we have managed to prepare all the components, we're ready to train the actual train. We start by creating TensorFlow session as follows:

```
sess = tf.Session()
sess.run(init_op)
for epoch in range(training_epochs):
    sess.run(training_step,feed_dict={X:train_x,Y:train_y})
    log_loss =
np.append(log_loss,sess.run(cost_op,feed_dict={X: train_x,Y:
train_y}))
```

Once the training is completed, we are able to make predictions on unseen data. However, it's even more exciting to see a visual representation of the completed training. So, let's plot the cost as a function of the number of iterations using Matplotlib:

```
plt.plot(range(len(log_loss)),log_loss)
plt.axis([0,training_epochs,0,np.max(log_loss)])
plt.show()
```

The following is the output of the preceding code:

>>>

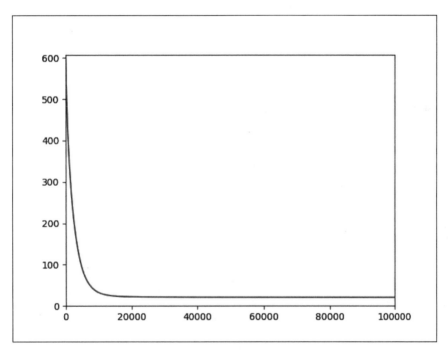

Figure 13: Cost as a function of the number of iterations

Make some predictions on the test dataset and calculate the mean squared error:

```
pred_y = sess.run(y_, feed_dict={X: test_x})
mse = tf.reduce_mean(tf.square(pred_y - test_y))
print("MSE: %.4f" % sess.run(mse))
```

The following is the output of the preceding code:

>>>
MSE: 27.3749

Finally, let's show the line of best fit:

```
fig, ax = plt.subplots()
ax.scatter(test_y, pred_y)
ax.plot([test_y.min(), test_y.max()], [test_y.min(),
test_y.max()], 'k--', lw=3)
ax.set_xlabel('Measured')
ax.set_ylabel('Predicted')
plt.show()
```

The following is the output of the preceding code:

>>>

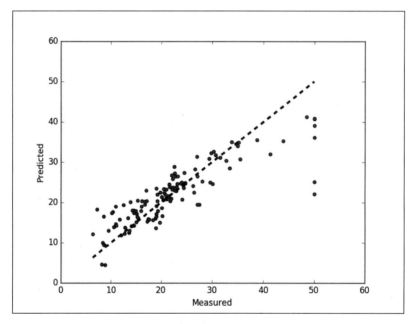

Figure 14: Predicted versus actual values

Summary

TensorFlow is designed to make predictive analytics through ML and DL easy for everyone, but using it does require a decent understanding of some general principles and algorithms. The latest release of TensorFlow comes with lots of exciting new features, so we have tried to cover them so that you can use them with ease. In summary, here is a brief recap of the key concepts of TensorFlow that have been explained in this chapter:

- **Graph**: Each TensorFlow computation can be represented as a data flow graph, where each graph is built as a set of operation objects. There are three core graph data structures: `tf.Graph` (https://www.tensorflow.org/api_docs/python/tf/Graph), `tf.Operation` (https://www.tensorflow.org/api_docs/python/tf/Operation), and `tf.Tensor` (https://www.tensorflow.org/api_docs/python/tf/Tensor).

- **Operation**: A graph node takes one or more tensors as input and produces one or more tensors as output. A node can be represented by an operation object for performing computational units such as addition, multiplication, division, subtraction, or more complex operations.

- **Tensors**: Like high-dimensional array objects. In other words, they can be represented as edges of a data flow graph and are the outputs of different operations.

- **Session**: They are like l session object is an entity that encapsulates the environment in which operation objects are executed for running calculations on the data flow graph. As a result, tensor objects are evaluated inside a `run()` or `eval()` invocation.

In a later section of the chapter, we introduced TensorBoard, which is a powerful tool for analyzing and debugging neural network models. Finally, we saw how to implement one of the simplest TensorFlow-based linear regression models on a fake and a real dataset.

In the next chapter, we will discuss the theoretical background of different **FFNN** architectures such as **Deep Belief Networks** (**DBNs**) and **Multilayer Perceptron** (**MLP**).

We will then show how to train and analyze the performance metrics that are needed to evaluate the models, followed by some ways to tune hyperparameters for FFNNs for optimized performance. Finally, we will provide two examples, using MLP and DBN, of how to build very robust and accurate predictive models for predictive analytics on a bank marketing dataset.

3
Feed-Forward Neural Networks with TensorFlow

ANNs are at the very core of DL. They are versatile, powerful, and scalable, making them ideal for tackling large and highly complex ML tasks. We can classify billions of images, power speech recognition services, and even recommend that hundreds of millions of users watch the best videos, by stacking multiple ANNs together. These multiple stacked ANNs are called **Deep Neural Networks** (**DNNs**). Using DNNs, we can build very robust and accurate models for predictive analytics.

The architectures of DNNs can be very different: they are often organized on different layers. The first layer receives the input signals and the last layer produces the output signals. Usually, these networks are identified as **Feed-Forward Neural Networks** (**FFNNs**). In this chapter, we will construct an FFNN that classifies an MNIST dataset. Later on, we will see two more implementations of FFNNs (for building very robust and accurate models for predictive analytics) called **Deep Belief Networks** (**DBNs**) and **Multilayer Perceptron** (**MLP**), on a bank marketing dataset. Finally, we will see how to tune the most important FFNN hyperparameters for optimized performance.

Concisely, the following topics will be covered throughout this chapter:

- Feed-forward neural networks
- Implementing a five-layer FFNN for digit classification
- Implementing a deep MLP for client-subscription prediction
- Revisiting client-subscription prediction: implementing a DBN
- Hyperparameter tuning and dropout optimization in an FFNN

Feed-forward neural networks (FFNNs)

An FFNN consists of a large number of neurons, organized in layers: one input layer, one or more hidden layers, and one output layer. Each neuron in a layer is connected to all the neurons of the previous layer, although the connections are not all the same because they have different weights. The weights of these connections encode the knowledge of the network. Data enters at the inputs and passes through the network, layer by layer until it arrives at the outputs. During this operation, there is no feedback between layers. Therefore, these types of networks are called feed-forward neural networks.

An FFNN with enough neurons in the hidden layer is able to approximate with arbitrary precision, and can model the linear, as well as non-linear, relationships in your data:

- Any continuous function, with one hidden layer
- Any function, even discontinuous, with two hidden layers

However, it is not possible to determine a priori, with adequate precision, the required number of hidden layers, or even the number of neurons that must be contained inside each hidden layer to compute a non-linear function. There is no straightforward answer to this, but we can try increasing the number of neurons in a hidden layer until the FFNN starts overfitting. We will discuss this later on. Despite some rules of thumb, setting the number of hidden layers relies on experience and on some heuristics to determine the structure of the network.

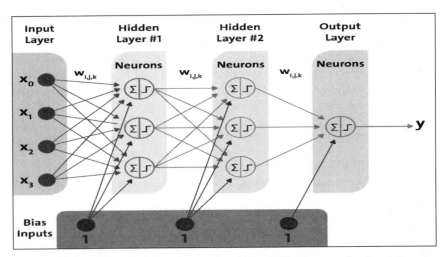

Figure 1: A feed-forward neural network with two hidden layers and an input bias

If a low number of hidden layers, or neurons, constitute the neural network architecture, then the network is not able to approximate with adequate precision the unknown function, for example. This could be because it is too complex, or because the backpropagation algorithm falls within a local minimum. If the network is composed of a high number of hidden layers, then we have an overfitting problem; namely a worsening of the network's generalization ability. One solution to this problem is regularization through dropout (we will discuss this later in the chapter).

Therefore, a complex network can consist of many neurons, hidden layers, and connections, but in general, an ANN with two or more hidden layers is called a **DNN**. From the implementation perspective, a DNN can be constructed by stacking multiple ANNs together.

Based on the types of layers used in DNNs and the corresponding learning method, DNNs can be classified as MLPs, **Staked Auto-Encoders** (**SAEs**), or DBNs. All of these are regular FFNNs, but they have a number of hidden layers and they are architecturally different. In this chapter we will mainly discuss MLPs and DBNs, using hands-on examples, but SAE will be covered in *Chapter 5, Optimizing TensorFlow Autoencoders*. However, first, let's focus on the feed-forward and the backpropagation mechanism.

Feed-forward and backpropagation

The backpropagation algorithm aims to minimize the error between the current and the desired output. Since the network is feed-forward, the activation flow always proceeds forward from the input units to the output units. The gradient of the cost function is backpropagated through the modification of weights.

This method is recursive and can be applied to any number of hidden layers. In such a method, the incorporation between two phases is important. The feed-forward learning models are:

- Forward pass
- Backward pass

In the forward pass, we do a bunch of operations and obtain some predictions or scores. Therefore, for each operation in the forward pass, we need to create a graph connecting operations top to bottom.

On the other hand, the backward pass is involved mainly with mathematical operations, such as creating derivatives for all the deferential operations (that is, auto differentiation methods) top to bottom (for example, loss function to weights update), for all the operations in the graph, and then using them in the chain rule. Note that there are two types of auto differentiation methods:

- **Reverse-mode**: Derivation of a single output with respect to all inputs
- **Forward-mode**: Derivation of all outputs with respect to one input

We will discuss how TensorFlow makes this easier for us. The backpropagation algorithm processes the information in such a way that the network decreases the global error during the learning iterations; however, this does not guarantee that the global minimum is reached. The presence of the hidden units and the non-linearity of the output function means that the behavior of the error is very complex and has many local minima.

The backpropagation algorithm, therefore, can stop at a local minimum, providing a suboptimal solution. Normally, the error always decreases on the training set, which improves the ability to represent the input-output relationship between the data supplied. Because the network is learning while on the testing set (which measures the predictive capabilities from a certain value), it can grow due to the over-fitting problem: the resulting network (or model) will have a high classification accuracy for the training samples, and a low classification accuracy for test samples.

Now let's see how TensorFlow performs the forward and backward pass. While developing deep learning applications using TensorFlow, we will only consider writing forward passcodes. Let's clarify the idea further by recapping some concepts from *Chapter 2, A First Look at TensorFlow*.

We have seen that a TensorFlow programme has the following two components:

- Graph creation (`https://www.tensorflow.org/api_docs/python/tf/Graph`)
- Session execution (`https://www.tensorflow.org/api_docs/python/tf/Session`)

The first one is for building the model and the second one is for feeding the data in and getting results. Each of these is executed on a C++ engine, which consists of the following components:

- Efficient implementations of different operations such as activation functions (for example, sigmoid, ReLU, tanh, softmax, cross entropy, and so on.)
- Derivatives of forward-mode operation

I am not kidding when I say that even a little addition or multiplication is not executed in Python since Python is just a wrapper. Anyway, let us return to our original discussion by introducing an example. Imagine that we want to perform the dropout op (we will see more details about this later in this chapter) to randomly turn off and on some neurons:

```
yi =dropout(Sigmoid(Wx+b))
```

Now, for this, even though we do not care about the backward pass, TensorFlow automatically creates derivatives for all the operations in a top to bottom fashion. When we start a session, TensorFlow automatically calculates gradients for all the deferential operations in the graph and uses them in the chain rule. Therefore, the forward pass consists of the following:

1. Variables and placeholders (weights W, input x, bias b)
2. Operations (nonlinear operations for example, ReLU, cross entropy loss and so on.)

Now the forward pass is what we create, but TensorFlow automatically creates a backward pass, which makes the training process run by transferring data when doing the chain rule.

Weights and biases

Besides considering the state of a neuron and the way that it is linked to others, we should consider the synaptic weight, which is the influence of that connection within the network. Each weight has a numerical value indicated by W_{ij}, which is the synaptic weight connecting the neuron i to neuron j.

Synaptic weight: This is evolved from Biology and refers to the strength or amplitude of a connection between two nodes; in Biology, this corresponds to the amount of influence the firing of one neuron has on another.

Depending on the point where a neuron is located, it will always have one or more links, which correspond to relative synaptic weights. The weights and output function determine the behavior of an individual neuron and the network in general. They should be correctly changed during the training phase to ensure the correct behavior of the model.

For each unit *i* an input vector can be defined by $x_i = (x_1, x_2, \ldots, x_n)$ and a weight vector can be defined by $w_i = (w_{i1}, w_{i2}, \ldots, w_{in})$. Then during the forward propagation, each unit in the hidden layer gets the following signal:

$$\text{net}_i = \sum_j w_{ij} x_j \quad \ldots\ldots\ldots\ldots\ldots\ldots(a)$$

The preceding equation signifies that each hidden unit gets the sum of inputs multiplied by the corresponding weight.

Among the weights, there is one special weight called a *bias*. It is not tied to any other unit of the network but is considered to have an input equal to one. This expedient allows for establishing a kind of *reference point*, or threshold for neurons, and formally, the bias performs a translation along the abscissa axis to the output function. The previous formula will become as follows:

$$\text{net}_i = \sum_j w_{ij} x_j + b_i \quad \ldots\ldots\ldots\ldots\ldots\ldots(b)$$

Now a tricky question: how do we initialize the weights? Well, if we initialize all weights to the same value (for example, zero or one), each hidden neuron will get exactly the same signal. Let's try to break it down:

- If all weights are initialized to one, then each unit gets a signal equal to the sum of the inputs
- If all weights are zeros, which is even worse, every neuron in a hidden layer will get zero signal

In short, no matter what the input was, if all the weights are the same then all the units in the hidden layer will be the same too. To get rid of this issue, one of the most common initialization techniques in training FNNs is the random initialization. The idea of using the random initialization is just to sample each weight from a normal distribution of the input dataset, with low deviation.

A low deviation allows you to bias the network towards the "simple" zero solutions. However, what does this mean? The thing is that the initialization can be completed without the bad repercussions of actually initializing the weights to zero.

Secondly, Xavier initialization is nowadays used for training neural networks. It is similar to random initialization but often turns out to work much better. Now let me explain the reason: imagine that you initialize the network weights randomly, but it turns out that you start too small. The signal will shrink as it passes through each layer until it is too tiny to be useful. On the other hand, if the weights in a network start too large, then the signal will grow as it passes through each layer until it is too massive to be useful.

The good thing when using Xavier initialization is that it makes sure the weights are "just right," keeping the signal in a reasonable range of values through many layers. In summary, it can automatically determine the scale of initialization based on the number of input and output neurons.

Interested readers should refer to this publication for detailed information: *Xavier Glorot, Yoshua Bengio, Understanding the difficulty of training deep feedforward neural networks, Proceedings of the 13th International Conference on Artificial Intelligence and Statistics (AISTATS) 2010, Chia Laguna Resort, Sardinia, Italy. Volume 9 of JMLR: W&CP.*

You may be wondering if you can get rid of the random initialization while training a regular DNN (for example, MLP or DBN). Well, recently, some researchers have been talking about random orthogonal matrix initializations that perform better than just any random initialization for training DNNs.

When it comes to initializing the biases, it is possible, and common, to initialize the biases to be zero since the asymmetry breaking is provided by the small random numbers in the weights. Setting the biases to a small constant value such as 0.01 for all biases ensures that all ReLU units can propagate some gradient. However, it neither performs well nor shows consistent improvement. Therefore, sticking with zero is recommended.

Activation functions

To allow a neural network to learn complex decision boundaries, we apply a non-linear activation function to some of its layers. Commonly used functions include tanh, ReLU, softmax, and variants of these. More technically, each neuron receives as input signal the weighted sum of the synaptic weights and the activation values of the neurons connected.

To allow the neuron to calculate its activation value, that is, what the neuron retransmits, the weighted sum must be passed as the argument of the activation function. The activation function allows the receiving neuron to transmit the received signal, modifying it. One of the most widely used functions for this purpose is the so-called sigmoid function. It is a special case of the logistic function, which is defined by the following formula:

$$out_i = \frac{1}{1+e^{-net_i}}$$

A sigmoid function is a bounded differentiable real function that is defined for all real input values and has a non-negative derivative at each point. In general, a sigmoid function is real-valued, monotonic, and differentiable, having a non-negative first derivative, which is bell-shaped.

The domain of this function, which includes all real numbers and the co-domain, is (0, 1). This means that any value obtained as an output from a neuron (as per the calculation of its activation state), will always be between 0 and 1. The sigmoid function, as represented in the following diagram, provides an interpretation of the saturation rate of a neuron: from not being active (= 0), to complete saturation, which occurs at a predetermined maximum value (=1).

When new data has to be analyzed, it is loaded by the input layer, which through (*a*) or (*b*) generates an output. This result, together with the output from neurons of the same layer, will form a new input to the neurons on the next layers. The process will be iterated until the last layer.

On the other hand, hyperbolic tangent, or tanh, is another form of the activation function. Tanh squashes a real-valued number to the range [-1, 1]. Like the sigmoid neuron, its activations saturate, but unlike the sigmoid neuron, its output is zero-centered. Therefore, in practice, the tanh non-linearity is always preferred over the sigmoid nonlinearity. Also, note that the tanh neuron is simply a scaled sigmoid neuron. In particular, the following holds true: `tanh(x)=2σ(2x)-1`, as shown in the following figure:

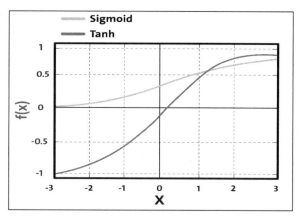

Figure 2: Sigmoid vs Tanh activation function

In general, in the last level of an FFNN, the softmax function is applied as the decision boundary. This is a common case, especially when solving a classification problem. Otherwise, we do not need to use any activation function for a regression problem at all.

In Mathematics, the **softmax function** is a generalization of the logistic function, that "squashes" a K-dimensional vector of arbitrary real values to a K-dimensional vector $\sigma(z)$ of real values in the range [0, 1] that add up to 1:

$$\sigma : \mathbb{R}^K \to [0,1]^K$$

Now the softmax function can be expressed as follows:

$$\sigma(z)_j = \frac{e^{z_j}}{\sum_{k=1}^{K} e^{z_k}} \text{ for } j = 1, \ldots, K$$

In the preceding equation, K represents the total number of outputs from the network. In probability theory, the output of the softmax function can be used to represent a categorical distribution – that is, a probability distribution over K different possible outcomes. In fact, it is the gradient-log-normalizer of the categorical probability distribution.

Nevertheless, the softmax function is used in various multiclass classification methods, such as multinomial logistic regression (also known as softmax regression), multiclass linear discriminant analyses, naive Bayes classifiers, and ANNs.

Now let us see how to use a few commonly-used activation functions in TensorFlow syntax. For providing different types of nonlinearities in neural networks, TensorFlow provides different activation ops. These include smooth nonlinearities such as sigmoid, tanh, elu, softplus, and softsign.

On the other hand, some continuous but not widespread differentiable functions that can be used are `ReLU`, `relu6`, `crelu`, and `relu_x`. All activation ops apply component-wise and produce a tensor of the same shape as the input tensor.

Using sigmoid

In TensorFlow, the signature `tf.sigmoid(x, name=None)` computes sigmoid of x element-wise using `y = 1 / (1 + exp(-x))` and returns a tensor with the same type x. Here is the parameter description:

- `x`: A tensor. Must be one of the following types: `float32`, `float64`, `int32`, `complex64`, `int64`, or `qint32`.
- `name`: A name for the operation (optional).

Using tanh

In TensorFlow, the signature `tf.tanh(x, name=None)` computes the hyperbolic tangent of x element-wise and returns a tensor with the same type x. Here is the parameter description:

- `x`: A tensor or sparse tensor with type `float`, `double`, `int32`, `complex64`, `int64`, or `qint32`.
- `name`: A name for the operation (optional).

Using ReLU

In TensorFlow, the signature `tf.nn.relu(features, name=None)` computes rectified linear using `max(features, 0)` and returns a tensor with the same type of features. Here is the parameter description:

- `features`: A tensor. Must be one of the following types: `float32`, `float64`, `int32`, `int64`, `uint8`, `int16`, `int8`, `uint16`, or `half`
- `name`: A name for the operation (optional)

Using softmax

In TensorFlow, the signature `tf.nn.softmax(logits, axis=None, name=None)` computes `softmax` activations and returns a tensor having the same type and shape as `logits`. Here is the parameter description:

- `logits`: A non-empty tensor. Must be one of the following types: `half`, `float32`, or `float64`
- `axis`: The dimension `softmax` would be performed on. The default is -1, which indicates the last dimension
- `name`: The name for the operation (optional)

This `softmax` function performs the equivalent of `softmax = tf.exp(logits) / tf.reduce_sum(tf.exp(logits), axis)`.

Implementing a feed-forward neural network

Automatic recognition of handwritten digits is an important problem, which can be found in many practical applications. In this section, we will implement a feed-forward network to address this.

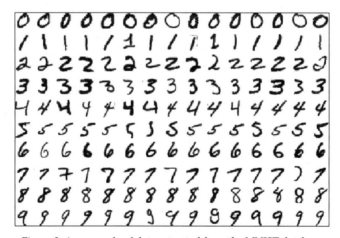

Figure 3: An example of data extracted from the MNIST database

To train, and test, the implemented models, we will be using one of the most famous datasets called MNIST of handwritten digits. The MNIST dataset is a training set of 60,000 examples and a test set of 10,000 examples. An example of the data, as it is stored in the files of the examples, is shown in the preceding figure.

The source images were originally in black and white. Later, to normalize them to the size of 20×20 pixels, intermediate brightness levels were introduced, due to the effect of the anti-aliasing filter for resizing. Subsequently, the images were focused in the center of mass of the pixels, in an area of 28×28 pixels, in order to improve the learning process. The entire database is stored in four files:

- `train-images-idx3-ubyte.gz`: Training set images (9912422 bytes)
- `train-labels-idx1-ubyte.gz`: Training set labels (28881 bytes)
- `t10k-images-idx3-ubyte.gz`: Test set images (1648877 bytes)
- `t10k-labels-idx1-ubyte.gz`: Test set labels (4542 bytes)

Each database consists of two files; the first contains the images, while the second contains the respective labels.

Exploring the MNIST dataset

Let's look at a short example of how to access the MNIST data, and how to display a selected image. For this just execute the `Explore_MNIST.py` script. First, we must import the numpy, because we have to do some image manipulation:

```
import numpy as np
```

The `pyplot` function in Matplotlib is used for drawing the images:

```
import matplotlib.pyplot as plt
```

We will use the `input_data` class from the `tensorflow.examples.tutorials.mnist` that allows us to download the MNIST database and build the dataset:

```
import tensorflow as tf
from tensorflow.examples.tutorials.mnist import input_data
```

Then we load the dataset using the `read_data_sets` method:

```
import os
dataPath = "temp/"
if not os.path.exists(dataPath):
    os.makedirs(dataPath)
input = input_data.read_data_sets(dataPath, one_hot=True)
```

The images will be saved in the `temp/` directory. Now let's see the shape of the images and labels:

```
print(input.train.images.shape)
print(input.train.labels.shape)
print(input.test.images.shape)
print(input.test.labels.shape)
```

Chapter 3

The following is the output of the preceding code:

```
>>>
(55000, 784)
(55000, 10)
(10000, 784)
(10000, 10)
```

Using the Python library, matplotlib, we want to visualize a single digit:

```
image_0 = input.train.images[0]
image_0 = np.resize(image_0,(28,28))
label_0 = input.train.labels[0]
print(label_0)
```

The following is the output of the preceding code:

```
>>>
[ 0.  0.  0.  0.  0.  0.  0.  1.  0.  0.]
```

The number 1 is the eighth position of the array. This means that the figure for our image is the digit 7. Finally, we must verify that the digit is really 7. We can use the imported plt function, to draw the image_0 tensor:

```
plt.imshow(image_0, cmap='Greys_r')
plt.show()
```

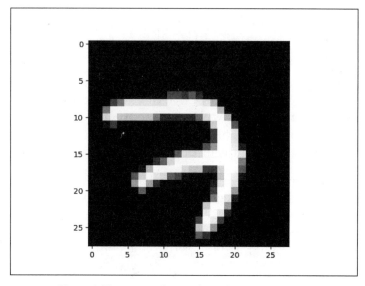

Figure 4: The extracted image from the MNIST dataset

Softmax classifier

In the previous section, we showed how to access and manipulate MNIST dataset. In this section, we will see how to use the preceding dataset to address the classification problem of handwritten digits with TensorFlow. We will apply the concepts learned to build more models of neural networks, in order to assess and compare the results of the different approaches followed.

The first feed-forward network architecture that will be implemented is represented in the following figure:

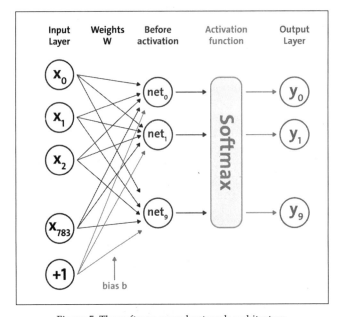

Figure 5: The softmax neural network architecture

We will construct a five-layer network: layers one to four are sigmoid and layer five is softmax activation function. Remember that this network is defined so that its activation is a set of positive values, with a total sum equal to 1. This means that the jth value of the output is the probability that j is the class that corresponds to the network input. Let's see how to implement our neural network model.

To determine the appropriate size of the network (that is, the number of neurons or units in a layer), that is, the number of hidden layers and the number of neurons per layer, typically we rely on general empirical criteria, the personal experience, or appropriate tests. These are a few hyperparameters to be tuned into. Later in this chapter, we will see some examples of hyperparameter optimization.

The following table summarizes the implemented network architecture. It shows the number of neurons per layer, and the respective activation functions:

Layer	Number of neurons	Activation function
First	L = 200	Sigmoid
Second	M = 100	Sigmoid
Third	N = 60	Sigmoid
Fourth	O = 30	Sigmoid
Fifth	10	Softmax

The activation function for the first four layers is the **sigmoid** function. The last layer of the activation function is always the **softmax** since the output of the network must express a probability for the input digit. In general, the number and the size, of the intermediate layers greatly affect the network performance:

- In a positive way because on these layers is based the ability of the network to generalize, and to detect peculiar characteristics of the input
- In a negative way because if the network is redundant then it unnecessarily weighs down the learning phase

For this just execute the `five_layers_sigmoid.py` script. First, we will start to implement the network by importing the following libraries:

```
import tensorflow as tf
from tensorflow.examples.tutorials.mnist import input_data
import math
from tensorflow.python.framework import ops
import random
import os
```

Next, we will set the following configuration parameters:

```
logs_path = 'log_sigmoid/' # logging path
batch_size = 100 # batch size while performing training
learning_rate = 0.003 # Learning rate
training_epochs = 10 # training epoch
display_epoch = 1
```

Feed-Forward Neural Networks with TensorFlow

We will then download images and labels, and prepare the dataset:

```
dataPath = "temp/"
if not os.path.exists(dataPath):
    os.makedirs(dataPath)
mnist = input_data.read_data_sets(dataPath, one_hot=True) # MNIST to be downloaded
```

Starting with the input layer, we will now see how to build the network's architecture. The input layer is now a tensor of the shape [1×784] –that is [1, 28 * 28], which represents the image to classify:

```
X = tf.placeholder(tf.float32, [None, 784], name='InputData') # image shape 28*28=784
XX = tf.reshape(X, [-1, 784]) # reshape input
Y_ = tf.placeholder(tf.float32, [None, 10], name='LabelData') # 0-9 digits => 10 classes
```

The first layer receives the pixels of the input image to be classified, combined with the W1 weight connections, and added to the respective values of the B1 biases tensor:

```
W1 = tf.Variable(tf.truncated_normal([784, L], stddev=0.1)) # Initialize random weights for the hidden layer 1
B1 = tf.Variable(tf.zeros([L])) # Bias vector for layer 1
```

The first layer sends its output to the second layer, through the sigmoid activation function:

```
Y1 = tf.nn.sigmoid(tf.matmul(XX, W1) + B1) # Output from layer 1
```

The second layer receives the Y1 output from the first layer, combines it with the W2 weight connections, and adds it to the respective values of the B2 biases tensor:

```
W2 = tf.Variable(tf.truncated_normal([L, M], stddev=0.1))
# Initialize random weights for the hidden layer 2
B2 = tf.Variable(tf.ones([M])) # Bias vector for layer 2
```

The second layer sends its output to the third layer, through the sigmoid activation function:

```
Y2 = tf.nn.sigmoid(tf.matmul(Y1, W2) + B2) # Output from layer 2
```

The third layer receives the Y2 output from the second layer, combines it with the W3 weight connections, and adds it to the respective values of the B3 biases tensor:

```
W3 = tf.Variable(tf.truncated_normal([M, N], stddev=0.1))
# Initialize random weights for the hidden layer 3
B3 = tf.Variable(tf.ones([N])) # Bias vector for layer 3
```

The third layer sends its output to the fourth layer, through the sigmoid activation function:

```
Y3 = tf.nn.sigmoid(tf.matmul(Y2, W3) + B3) # Output from layer 3
```

The fourth layer receives the Y3 output from the third layer, combines it with the W4 weight connections, and adds it to the respective values of the B4 biases tensor:

```
W4 = tf.Variable(tf.truncated_normal([N, O], stddev=0.1))
# Initialize random weights for the hidden layer 4
B4 = tf.Variable(tf.ones([O])) # Bias vector for layer 4
```

The output of fourth layer is then propagated to the fifth layer, through the sigmoid activation function:

```
Y4 = tf.nn.sigmoid(tf.matmul(Y3, W4) + B4) # Output from layer 4
```

The fifth layer will receive in input the O = 30 stimuli, coming from the fourth layer that will be converted in the respective classes of probability for each number, through the softmax activation function:

```
W5 = tf.Variable(tf.truncated_normal([O, 10], stddev=0.1))
# Initialize random weights for the hidden layer 5
B5 = tf.Variable(tf.ones([10])) # Bias vector for layer 5
Ylogits = tf.matmul(Y4, W5) + B5 # computing the logits
Y = tf.nn.softmax(Ylogits)
# output from layer 5
```

Here, our loss function is the cross-entropy between the target and the softmax activation function, applied to the model's prediction:

```
cross_entropy =
tf.nn.softmax_cross_entropy_with_logits_v2(logits=Ylogits,
labels=Y) # final outcome using softmax cross entropy
cost_op = tf.reduce_mean(cross_entropy)*100
```

In addition, we define the correct_prediction and the model's accuracy:

```
correct_prediction = tf.equal(tf.argmax(Y, 1), tf.argmax(Y_, 1))
accuracy = tf.reduce_mean(tf.cast(correct_prediction, tf.float32))
```

Now we need to use an optimizer to reduce the training error. The AdamOptimizer offers several advantages over the simple GradientDescentOptimizer. In fact, it uses a larger effective step size, and the algorithm will converge to this step size without fine-tuning:

```
# Optimization op (backprop)
train_op =
tf.train.AdamOptimizer(learning_rate).minimize(cost_op)
```

The `Optimizer` base class provides methods to compute gradients for a loss and apply gradients to variables. A collection of subclasses implements classic optimization algorithms, such as `GradientDescent` and `Adagrad`. While training a NN model in TensorFlow, we never instantiate the `Optimizer` class itself, but instead instantiate one of the following subclasses:

- `tf.train.Optimizer` (https://www.tensorflow.org/api_docs/python/tf/train/Optimizer)
- `tf.train.GradientDescentOptimizer` (https://www.tensorflow.org/api_docs/python/tf/train/GradientDescentOptimizer)
- `tf.train.AdadeltaOptimizer` (https://www.tensorflow.org/api_docs/python/tf/train/AdadeltaOptimizer)
- `tf.train.AdagradOptimizer` (https://www.tensorflow.org/api_docs/python/tf/train/AdagradOptimizer)
- `tf.train.AdagradDAOptimizer` (https://www.tensorflow.org/api_docs/python/tf/train/AdagradDAOptimizer)
- `tf.train.MomentumOptimizer` (https://www.tensorflow.org/api_docs/python/tf/train/MomentumOptimizer)
- `tf.train.AdamOptimizer` (https://www.tensorflow.org/api_docs/python/tf/train/AdamOptimizer)
- `tf.train.FtrlOptimizer` (https://www.tensorflow.org/api_docs/python/tf/train/FtrlOptimizer)
- `tf.train.ProximalGradientDescentOptimizer` (https://www.tensorflow.org/api_docs/python/tf/train/ProximalGradientDescentOptimizer)
- `tf.train.ProximalAdagradOptimizer` (https://www.tensorflow.org/api_docs/python/tf/train/ProximalAdagradOptimizer)
- `tf.train.RMSPropOptimizer` (https://www.tensorflow.org/api_docs/python/tf/train/RMSPropOptimizer)

See https://www.tensorflow.org/api_guides/python/train and `tf.contrib.opt` (https://www.tensorflow.org/api_docs/python/tf/contrib/opt) for more optimizers.

Then let's construct a model encapsulating all ops into scopes, making TensorBoard's graph visualization more convenient:

```
# Create a summary to monitor cost tensor
tf.summary.scalar("cost", cost_op)
# Create a summary to monitor accuracy tensor
tf.summary.scalar("accuracy", accuracy)
# Merge all summaries into a single op
summary_op = tf.summary.merge_all()
```

Finally, we'll start the training:

```
with tf.Session() as sess:
        # Run the initializer
    sess.run(init_op)

    # op to write logs to TensorBoard
    writer = tf.summary.FileWriter(logs_path,
graph=tf.get_default_graph())

    for epoch in range(training_epochs):
        batch_count = int(mnist.train.num_examples/batch_size)
        for i in range(batch_count):
            batch_x, batch_y = mnist.train.next_batch(batch_size)
            _, summary = sess.run([train_op, summary_op],
feed_dict={X: batch_x, Y_: batch_y})
            writer.add_summary(summary, epoch * batch_count + i)

        print("Epoch: ", epoch)
    print("Optimization Finished!")

    print("Accuracy: ", accuracy.eval(feed_dict={X:
mnist.test.images, Y_: mnist.test.labels}))
```

The source code for the definition of the summaries and the running of the session is almost identical to the previous one. We can move directly to evaluating the implemented model. When running the model, we have the following output:

The final test set accuracy after running this code should be approximately 97%:

```
Extracting temp/train-images-idx3-ubyte.gz
Extracting temp/train-labels-idx1-ubyte.gz
Extracting temp/t10k-images-idx3-ubyte.gz
Extracting temp/t10k-labels-idx1-ubyte.gz
Epoch:   0
```

```
Epoch:   1
Epoch:   2
Epoch:   3
Epoch:   4
Epoch:   5
Epoch:   6
Epoch:   7
Epoch:   8
Epoch:   9
Optimization Finished!
Accuracy:    0.9715
```

We can now move onto TensorBoard by simply opening Terminal in the running folder, then performing this command:

```
$> tensorboard --logdir='log_sigmoid/' # if required, provide
absolute path
```

We then open our browser at localhost. In the following figure, we show the trend of the cost function, as a function of the number of examples, over the training set, and the accuracy on the test set:

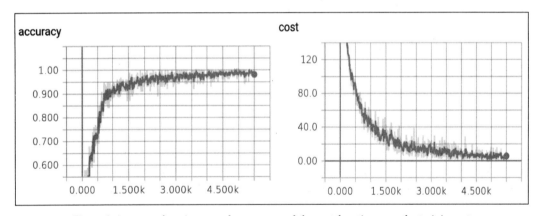

Figure 6: Accuracy function over the test set, and the cost function over the training set

The cost function decreases with increasing iterations. If this does not happen, it means that something went wrong. In the best-case scenario, this could simply be because some parameters have not been set properly. At worst, there could be a problem in the constructed dataset, for example, there could be too little information or poor-quality images. If this happens, we must directly fix the dataset.

So far, we have seen an implementation of an FFNN. However, it would be great to explore the more useful implementations of FFNNs using real-life datasets. We will start with MLP.

Implementing a multilayer perceptron (MLP)

A perceptron is composed of a single layer of LTUs, with each neuron connected to all the inputs. These connections are often represented using special pass-through neurons called **input neurons**: they just output whatever input they are fed. Moreover, an extra bias feature is generally added ($x0 = 1$).

This bias feature is typically represented using a special type of neuron called a **bias neuron**, which just outputs 1 all the time. A perceptron with two inputs and three outputs is represented in Figure 7. This perceptron can simultaneously classify instances into three different binary classes, which makes it a multioutput classifier:

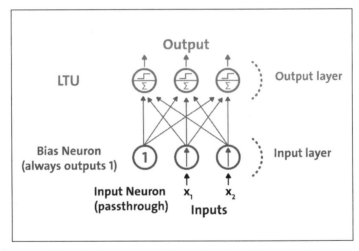

Figure 7: A perceptron with two inputs and three outputs

Since the decision boundary of each output neuron is linear, perceptrons are incapable of learning complex patterns. However, if the training instances are linearly separable, research has shown that this algorithm will converge to a solution called "perceptron convergence theorem."

An MLP is an FFNN, which means that it is the only connection between neurons from different layers. More specifically, an MLP is composed of one (pass through) input layer, one or more layers of LTUs (called **hidden layers**), and one final layer of LTUs called the output layer. Every layer, except the output layer, includes a bias neuron, and is connected to the next layer as a fully connected bipartite graph:

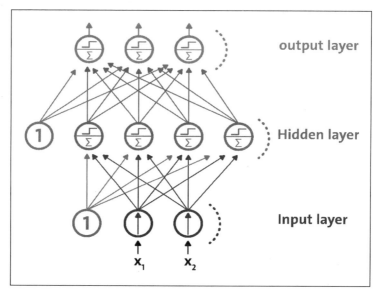

Figure 8: An MLP is composed of one input layer, one hidden layer, and an output layer

Training an MLP

An MLP was trained successfully using the backpropagation training algorithm for the first time in 1986. However, nowadays the optimized version of this algorithm is called gradient descent. During the training phase, for each training instance, the algorithm feeds it to the network and computes the output of every neuron, in each consecutive layer.

Training the algorithm measures the network's output error (that is, the difference between the desired output and the actual output of the network), and it computes how much each neuron in the last hidden layer contributed to each output neuron's error. It then proceeds to measure how much of these error contributions came from each neuron in the previously hidden layer, and so on until the algorithm reaches the input layer. This reverse pass efficiently measures the error gradient across all the connection weights in the network, by propagating the error gradient backward in the network.

More technically, the calculation of the gradient of the cost function for each layer is done by the backpropagation method. The idea of gradient descent is to have a cost function that shows the difference between the predicted outputs of some neural network, with the actual output:

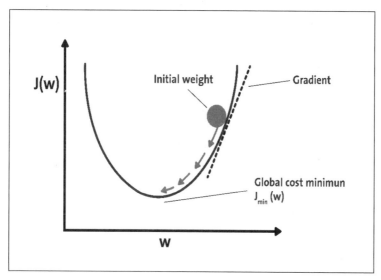

Figure 9: Sample implementation of an ANN for unsupervised learning

There are several known types of the cost function, such as the squared error function and the log-likelihood function. The choice for this cost function can be based on many factors. The gradient descent method optimizes the network's weight, by minimizing this cost function. The steps are as follows:

1. Weight initialization
2. Calculation of a neural network's predicted output, which is usually called the **forwarding propagation step**
3. Calculation of cost/loss function. Some common cost/loss functions include the **log-likelihood function** and the **squared error function**
4. Calculation of the gradient of the cost/lost function. For most DNN architecture, the most common method is backpropagation
5. Weight update based on the current weight, and the gradient of the cost/loss function
6. Iteration of steps two to five, until the cost function, reaches a certain threshold or after a certain amount of iteration

An illustration of gradient descent can be seen in Figure 9. The graph shows a neural network's cost function based on the network's weight. In the first iteration of gradient descent, we apply the cost function on some random initial weight. With each iteration, we update the weight in the direction of the gradient, which corresponds to the arrows in Figure 9. The weight update is repeated until a certain number of iterations or until the cost function reaches a certain threshold.

Using MLPs

Multilayer perceptrons are commonly used for solving classification and regression problems in a supervised way. Although CNNs have gradually replaced their implementation in image and video data, a low dimensional and numerical feature MLP still can be used effectively: both the binary and multiclass classification problems can be solved.

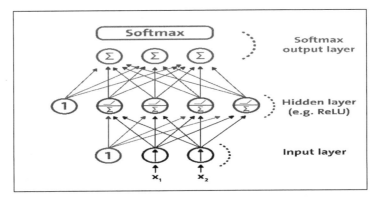

Figure 10: A modern MLP (including ReLU and softmax) for classification

However, for multiclass classification tasks and training, the output layer is typically modified, by replacing the individual activation functions with a shared softmax function. The output of each neuron corresponds to the estimated probability of the corresponding class. Note that the signal flows from the input to output in one direction only, so this architecture is an example of an FFNN.

As a case study, we will be using bank-marketing datasets. The data is related to the direct marketing campaigns of a Portuguese banking institution. The marketing campaigns were based on phone calls. Often, the same client was contacted more than once, in order to assess whether the product (bank term deposit) would be (yes) or would not be (no) subscribed. The target is to use MLP to predict whether the client will subscribe a term deposit (variable y), that is, a binary classification problem.

Dataset description

There are two sources that I would like to acknowledge here. This dataset was used in a research paper published by *Moro and others*: *A data-driven approach to predict the success of bank telemarketing, Decision support systems, Elsevier, June 2014*. Later on, it was donated to the UCI Machine Learning Repository, which can be downloaded from `https://archive.ics.uci.edu/ml/datasets/bank+marketing`.

According to the dataset description, there are four datasets:

- `bank-additional-full.csv`: This includes all examples (41,188) and 20 inputs, which are ordered by date (from May 2008 to November 2010). This data is very close to the data analyzed by Moro and others
- `bank-additional.csv`: This includes 10% of the examples (4119), randomly selected from 1, and 20 inputs
- `bank-full.csv`: This includes all the examples and 17 inputs, ordered by date (an older version of this dataset with fewer inputs)
- `bank.csv`: This includes 10% of the examples and 17 inputs, randomly selected from 3 (the older version of this dataset with fewer inputs)

There are 21 attributes in the dataset. The independent variables can be further categorized as bank-client-related data (attributes 1 to 7), related to the last contact from the current campaign (attributes 8 to 11). Other attributes (attributes 12 to 15), and social and economic context attributes (attributes 16 to 20) are categorized. The dependent variable is specified by y, the last attribute (21):

ID	Attribute	Explanation
1	`age`	Age in numbers.
2	`job`	This is the type of job in a categorical format with these possible values: `admin`, `blue-collar`, `entrepreneur`, `housemaid`, `management`, `retired`, `self-employed`, `services`, `student`, `technician`, `unemployed`, and `unknown`.
3	`marital`	This is the marital status in a categorical format with possible values: `divorced` (or `widowed`), `married`, `single`, and `unknown`.
4	`education`	This is the educational background in a categorical format with possible values as follows: `basic.4y`, `basic.6y`, `basic.9y`, `high.school`, `illiterate`, `professional.course`, `university.degree`, and `unknown`.
5	`default`	This is a categorical format with possible values in credit in default: `no`, `yes` and `unknown`.

ID	Attribute	Explanation
6	housing	Has the customer had a housing loan?
7	loan	The personal loan in a categorical format with possible values: no, yes, and unknown.
8	contact	This is the communication type in a categorical format with possible values: cellular or telephone.
9	month	This is the last contact month of the year in a categorical format with possible values: jan, feb, mar, ... nov, and dec.
10	day_of_week	This is the last contact day of the week in a categorical format with possible values: mon, tue, wed, thu, and fri.
11	duration	This is the last contact duration in seconds (numerical value). This attribute highly affects the output target (for example, if duration=0, then y=no). Yet, the duration is not known before a call is performed. In addition, after the end of the call, y is obviously known. Thus, this input should only be included for benchmark purposes and should be discarded if the intention is to have a realistic predictive model.
12	campaign	This is the number of contacts made during this campaign and for this client.
13	pdays	This is the number of days that passed by after the client was last contacted by a previous campaign (numeric - 999 means the client was not previously contacted).
14	previous	This is the number of contacts performed before this campaign and for this client (numeric).
15	poutcome	The outcome of the previous marketing campaign (categorical: failure, nonexistent, and success).
16	emp.var.rate	This is the employment variation rate and quarterly indicator (numeric).
17	cons.price.idx	This is the consumer price index and monthly indicator (numeric).
18	cons.conf.idx	This is the consumer confidence index and monthly indicator (numeric).
19	euribor3m	This is the euribor 3-month rate and daily indicator (numeric).
20	nr.employed	This is the number of employees and quarterly indicator (numeric).
21	y	Signifies if the client subscribed a term deposit, with the possible binary: yes and no values.

Preprocessing

You can see that the dataset is not ready to feed to your MLP, or DBN classifier, directly since the feature is mixed with numerical and categorical values. In addition, the outcome variable is in categorical value. Therefore, we need to convert the categorical values into numerical values, so that the feature and the outcome variable are in numerical form. The next step shows this process. Refer to the `preprocessing_b.py` file for this preprocessing.

Firstly, we must load the required packages and libraries needed for the preprocessing:

```
import pandas as pd
import numpy as np
from sklearn import preprocessing
```

Then download the data file from the afore-mentioned URL and place it in your convenient place – say `input`:

Then, we'll load and parse the dataset:

```
data = pd.read_csv('input/bank-additional-full.csv', sep = ";")
```

Next, we'll extract variable names:

```
var_names = data.columns.tolist()
```

Now, based on the dataset description in Table 1, we'll extract the categorical variables:

```
categs =
['job','marital','education','default','housing','loan','contact',
'month','day_of_week','duration','
poutcome','y']
```

Then, we'll extract the quantitative variables:

```
# Quantitative vars
quantit = [i for i in var_names if i not in categs]
```

Then let's get the dummy variables for categorical variables:

```
job = pd.get_dummies(data['job'])
marital = pd.get_dummies(data['marital'])
education = pd.get_dummies(data['education'])
default = pd.get_dummies(data['default'])
housing = pd.get_dummies(data['housing'])
loan = pd.get_dummies(data['loan'])
contact = pd.get_dummies(data['contact'])
month = pd.get_dummies(data['month'])
```

```
day = pd.get_dummies(data['day_of_week'])
duration = pd.get_dummies(data['duration'])
poutcome = pd.get_dummies(data['poutcome'])
```

Now, it's time to map variables to predict:

```
dict_map = dict()
y_map = {'yes':1,'no':0}
dict_map['y'] = y_map
data = data.replace(dict_map)
label = data['y']
df_numerical = data[quantit]
df_names = df_numerical .keys().tolist()
```

Once we have converted the categorical variables into numerical variables, the next task is to normalize the numerical variables too. So, using the normalization, we scale an individual sample to have unit norm. This process can be useful if you plan to use a quadratic form such as the dot product, or any other kernel, to quantify the similarity of any pair of samples. This assumption is the basis of the vector space model (https://en.wikipedia.org/wiki/Vector_space_model) often used in text classification and clustering contexts.

So, let's scale the quantitative variables:

```
min_max_scaler = preprocessing.MinMaxScaler()
x_scaled = min_max_scaler.fit_transform(df_numerical)
df_temp = pd.DataFrame(x_scaled)
df_temp.columns = df_names
```

Now that we have the temporary data frame for the (original) numerical variables, the next task is to combine all the data frames together and generate the normalized data frame. We will use pandas for this:

```
normalized_df = pd.concat([df_temp,
                job,
                marital,
                education,
                default,
                housing,
                loan,
                contact,
                month,
                day,
                poutcome,
                duration,
                label], axis=1)
```

Finally, we need to save the resulting data frame in a CSV file as follows:

```
normalized_df.to_csv('bank_normalized.csv', index = False)
```

A TensorFlow implementation of MLP for client-subscription assessment

For this example, we will be using the bank marketing dataset that we have normalized in the previous example. There are several steps to follow. To begin with, we need to import TensorFlow, and the other necessary packages and modules:

```
import tensorflow as tf
import pandas as pd
import numpy as np
import os
from sklearn.cross_validation import train_test_split # for random split of train/test
```

Now, we need to load the normalized bank marketing dataset, where all the features and the labels are numeric. For this we use the `read_csv()` method from the pandas library:

```
FILE_PATH = 'bank_normalized.csv'          # Path to .csv dataset
raw_data = pd.read_csv(FILE_PATH)          # Open raw .csv
print("Raw data loaded successfully...\n")
```

The following is the output of the preceding code:

```
>>>
Raw data loaded successfully...
```

As mentioned in the previous section, tuning the hyperparameters for DNNs is not straightforward. However, it often depends on the dataset that you are handling. For some datasets, a possible workaround is setting these values based on dataset-related statistics, for example, the number of training instances, the input size, and the number of classes.

DNNs are not suitable for very small and low-dimensional datasets. In these cases, a better option is to use the linear models instead. To get started, let us put a pointer to the label column itself, compute the number of instances and number of classes, and define the train/test split ratio as follows:

```
Y_LABEL = 'y'     # Name of the variable to be predicted
KEYS = [i for i in raw_data.keys().tolist() if i != Y_LABEL]
    # Name of predictors
N_INSTANCES = raw_data.shape[0]           # Number of instances
```

```python
N_INPUT = raw_data.shape[1] - 1          # Input size
N_CLASSES = raw_data[Y_LABEL].unique().shape[0] # Number of classes
TEST_SIZE = 0.25          # Test set size (% of dataset)
TRAIN_SIZE = int(N_INSTANCES * (1 - TEST_SIZE))  # Train size
```

Now, let's see the statistics of the dataset that we are going to use to train the MLP model:

```python
print("Variables loaded successfully...\n")
print("Number of predictors \t%s" %(N_INPUT))
print("Number of classes \t%s" %(N_CLASSES))
print("Number of instances \t%s" %(N_INSTANCES))
print("\n")
```

The following is the output of the preceding code:

```
>>>
Variables loaded successfully...
Number of predictors     1606
Number of classes        2
Number of instances      41188
```

The next task is to define the other parameters such as learning rate, training epochs, batch size, and the standard deviation for the weights. Usually, a low value of training rate will help your DNN to learn more slowly, but intensively. Note that we need to define more parameters, such as the number of hidden layers, and the activation function.

```python
LEARNING_RATE = 0.001     # learning rate
TRAINING_EPOCHS = 1000    # number of training epoch for the forward pass
BATCH_SIZE = 100     # batch size to be used during training
DISPLAY_STEP = 20    # print the error etc. at each 20 step
HIDDEN_SIZE = 256    # number of neurons in each hidden layer
# We use tanh as the activation function, but you can try using ReLU as well
ACTIVATION_FUNCTION_OUT = tf.nn.tanh
STDDEV = 0.1         # Standard Deviations
RANDOM_STATE = 100
```

The preceding initialization is set on a trial-and-error basis. Therefore, depending on your use case and data types, set them wisely but we will provide some guidelines later in this chapter. In addition, for the preceding code, RANDOM_STATE is used to signify random state for the train and test split. At first, we separate the raw features and the labels:

```
data = raw_data[KEYS].get_values()      # X data
labels = raw_data[Y_LABEL].get_values()  # y data
```

Now that we have the labels, they have to be coded:

```
labels_ = np.zeros((N_INSTANCES, N_CLASSES))
labels_[np.arange(N_INSTANCES), labels] = 1
```

Finally, we must split the training and test sets. As mentioned earlier, we'll keep 75% of the input for training and the remaining 25% for the test set:

```
data_train, data_test, labels_train, labels_test =
train_test_split(data,labels_,test_size = TEST_SIZE,random_state =
RANDOM_STATE)
print("Data loaded and splitted successfully...\n")
```

The following is the output of the preceding code:

```
>>>
Data loaded and splitted successfully
```

Since this is a supervised classification problem, we should have placeholders for the features and the labels:

As mentioned previously, an MLP is composed of one input layer, several hidden layers, and one final layer of LTUs called the output layer. For this example, I am going to incorporate the training with four hidden layers. Thus, we are calling our classifier a deep feed-forward MLP. Note that we also need to have the weight in each layer (except in the input layer), and the bias in each layer (except the output layer). Usually, each hidden layer includes a bias neuron, and is fully connected to the next layer as a fully-connected bipartite graph (feed-forward) from one hidden layer to another. So, let's define the size of the hidden layers:

```
n_input = N_INPUT                  # input n labels
n_hidden_1 = HIDDEN_SIZE           # 1st layer
n_hidden_2 = HIDDEN_SIZE           # 2nd layer
n_hidden_3 = HIDDEN_SIZE           # 3rd layer
n_hidden_4 = HIDDEN_SIZE           # 4th layer
n_classes = N_CLASSES              # output m classes
```

Feed-Forward Neural Networks with TensorFlow

Since this is a supervised classification problem, we should have placeholders for the features and the labels:

```
# input shape is None * number of input
X = tf.placeholder(tf.float32, [None, n_input])
```

The first dimension of the placeholder is None, meaning we can have any number of rows. The second dimension is fixed at number of features, meaning each row needs to have that number of columns of features.

```
# label shape is None * number of classes
y = tf.placeholder(tf.float32, [None, n_classes])
```

Additionally, we need another placeholder for dropout, which is implemented by only keeping a neuron active with some probability (say p < 1.0, or setting it to zero otherwise). Note that this is also hyperparameters to be tuned and the training time, but not the test time:

```
dropout_keep_prob = tf.placeholder(tf.float32)
```

Using the scaling given here enables the same network to be used for training (with dropout_keep_prob < 1.0) and evaluation (with dropout_keep_prob == 1.0). Now, we can define a method that implements the MLP classifier. For this, we are going to provide four parameters such as input, weight, biases, and the drop out probability as follows:

```
def DeepMLPClassifier(_X, _weights, _biases, dropout_keep_prob):
    layer1 = tf.nn.dropout(tf.nn.tanh(tf.add(tf.matmul(_X, _weights['h1']), _biases['b1'])), dropout_keep_prob)
    layer2 = tf.nn.dropout(tf.nn.tanh(tf.add(tf.matmul(layer1, _weights['h2']), _biases['b2'])), dropout_keep_prob)
    layer3 = tf.nn.dropout(tf.nn.tanh(tf.add(tf.matmul(layer2, _weights['h3']), _biases['b3'])), dropout_keep_prob)
    layer4 = tf.nn.dropout(tf.nn.tanh(tf.add(tf.matmul(layer3, _weights['h4']), _biases['b4'])), dropout_keep_prob)
    out = ACTIVATION_FUNCTION_OUT(tf.add(tf.matmul(layer4, _weights['out']), _biases['out']))
    return out
```

The return value of the preceding method is the output of the activation function. The preceding method is a stub implementation that did not tell anything concrete about the weights and biases, so before we start the training, we should have them defined:

```
weights = {
    'w1': tf.Variable(tf.random_normal([n_input, n_hidden_1], stddev=STDDEV)),
```

```
        'w2': tf.Variable(tf.random_normal([n_hidden_1,
n_hidden_2],stddev=STDDEV)),
        'w3': tf.Variable(tf.random_normal([n_hidden_2, n_
hidden_3],stddev=STDDEV)),
        'w4': tf.Variable(tf.random_normal([n_hidden_3, n_
hidden_4],stddev=STDDEV)),
        'out': tf.Variable(tf.random_normal([n_hidden_4, n_
classes],stddev=STDDEV)),
    }
    biases = {
        'b1': tf.Variable(tf.random_normal([n_hidden_1])),
        'b2': tf.Variable(tf.random_normal([n_hidden_2])),
        'b3': tf.Variable(tf.random_normal([n_hidden_3])),
        'b4': tf.Variable(tf.random_normal([n_hidden_4])),
        'out': tf.Variable(tf.random_normal([n_classes]))
    }
```

Now we can invoke the preceding implementation of the MLP with real arguments (an input layer, weights, biases, and the drop out) keeping probability as follows:

```
pred = DeepMLPClassifier(X, weights, biases, dropout_keep_prob)
```

We have built the MLP model and it's time to train the network itself. At first, we need to define the cost op and then we will use Adam optimizer, which will learn slowly and try to reduce the training loss as much as possible:

```
cost =
tf.reduce_mean(tf.nn.softmax_cross_entropy_with_logits_v2(logits=
pred, labels=y))

# Optimization op (backprop)
optimizer = tf.train.AdamOptimizer(learning_rate =
LEARNING_RATE).minimize(cost_op)
```

Next, we need to define additional parameters for computing the classification accuracy:

```
correct_prediction = tf.equal(tf.argmax(pred, 1), tf.argmax(y, 1))
accuracy = tf.reduce_mean(tf.cast(correct_prediction, tf.float32))
print("Deep MLP networks has been built successfully...")
print("Starting training...")
```

After that, we need to initialize all the variables and placeholders, before launching a TensorFlow session:

```
init_op = tf.global_variables_initializer()
```

Feed-Forward Neural Networks with TensorFlow

Now, we are very closer to starting the training, but before that, the last step is to create a TensorFlow session and launch it as follows:

```
sess = tf.Session()
sess.run(init_op)
```

Finally, we are ready to start training our MLP on the training set. We'll iterate over all the batches and fit using the batched data to compute the average training cost. Nevertheless, it would be great to show the training cost and accuracy for each epoch:

```
for epoch in range(TRAINING_EPOCHS):
    avg_cost = 0.0
    total_batch = int(data_train.shape[0] / BATCH_SIZE)
    # Loop over all batches
    for i in range(total_batch):
        randidx = np.random.randint(int(TRAIN_SIZE), size = BATCH_SIZE)
        batch_xs = data_train[randidx, :]
        batch_ys = labels_train[randidx, :]
        # Fit using batched data
        sess.run(optimizer, feed_dict={X: batch_xs, y: batch_ys, dropout_keep_prob: 0.9})
        # Calculate average cost
        avg_cost += sess.run(cost, feed_dict={X: batch_xs, y: batch_ys, dropout_keep_prob:1.})/total_batch
    # Display progress
    if epoch % DISPLAY_STEP == 0:
        print("Epoch: %3d/%3d cost: %.9f" % (epoch, TRAINING_EPOCHS, avg_cost))
        train_acc = sess.run(accuracy, feed_dict={X: batch_xs, y: batch_ys, dropout_keep_prob:1.})
        print("Training accuracy: %.3f" % (train_acc))
print("Your MLP model has been trained successfully.")
```

The following is the output of the preceding code:

```
>>>
Starting training...
Epoch:   0/1000 cost: 0.356494816
Training accuracy: 0.920

...

Epoch: 180/1000 cost: 0.350044933
Training accuracy: 0.860

….
```

[108]

```
Epoch: 980/1000 cost: 0.358226758
Training accuracy: 0.910
```

Well done, our MLP model has been trained successfully! Now, what if we see the cost and the accuracy graphically? Let's try it out:

```
# Plot loss over time
plt.subplot(221)
plt.plot(i_data, cost_list, 'k--', label='Training loss',
linewidth=1.0)
plt.title('Cross entropy loss per iteration')
plt.xlabel('Iteration')
plt.ylabel('Cross entropy loss')
plt.legend(loc='upper right')
              plt.grid(True)
```

The following is the output of the preceding code:

```
>>>
```

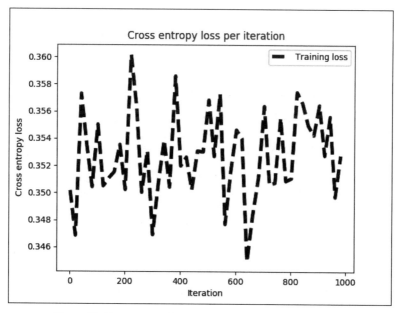

Figure 11: Cross entropy loss per iteration in the training phase

Feed Forward Neural Networks with TensorFlow

The preceding figure shows that the cross-entropy loss is more or less stable between **0.34** and **0.36**, but with a little fluctuation. Now, let's see how this affects the training accuracy overall:

```
# Plot train and test accuracy
plt.subplot(222)
plt.plot(i_data, acc_list, 'r--', label='Accuracy on the training
set', linewidth=1.0)
plt.title('Accuracy on the training set')
plt.xlabel('Iteration')
plt.ylabel('Accuracy')
plt.legend(loc='upper right')
plt.grid(True)
plt.show()
```

The following is the output of the preceding code:

>>>

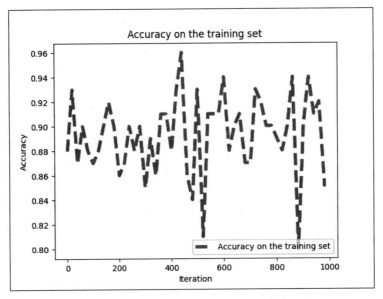

Figure 12: Accuracy on the training set on each iteration

We can see that the training accuracy fluctuates between 79% and 96%, but does not increase or decrease uniformly. One possible way around this is to add more hidden layers and use different optimizers, such as gradient descent, which was discussed earlier in this chapter. We will increase the dropout probability to 100%, that is, 1.0. The reason is to have the same network used for testing as well:

```
print("Evaluating MLP on the test set...")
test_acc = sess.run(accuracy, feed_dict={X: data_test, y:
labels_test, dropout_keep_prob:1.})
print ("Prediction/classification accuracy: %.3f" % (test_acc))
```

The following is the output of the preceding code:

```
>>>
Evaluating MLP on the test set...
Prediction/classification accuracy: 0.889
Session closed!
```

Thus, the classification accuracy is about 89%. Not bad at all! Now, if higher accuracy is desired, we can use another architecture of DNNs called **Deep Belief Networks (DBNs)**, that can be trained either in a supervised or unsupervised way.

This is the easiest way to observe DBN in its application as a classifier. If we have a DBN classifier, then the pre training method is done in an unsupervised way similar to an autoencoder, which will be described in *Chapter 5, Optimizing TensorFlow Autoencoders*, and the classifier is trained (fine-tuned) in a supervised way, exactly like the one in MLP.

Deep Belief Networks (DBNs)

To overcome the overfitting problem in MLP, we set up a DBN, do unsupervised pre training to get a decent set of feature representations for the inputs, then fine-tune on the training set to get predictions from the network.

While weights of an MLP are initialized randomly, a DBN uses a greedy layer-by-layer pre training algorithm to initialize the network weights, through probabilistic generative models. These models are composed of a visible layer, and multiple layers of stochastic, latent variables, which are called hidden units or feature detectors.

RBMs in the DBN are stacked, forming an undirected probabilistic graphical model, similar to **Markov Random Fields** (**MRF**): the two layers are composed of visible neurons and then hidden neurons.

The top two layers in a stacked RBM have undirected, symmetric connections between them and form an associative memory, whereas lower layers receive top-down, directed connections from the layer above:

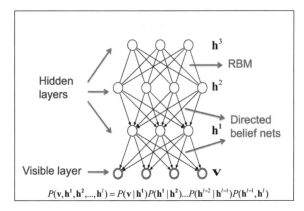

Figure 13: A high-level view of a DBN with RBM as the building block

The top two layers have undirected, symmetric connections between them and form an associative memory, whereas lower layers receive top-down, directed connections from the preceding layers. Several RBMs are stacked one after another to form DBNs.

Restricted Boltzmann Machines (RBMs)

An RBM is an undirected probabilistic graphical model called Markov random fields. It consists of two layers. The first layer is composed of visible neurons and second layer consist of hidden neurons. Figure 14 shows the structure of a simple RBM. Visible units accept inputs, and hidden units are nonlinear feature detectors. Each visible neuron is connected to all the hidden neurons, but there is no internal connection among neurons in the same layer:

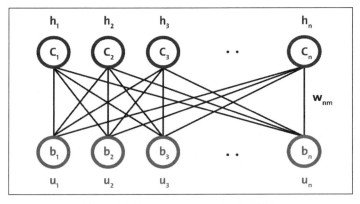

Figure 14: The structure of a simple RBM

The RBM in Figure 14 consists of m visible units, $V = (v_1, \ldots v_m)$ and n hidden units, $H = (h_1, \ldots h_n)$. Visible units accept values between 0 and 1 and generated values of hidden units are between 0 and 1. The joint probability of the model is an energy function given by the following equation:

$$E(v,h) = -\sum_{i=1}^{m} b_i v_i - \sum_{j=1}^{n} c_j h_j - \sum_{i=1}^{m}\sum_{j=1}^{n} v_i h_j w_i j \quad (1)$$

In the preceding equation, $i = 1\ldots m$, $j = 1\ldots n$, b_i, and c_j are biases of visible and hidden units respectively, and w_{ij} is the weight between v_i and h_j. The probability assigned by the model to a visible vector v is given by the following equation:

$$p(v) = \frac{1}{Z}\sum_{h} e^{-E(v,h)} \quad (2)$$

In the second equation, Z is a partition function defined as follows:

$$Z = \sum_{v,h} e^{-E(v,h)} \quad (3)$$

The learning of weight can be attained by the following equation:

$$\Delta w_{ij} = \epsilon \left(v_i h_{j_{data}} - v_i h_{j_{model}} \right) \quad (4)$$

In equation 4, the learning rate is defined by ϵ. In general, a smaller value of ϵ ensures that training is more intensive. However, if you want your network to learn quickly, you can set this value higher.

It is easy to calculate the first term since there are no connections among units in the same layer. Conditional distributions of $p(h\,|\,v)$ and $p(v\,|\,h)$ are factorial, and given by the logistic function in the following equations:

$$p(h_j = 1\,|\,v) = g\left(c_j + \sum_i v_i w_{ij}\right) \quad (5)$$

$$p(v_j = 1\,|\,h) = g\left(b_i + \sum_j h_i w_{ij}\right) \quad (6)$$

$$g(x) = \frac{1}{1+\exp(-x)} \quad (7)$$

Hence, the sample $v_i h_j$ is unbiased. However, calculating the log-likelihood of the second term is exponentially expensive to compute. Although it is possible to get unbiased samples of the second term, using Gibbs sampling by running **Markov Chain Monte Carlo** (**MCMC**), this process is not cost-effective either. Instead, RBM uses an efficient approximation method called **contrastive divergence**.

In general, MCMC requires many sampling steps to reach convergence to stationary. Running Gibbs sampling for few steps (usually one) is enough to train a model, which is called contrastive divergence learning. The first step of contrastive divergence is to initialize the visible units with a training vector.

The next step is to compute all hidden units, using visible units at the same time, with equation five, then reconstruct visible units from hidden units using equation four. Lastly, the hidden units are updated with the reconstructed visible units. Therefore, instead of equation four, we get the following weight-learning model in the end:

$$\Delta w_{ij} = \epsilon \left(v_i h_{j_{data}} - v_i h_{j_{recons}}\right) \quad (8)$$

In short, this process tries to reduce the reconstruction error between input data and reconstructed data. Several iterations of parameter update are required for the algorithm to converge. Iterations are called epoch. Input data is divided into mini batches, and parameters are updated after each mini batch, with the average values of the parameters.

Finally, as stated earlier, RBM maximizes the probability of visible units $p(v)$, which is defined by the mode and overall training data. It is equivalent to minimizing the KL-divergence (https://en.wikipedia.org/wiki/Kullback%E2%80%93Leibler_divergence) between the model distribution and the empirical data distribution.

Contrastive divergence is only a crude approximation of this objective function, but it works very well in practice. Although it is convenient, the reconstruction error is actually a very poor measure of the progress of learning. Considering these aspects, RBM requires some time to converge, but if you see that the reconstruction is decent, then your algorithm works well.

Construction of a simple DBN

A single hidden layer RBM cannot extract all the features from the input data, due to its inability to model the relationship between variables. Hence, multiple layers of RBMs are used one after another to extract non-linear features. In DBNs, an RBM is trained first with input data, and the hidden layer represents learned features in a greedy learning approach.

These learned features of the first RBM are used as the input of the second RBM, as another layer in the DBN, which is shown in Figure 15. Similarly, learned features of the second layer are used as input for another layer.

This way, DBNs can extract deep and non-linear features from input data. The hidden layer of the last RBM represents the learned features of the whole network. The process of learning features, described earlier for all RBM layers, is called pre-training.

Unsupervised pre-training

Suppose you want to tackle a complex task, for which you do not have much-labeled training data. It will be difficult to find a suitable DNN implementation, or architecture, to be trained and used for predictive analytics. Nevertheless, if you have plenty of unlabeled training data, you can try to train the layers one by one, starting with the lowest layer and then going up, using an unsupervised feature detector algorithm. This is how exactly RBMs (Figure 15) or autoencoders (Figure 16) work.

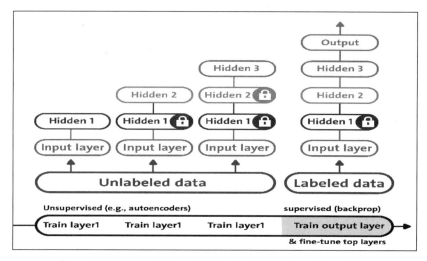

Figure 15: Unsupervised pre-training in a DBN using autoencoders

Unsupervised pre-training is still a good option when you have a complex task to solve, no similar model you can reuse, and little-labeled training data, but plenty of unlabeled training data. The current trend is using autoencoders rather than RBMs; however, for the example in the next section, RBMs will be used for simplicity. Readers can also try using autoencoders rather than RBMs.

Pre-training is an unsupervised learning process. After pre-training, fine-tuning of the network is carried out, by adding a labeled layer at the top of the last RBM layer. This step is a supervised learning process. The unsupervised pre-training step tries to find network weights:

Chapter 3

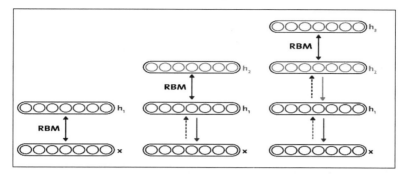

Figure 16: Unsupervised pre-training in a DBN by constructing a simple a DBN with a stack of RBMs

Supervised fine-tuning

In the supervised learning stage (also called supervised fine-tuning), instead of randomly initializing network weights, they are initialized with the weights computed in the pre-training step. This way, DBNs can avoid converging to a local minimum when a supervised gradient descent is used.

As stated earlier, using a stack of RBMs, a DBN can be constructed as follows:

- Train the bottom RBM (first RBM) with parameter $\mathbf{W}^1$
- Initialize the second layer weights to $\mathbf{W}^2 = \mathbf{W}^{1T}$, which ensures that the DBN is at least as good as our base RBM

Therefore, putting these steps together, Figure 17 shows the construction of a simple DBN, consisting of three RBMs:

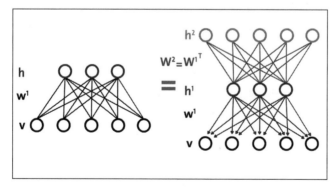

Figure 17: Construction of a simple DBN using several RBMs

Now, when it comes to tuning a DBN for better predictive accuracy, we should tune several hyper-parameters, so that DBNs fit the training data by untying and refining W^2. Putting this all together, we have the conceptual workflow for creating a DBN-based classifier or regressor.

Now that we have enough theoretical background on how to construct a DBN using several RBMs, it is time to apply our theory in practice. In the next section, we will see how to develop a supervised DBN classifier for predictive analytics.

Implementing a DBN with TensorFlow for client-subscription assessment

In the previous example of the bank marketing dataset, we observed about 89% classification accuracy using MLP. We also normalized the original dataset, before feeding it to the MLP. In this section, we will see how to use the same datasets for the DBN-based predictive model.

We will use the DBN implementation of the recently published book *Predictive Analytics with TensorFlow*, by *Md. Rezaul Karim*, that can be downloaded from GitHub at https://github.com/PacktPublishing/Predictive-Analytics-with-TensorFlow/tree/master/Chapter07/DBN.

The aforementioned implementation is a simple, clean, fast Python implementation of DBNs, based on RBMs, and built upon NumPy and TensorFlow libraries, in order to take advantage of GPU computation. This library is implemented based on the following two research papers:

- *A fast learning algorithm for deep belief nets*, by *Geoffrey E. Hinton, Simon Osindero*, and *Yee-Whye Teh. Neural Computation 18.7 (2006): 1527-1554.*
- *Training Restricted Boltzmann Machines: An Introduction*, *Asja Fischer*, and *Christian Igel. Pattern Recognition 47.1 (2014): 25-39.*

We will see how to train the RBMs in an unsupervised way and then we will train the network in a supervised way. In short, there are several steps to be followed. The main classifier is `classification_demo.py`.

Although the dataset is not that big or high dimensional when training a DBN in both a supervised and unsupervised way, there will be so many computations in the training time and this requires huge resources. Nevertheless, RBM requires a lot of time to converge. Therefore, I would suggest that readers perform the training on GPU, having at least 32 GB of RAM and a corei7 processor.

We will start by loading required modules and libraries:

```
import numpy as np
import pandas as pd
from sklearn.datasets import load_digits
from sklearn.model_selection import train_test_split
from sklearn.metrics.classification import accuracy_score
from sklearn.metrics import precision_recall_fscore_support
from sklearn.metrics import confusion_matrix
import itertools
from tf_models import SupervisedDBNClassification
import matplotlib.pyplot as plt
```

We then load the already normalized dataset used in the previous MLP example:

```
FILE_PATH = FILE_PATH = '../input/bank_normalized.csv'
raw_data = pd.read_csv(FILE_PATH)
```

In the preceding code, we have used pandas `read_csv()` method and have created a `DataFrame`. Now, the next task is to spate the features and labels as follows:

```
Y_LABEL = 'y'
KEYS = [i for i in raw_data.keys().tolist() if i != Y_LABEL]
X = raw_data[KEYS].get_values()
Y = raw_data[Y_LABEL].get_values()
class_names = list(raw_data.columns.values)
print(class_names)
```

In the preceding lines, we have separated the features and labels. The features are stored in X and the labels are in Y. The next task is to split them into the train (75%) and the test set (25%) as follows:

```
X_train, X_test, Y_train, Y_test = train_test_split(X, Y,
test_size=0.25, random_state=100)
```

Now that we have the training and test set, we can go to the DBN training step directly. However, first we need to instantiate the DBN. We will do it in a supervised way for classification, but we need to provide the hyperparameters for this DNN architecture:

```
classifier = SupervisedDBNClassification(hidden_layers_structure=[64,
64],
                        learning_rate_rbm=0.05,
                        learning_rate=0.01,
                        n_epochs_rbm=10,
                        n_iter_backprop=100,
                        batch_size=32,
                        activation_function='relu',
                        dropout_p=0.2)
```

In the preceding code segment, `n_epochs_rbm` is the number of epoch for the pre-training (unsupervised) and `n_iter_backprop` for the supervised fine-tuning. Nevertheless, we have defined two separate learning rates for these two phases, as well using `learning_rate_rbm` and `learning_rate` respectively.

Nevertheless, we will describe this class implementation for `SupervisedDBNClassification` later in this section.

This library has an implementation to support sigmoid, ReLU, and tanh activation functions. In addition, it utilizes the l2 regularization to avoid overfitting. We will do the actual fitting as follows:

```
classifier.fit(X_train, Y_train)
```

If everything goes fine, you should observe the following progress on the console:

```
[START] Pre-training step:
>> Epoch 1 finished      RBM Reconstruction error 1.681226
…..
>> Epoch 3 finished      RBM Reconstruction error 4.926415
>> Epoch 5 finished      RBM Reconstruction error 7.185334
…
>> Epoch 7 finished      RBM Reconstruction error 37.734962
>> Epoch 8 finished      RBM Reconstruction error 467.182892
…..
>> Epoch 10 finished     RBM Reconstruction error 938.583801
[END] Pre-training step
[START] Fine tuning step:
>> Epoch 0 finished      ANN training loss 0.316619
>> Epoch 1 finished      ANN training loss 0.311203
>> Epoch 2 finished      ANN training loss 0.308707
…..
>> Epoch 98 finished     ANN training loss 0.288299
>> Epoch 99 finished     ANN training loss 0.288900
```

Chapter 3

Since the weights of the RBM are randomly initialized, the difference between the reconstructions and the original input is often large.

More technically, we can think of reconstruction error as the difference between the reconstructed values and the input values. This error is then backpropagated against the RBM's weights several times that is, in an iterative learning process until an error minimum is reached.

Nevertheless, in our case, the reconstruction reaches up to 938, which is not that big (that is, not infinity) so we can still expect good accuracy. Anyway, after 100 iterations, the fine-tuning graph showing training gloss per epoch is as follows:

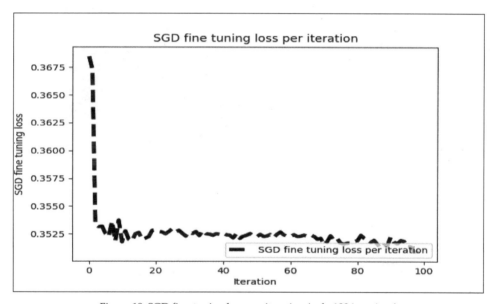

Figure 18: SGD fine-tuning loss per iteration (only 100 iterations)

However, when I iterated the preceding training and fine-tuning up to 1000 epochs, I did not see any significant improvement in the training loss:

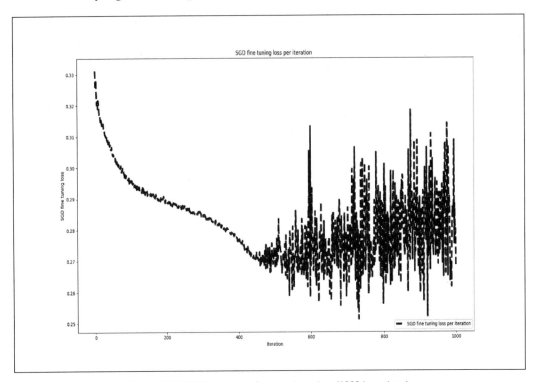

Figure 19: SGD fine-tuning loss per iteration (1000 iterations)

Here is the implementation of supervised DBN classifiers. This class implements a DBN for classification problems. It converts network output to original labels. It also takes network parameters and returns a list, after performing index to label mapping.

This class then predicts the probability distribution of classes for each sample in the given data and returns a list of dictionaries (one per sample). Finally, it appends a softmax linear classifier as an output layer:

```
class SupervisedDBNClassification(TensorFlowAbstractSupervisedDBN,
ClassifierMixin):
    def _build_model(self, weights=None):
        super(SupervisedDBNClassification, self)._build_model(weights)
        self.output = tf.nn.softmax(self.y)
        self.cost_function = tf.reduce_mean(tf.nn.softmax_cross_entropy_with_logits_v2(logits=self.y, labels=self.y_))
```

```python
            self.train_step =
self.optimizer.minimize(self.cost_function)
    @classmethod

    def _get_param_names(cls):
        return super(SupervisedDBNClassification,
cls)._get_param_names() + ['label_to_idx_map', 'idx_to_label_map']

    @classmethod
    def from_dict(cls, dct_to_load):
        label_to_idx_map = dct_to_load.pop('label_to_idx_map')
        idx_to_label_map = dct_to_load.pop('idx_to_label_map')
        instance = super(SupervisedDBNClassification,
cls).from_dict(dct_to_load)
        setattr(instance, 'label_to_idx_map', label_to_idx_map)
        setattr(instance, 'idx_to_label_map', idx_to_label_map)
        return instance

    def _transform_labels_to_network_format(self, labels):
        """
        Converts network output to original labels.
        :param indexes: array-like, shape = (n_samples, )
        :return:
        """
        new_labels, label_to_idx_map, idx_to_label_map =
to_categorical(labels, self.num_classes)
        self.label_to_idx_map = label_to_idx_map
        self.idx_to_label_map = idx_to_label_map
        return new_labels

    def _transform_network_format_to_labels(self, indexes):
        return list(map(lambda idx: self.idx_to_label_map[idx],
indexes))

    def predict(self, X):
        probs = self.predict_proba(X)
        indexes = np.argmax(probs, axis=1)
        return self._transform_network_format_to_labels(indexes)

    def predict_proba(self, X):
        """
        Predicts probability distribution of classes for each
sample in the given data.
        :param X: array-like, shape = (n_samples, n_features)
        :return:
```

```
        """
        return super(SupervisedDBNClassification,
self)._compute_output_units_matrix(X)

    def predict_proba_dict(self, X):
        """
        Predicts probability distribution of classes for each
sample in the given data.
        Returns a list of dictionaries, one per sample. Each dict
contains {label_1: prob_1, ..., label_j: prob_j}
        :param X: array-like, shape = (n_samples, n_features)
        :return:
        """
        if len(X.shape) == 1:  # It is a single sample
            X = np.expand_dims(X, 0)
        predicted_probs = self.predict_proba(X)
        result = []
        num_of_data, num_of_labels = predicted_probs.shape
        for i in range(num_of_data):
            # key : label
            # value : predicted probability
            dict_prob = {}
            for j in range(num_of_labels):
                dict_prob[self.idx_to_label_map[j]] = predicted_probs[i][j]
            result.append(dict_prob)
        return result
    def _determine_num_output_neurons(self, labels):
        return len(np.unique(labels))
```

As we mentioned in the previous example, and the running section, fine-tuning the parameters of a neural network is a tricky process. There are many different approaches out there, but there is no one-size-fits-all approach to my knowledge. Nevertheless, with the preceding combination, I have received better classification results. Another important parameter to select is the learning rate. Adapting the learning rate as your model goes is an approach that can be taken in order to reduce training time while avoiding local minimums. Here, I would like to discuss some tips that really helped me to get better predictive accuracy, not only for this application but for others as well.

Now that we have our model built, it is time to evaluate its performance. To evaluate the classification accuracy, we will use several performance metrics such as `precision`, `recall`, and `f1 score`. Moreover, we will draw the confusion matrix, to observe the predicted labels against the true labels. First, let us compute the prediction accuracy as follows:

```
Y_pred = classifier.predict(X_test)
print('Accuracy: %f' % accuracy_score(Y_test, Y_pred))
```

Next, we need to compute the `precision`, `recall`, and `f1 score` of the classification:

```
p, r, f, s = precision_recall_fscore_support(Y_test, Y_pred,
average='weighted')
print('Precision:', p)
print('Recall:', r)
print('F1-score:', f)
```

The following is the output of the preceding code:

```
>>>
Accuracy: 0.900554
Precision: 0.8824140209830381
Recall: 0.9005535592891133
F1-score: 0.8767190584424599
```

Fantastic! Using our DBN implementation we have solved the same classification problem that we did using MLP. Nevertheless, we have managed to achieve a slightly better accuracy compared to MLP.

Now, if you want to solve a regression problem, where the labels to be predicted are continuous, you will have to use the `SupervisedDBNRegression()` function for this implementation. The regression script (that is `regression_demo.py`) in the DBN folder can be used to perform the regression operation too.

However, using another dataset specially prepared for regression y would be the better idea. All you need to do is to prepare your dataset so that it can be consumed by the TensorFlow-based DBN. So, for minimal demonstration, I used the *House Prices: Advanced Regression Techniques dataset* to predict the housing price.

Tuning hyperparameters and advanced FFNNs

The flexibility of neural networks is also one of their main drawbacks: there are many hyperparameters to tweak. Even in a simple MLP, you can change the number of layers, the number of neurons per layer, and the type of activation function to use in each layer. You can also change the weight initialization logic, the drop out keep probability, and so on.

Additionally, some common problems in FFNNs, such as the gradient vanishing problem, and selecting the most suitable activation function, learning rate, and optimizer, are of prime importance.

Tuning FFNN hyperparameters

Hyperparameters are parameters that are not directly learned within estimators. It is possible and recommended that you search the hyperparameter space for the best cross-validation (http://scikit-learn.org/stable/modules/cross_validation.html#cross-validation) score. Any parameter provided when constructing an estimator may be optimized in this manner. Now, the question is: how you do know what combination of hyperparameters is best for your task? Of course, you can use grid search, with cross-validation, to find the right hyperparameters for linear machine learning models.

However, for the DNNs, there are many hyperparameters to tune. Since training a neural network on a large dataset takes a lot of time, you will only be able to explore a tiny part of the hyperparameter space in a reasonable amount of time. Here are some insights that can be followed.

Moreover, of course, as I said, you can use grid search or randomized searches, with cross-validation, to find the right hyperparameters for linear machine learning models. We will see some possible ways of exhaustive and randomized grid searching and cross-validation later in this section.

Number of hidden layers

For many problems, you can start with just one or two hidden layers and this setting will work just fine using two hidden layers, with the same total amount of neurons (see below to get an idea about a number of neurons), in roughly the same amount of training time. Now let's see some naïve estimation about setting the number of hidden layers:

- **0** : This is only capable of representing linear separable functions or decisions

- **1** : This can approximate any function that contains a continuous mapping from one finite space to another
- **2** : This can represent an arbitrary decision boundary to arbitrary accuracy, with rational activation functions, and can approximate any smooth mapping to any accuracy

However, for a more complex problem, you can gradually ramp up the number of hidden layers, until you start overfitting the training set. Very complex tasks, such as large image classification or speech recognition, typically require networks with dozens of layers, and they need a large amount of training data.

Nevertheless, you can try increasing the number of neurons gradually until the network starts overfitting. This means the upper bound on the number of hidden neurons that will not result in overfitting is:

$$N_h = \frac{N_s}{(\alpha * (N_i + N_o))}$$

In the preceding equation:

N_i = number of input neurons

N_o = number of output neurons

N_s = number of samples in training dataset

α = an arbitrary scaling factor usually 2-10.

Note that the above equation does not come from any research, but from my personal working experience. However, for an automated procedure, you would start with an alpha of 2, that is twice as many degrees of freedom in your training data as your model, and work your way up to 10, if the error for training data is significantly smaller than for the cross-validation data set.

Number of neurons per hidden layer

Obviously, the number of neurons in the input and output layers is determined by the type of input and output your task requires. For example, if your dataset has the shape of *28x28*, it should have input neurons of size 784, and the output neurons should be equal to the number of classes to be predicted.

We will see how this works in practice in the next example, using MLP, where there will be four hidden layers with 256 neurons (just one hyperparameter to tune, instead of one per layer). Just like for the number of layers, you can try increasing the number of neurons gradually until the network starts overfitting.

There are some empirically derived rules-of-thumb, of which the most commonly relied on is: "The optimal size of the hidden layer is usually between the size of the input and size of the output layers."

In summary, for most problems, you could probably get decent performance (even without a second optimization step) by setting the hidden layer configuration using just two rules:

- The number of hidden layers equals one
- The number of neurons in that layer is the mean of the neurons in the input and output layers

Nevertheless, just like for the number of layers, you can try increasing the number of neurons gradually until the network starts overfitting.

Weight and biases initialization

As we will see in the next example, initializing weight and biases for the hidden layers is an important hyperparameter to be taken care of:

- **Do not do all zero initialization**: A reasonable-sounding idea might be to set all the initial weights to zero, but it does not work in practice. This is because if every neuron in the network computes the same output, there will be no source of asymmetry between neurons if their weights are initialized to be the same.
- **Small random numbers**: It is also possible to initialize the weights of the neurons to small numbers, but not identically zero. Alternatively, it is possible to use small numbers drawn from a uniform distribution.
- **Initializing the biases**: It is common to initialize the biases to be zero since the small random numbers in the weights provide the asymmetry breaking. Setting the biases to a small constant value, such as 0.01 for all biases, ensures that all ReLU units can propagate some gradient. However, it neither performs well nor shows consistent improvement. Therefore, sticking with zero is recommended.

Selecting the most suitable optimizer

Since, in FFNNs, one of the objective functions is to minimize the evaluated cost, we must define an optimizer. We have already seen how to use `tf.train.AdamOptimizer` (https://www.tensorflow.org/api_docs/python/tf/train/AdamOptimizer). Tensorflow `tf.train` (https://www.tensorflow.org/api_docs/python/tf/train) provides a set of classes and functions that help to train models. Personally, I have found that Adam optimizer works well for me in practice, without having to think much about learning rates and so on.

For most of the cases, we can utilize Adam, but sometimes we can adopt the implemented `RMSPropOptimizer` function, which is an advanced form of gradient descent. The `RMSPropOptimizer` function implements the `RMSProp` algorithm.

The `RMSPropOptimizer` function also divides the learning rate by an exponentially decaying average of squared gradients. The suggested setting value of the decay parameter is `0.9`, while a good default value for the learning rate is `0.001`:

```
optimizer = tf.train.RMSPropOptimizer(0.001,
0.9).minimize(cost_op)
```

Using the most common optimizer **SGD**, the learning rates must scale with `1/T` to get convergence, where `T` is the number of iterations. `RMSProp` tries to overcome this limitation automatically by adjusting the step size so that the step is on the same scale as the gradients.

So if you're training a neural network, but computing the gradients is mandatory, using `tf.train.RMSPropOptimizer()` would be the faster way of learning in a mini-batch setting. Researchers also recommend using Momentum optimizer while training a deep network such as CNN.

Finally, if you want to play around by setting these optimizers, you just need to change one line. Due to time constraints, I have not tried all of these. However, according to a recent research paper by *Sebastian Ruder* (see at https://arxiv.org/abs/1609.04747), optimizers with adaptive learning-rate methods that is, `Adagrad`, `Adadelta`, `RMSprop`, and `Adam` are most suitable and provide the best convergence for these scenarios.

GridSearch and randomized search for hyperparameters tuning

Two generic approaches to sampling search candidates are provided in other Python-based machine-learning libraries such as Scikit-learn. For given values, `GridSearchCV` (http://scikit-learn.org/stable/modules/generated/sklearn.model_selection.GridSearchCV.html#sklearn.model_selection.GridSearchCV) exhaustively considers all parameter combinations, while `RandomizedSearchCV` (http://scikit-learn.org/stable/modules/generated/sklearn.model_selection.RandomizedSearchCV.html#sklearn.model_selection.RandomizedSearchCV) can sample a given number of candidates from a parameter space with a specified distribution.

`GridSearchCV` is a great way to test and optimize hyperparameters automatically. I often use it with Scikit-learn. However, it is not yet so straightforward with `TensorFlowEstimator` to optimize `learning_rate`, `batch_size`, and so on. Moreover, as I said, we often have so many hyperparameters to tune to get the best result. Nevertheless, I found this article quite useful to learn how to tune aforementioned hyperparameters:https://machinelearningmastery.com/grid-search-hyperparameters-deep-learning-models-python-keras/

The randomized search and the grid search explore exactly the same space of parameters. The result in parameter settings is quite similar, while the runtime for the randomized search is drastically lower.

Some benchmarks (for example, http://scikit-learn.org/stable/auto_examples/model_selection/) have reported that the performance is slightly worse for the randomized search, though this is most likely a noise effect and would not carry over to a held-out test set.

Regularization

There are several ways of controlling the training of DNNs to prevent overfitting in the training phase, for example, L2/L1 regularization, max norm constraints, and drop out:

- **L2 regularization**: This is probably the most common form of regularization. Using the gradient descent parameter update, L2 regularization signifies that every weight will be decayed linearly towards zero.
- **L1 regularization**: For each weight w, we add the term $\lambda |w|$ to the objective. However, it is also possible to combine L1 and L2 regularization *to achieve* elastic net regularization.

- **Max-norm constraints**: This is issued to enforce an absolute upper bound on the magnitude of the weight vector for each hidden layer neuron. Projected gradient descent can be used further to enforce the constraint.

The vanishing gradient problem arises in very deep neural networks (typically RNNs, which will have a dedicated chapter on), that use activation functions, whose gradients tend to be small (in the range of **0** from **1**).

Since these small gradients are further multiplied during the backpropagation, they tend to "vanish" throughout the layers, preventing the network from learning long-range dependencies. A common way to counter this problem is to use activation functions like **Linear Unit** (aka. **ReLU**), which does not suffer from small gradients. We will see an improved variant of an RNN, called **Long Short-Term Memory** (aka. **LSTM**), which can combat this problem. We will see a more detailed discussion on this topic in *Chapter 5, Optimizing TensorFlow Autoencoders*.

Nevertheless, we have seen that the last architectural change improved the accuracy of our model, but we can do even better by changing the sigmoid activation function with the ReLU, shown as follows:

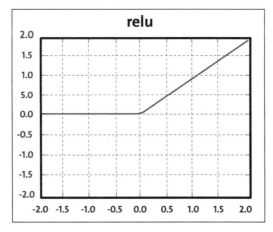

Figure 20: ReLU function

A **ReLU** unit computes the function `f(x) = max(0, x)`. ReLU is *computationally* fast because it does not require any exponential computation, such as that required in sigmoid or tanh activation. Furthermore, it was found to accelerate the convergence of stochastic gradient descent greatly, compared to the sigmoid/tanh functions. To use the ReLU function, we simply change, in the previously implemented model, the following definitions of the first four layers:

First layer output:

```
Y1 = tf.nn.relu(tf.matmul(XX, W1) + B1) # Output from layer 1
```

Second layer output:

```
Y2 = tf.nn.relu(tf.matmul(Y1, W2) + B2) # Output from layer 2
```

Third layer output:

```
Y3 = tf.nn.relu(tf.matmul(Y2, W3) + B3) # Output from layer 3
```

Fourth layer output:

```
Y4 = tf.nn.relu(tf.matmul(Y3, W4) + B4) # Output from layer 4
```

Output layer:

```
Ylogits = tf.matmul(Y4, W5) + B5 # computing the logits
Y = tf.nn.softmax(Ylogits) # output from layer 5
```

Of course, `tf.nn.relu` is TensorFlow's implementation of ReLU. The accuracy of the model is almost 98%, as you could see running the network:

```
>>>
Loading data/train-images-idx3-ubyte.mnist
Loading data/train-labels-idx1-ubyte.mnist Loading data/t10k-images-idx3-ubyte.mnist
Loading data/t10k-labels-idx1-ubyte.mnist
Epoch: 0
Epoch: 1
Epoch: 2
Epoch: 3
Epoch: 4
Epoch: 5
Epoch: 6
Epoch: 7
Epoch: 8
```

```
Epoch: 9
Accuracy:0.9789
done
>>>
```

As concerns, the TensorBoard analysis, from the folder where the source has been executed, you should digit:

```
$> Tensorboard --logdir = 'log_relu' # Don't put space before or after '='
```

Then open the browser at localhost to visualize TensorBoard's starting page. In the following figure, we show the trend's accuracy over the number of examples of the training set:

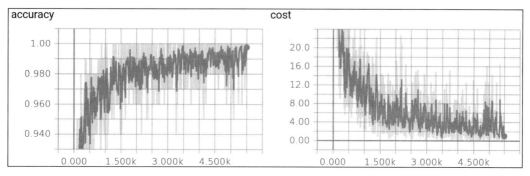

Figure 21: Accuracy function over the training set

You can easily see how the accuracy, after a bad initial trend, begins a rapid progressive improvement after about 1000 examples.

Dropout optimization

While working with a DNN, we need another placeholder for dropout, which is a hyperparameter to be tuned. It is implemented by only keeping a neuron active with some probability (say *p<1.0*) or setting it to zero otherwise. The idea is to use a single neural net at test time without dropout. The weights of this network are scaled-down versions of the trained weights. If a unit is retained with *dropout_keep_prob* < 1.0 during training, the outgoing weights of that unit are multiplied by *p* at test time.

During the learning phase, the connections with the next layer can be limited to a subset of neurons, to reduce the weights to be updated. This learning optimization technique is called **dropout**. The dropout is, therefore, a technique used to decrease the overfitting within a network with many layers and/or neurons. In general, the dropout layers are positioned after the layers that possess a large number of trainable neurons.

This technique allows the setting to 0, and then excluding the activation, of a certain percentage of the neurons of the preceding layer. The probability that the neuron's activation is set to 0 is indicated by the dropout ratio parameter within the layer, via a number between 0 and 1. In practice, the activation of a neuron is held with probability equal to the dropout ratio; otherwise, it is discarded, that is, set to 0.

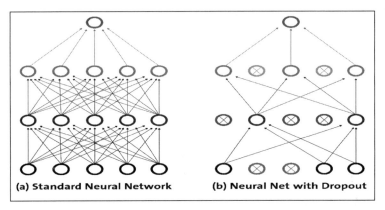

Figure 22: Dropout representation

In this way, for each input, the network owns an architecture slightly different from the previous one. Some connections are active and some are not, in a different way, every time, even if these architectures possess the same weights. The preceding figure shows how the dropout works: each hidden unit is randomly omitted from the network with a probability of p.

One thing to notice, though, is that selected dropout units are different for each training instance; that is why this is more of a training problem. Dropout can be seen as an efficient way to perform model averaging, across a large number of different neural networks, where overfitting can be avoided with much less cost of computation than an architecture problem. The dropout reduces the possibility that a neuron relies on the presence of other neurons. In this way, it is forced to learn more about robust features, and that they are useful with linkages to other different neurons.

The TensorFlow function that allows building a dropout layer is `tf.nn.dropout`. The input of this function is the output of the previous layer, and a dropout parameter, `tf.nn.dropout`, returns an output tensor of the same size as the input tensor. The implementation of this model follows the same rules used for the five-layer network. In this case, we must insert the dropout function between one layer and another layer:

```
pkeep = tf.placeholder(tf.float32)


Y1 = tf.nn.relu(tf.matmul(XX, W1) + B1) # Output from layer 1
Y1d = tf.nn.dropout(Y1, pkeep)

Y2 = tf.nn.relu(tf.matmul(Y1, W2) + B2) # Output from layer 2
Y2d = tf.nn.dropout(Y2, pkeep)

Y3 = tf.nn.relu(tf.matmul(Y2, W3) + B3) # Output from layer 3
Y3d = tf.nn.dropout(Y3, pkeep)

Y4 = tf.nn.relu(tf.matmul(Y3, W4) + B4) # Output from layer 4
Y4d = tf.nn.dropout(Y4, pkeep)

Ylogits = tf.matmul(Y4d, W5) + B5 # computing the logits
Y = tf.nn.softmax(Ylogits) # output from layer 5
```

The dropout optimization produces the following results:

```
>>>
Loading data/train-images-idx3-ubyte.mnist Loading data/train-labels-idx1-ubyte.mnist Loading data/t10k-images-idx3-ubyte.mnist Loading data/t10k-labels-idx1-ubyte.mnist Epoch:     0
Epoch: 1
Epoch: 2
Epoch: 3
Epoch: 4
Epoch: 5
Epoch: 6
Epoch: 7
Epoch: 8
Epoch: 9
Accuracy:      0.9666 done
>>>
```

Despite this implementation, the previous ReLU network is still better, but you can try to change the network parameters to improve the model's accuracy. Also, since this is a tiny network and we dealt with a small-scale dataset, when you handle a large-scale high-dimensional dataset with a more complex network, you will realize that the dropout could be really important. We will see a few hands-on examples in the next chapter.

Now, to see the effect of the dropout optimization, let's start the TensorBoard analysis. Just type the following:

```
$> Tensorboard --logdir=' log_softmax_relu_dropout/'
```

The following graph shows the accuracy cost function as a function of the training examples:

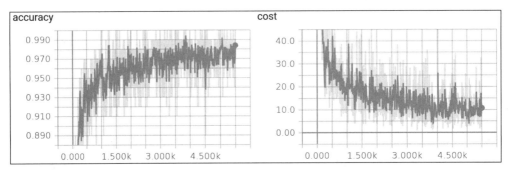

Figure 23: a) accuracy in dropout optimization, b) the cost function over the training set

In the preceding chart, we display the cost function as a function of the training examples. Both trends are what we expected: the accuracy increases with training examples, while the cost function decreases with increasing iterations.

Summary

We have seen how to implement FFNN architectures that are characterized by a set of input units, a set of output units, and one or more hidden units that connect the input level from that output. We have seen how to organize the network layers so that the connections between the levels are total and in a single direction: each unit receives a signal from all the units of the previous layer and transmits its output value, suitably weighed to all units of the next layer.

We have also seen how to define an activation function (for example, sigmoid, ReLU, tanh, and softmax) for each layer, where the choice of an activation function depends on the architecture and the problem being addressed.

We then implemented four different FFNN models. The first model had a single hidden layer, with a softmax activation function. The three other more complex models had five hidden layers in total, but with different activation function. We have also seen how to implement a deep MLP and DBN with TensorFlow, for solving a classification task. Using these implementations, we managed to achieve above 90% accuracy. Finally, we have discussed how to tune the hyperparameters for DNNs for better and more optimized performance.

Although a regular FFNN, such as an MLP, works fine for small images (for example, MNIST or CIFAR-10), it breaks down for larger images because of the huge number of parameters required. For example, a 100×100 image has 10,000 pixels, and if the first layer has just 1,000 neurons (which already severely restricts the amount of information transmitted to the next layer), this means 10 million connections. In addition, that is just for the first layer.

Importantly, a DNN has no prior knowledge of how pixels are organized, so it does not know that nearby pixels are close. The architecture of a CNN embeds this prior knowledge. Lower layers typically identify features in small areas of the images, while higher layers combine the lower-level features into larger features. This works well with most natural images, giving CNNs a decisive head start compared to DNNs.

In the next chapter, we will look further into the complexity of neural network models, introducing CNNs, which may have a big impact on deep learning techniques. We will study the main features and see some implementation examples.

4
Convolutional Neural Networks

In this chapter, we will talk about CNNs, which are a feather in the cap of deep learning. CNNs have achieved excellent results in many practical applications, particularly in the field of *object recognition in images*. We will explain and implement the **LeNet** architecture (LeNet5), which was the first CNN to have great success with the classic **MNIST** digit classification system. We will also analyze **AlexNet**, which is a deep CNN that was invented by *Alex Krizhevsky*. We'll use these networks to introduce *transfer learning*, which is a machine learning method that utilizes a pre-trained neural network. We will also introduce the VGG architecture, which is usually used as a deep CNN for object recognition. This was developed by Oxford University's renowned **Visual Geometry Group** (**VGG**), which performed very well with the **ImageNet** dataset. This architecture gives us the opportunity to show how to use a neural network to draw a picture in a certain artistic style (*artistic style learning*).

We will move on to the **Inception-v3** model, which was created for the **ImageNet Large-Scale Visual Recognition Challenge** (**ILSVRC**) using the data from the 2012 competition. This is a standard task in computer vision, in which models try to classify 1.2M images of 1000 different categories. We'll demonstrate how to train your own image classifier with Inception in TensorFlow. The last example is taken from the **Kaggle** platform. The purpose here is to train a network on a series of *facial images* to classify their *emotional stretch*. We'll evaluate the accuracy of the model and then test it on a single image that does not belong to the original dataset. The topics covered in this chapter are as follows:

- Main concepts of CNNs
- CNNs in action
- LeNet and the MNIST classification problem

- AlexNet and transfer learning
- VGG and artistic style learning
- Inception-v3 model
- Emotion recognition

Main concepts of CNNs

Recently, **Deep Neural Networks** (**DNNs**) have given fresh impetus to research and therefore they are being used widely. CNNs are a special type of DNN, and they have been used with great success in image classification problems. Before diving into the implementation of an image classifier based on CNNs, we'll introduce some basic concepts in image recognition, such as *feature detection* and *convolution*.

In computer vision, it is well known that a real image is associated with a grid composed of a high number of small squares called **pixels**. The following figure represents a *black and white* image related to a 5×5 grid of pixels:

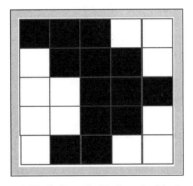

Figure 1: Pixel view of a black and white image.

Each element in the grid corresponds to a pixel. In the case of a black and white image, a value of 1 is associated with *black* and a value of 0 is associated with *white*. Alternatively, for a grayscale image, the allowed values for each grid element are in the range [0, 255], where 0 is associated with black and 255 is associated with white.

Finally, a *color image* is represented by a group of three matrices, each corresponding to one color channel (*red*, *green*, and *blue*). Each element of each matrix can vary over an interval of 0 to 255 that specifies the brightness of the fundamental color (or base color). This is shown in the following figure, in which each matrix is 4×4 and the number of color channels is three:

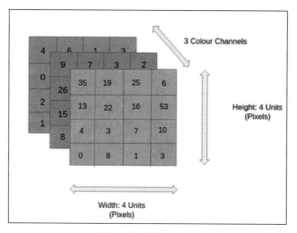

Figure 2: Color image

Let's focus now on the black and white image 5×5 matrix. Suppose we slide a second matrix of lower dimensions, for example a 3×3 matrix (see the figure below), across the width and height of the image matrix:

$$\begin{bmatrix} 1 & 0 & 1 \\ 0 & 1 & 0 \\ 1 & 0 & 1 \end{bmatrix}$$

Figure 3: Kernel filter

This flowing matrix is called a *kernel filter* or a *feature detector*. While a kernel filter moves along the *input matrix* (or input image), it performs a scalar product of the kernel values and the values of the matrix portion to which it is applied. The result is a new matrix called a **convolution matrix**.

The next figure displays the convolution procedure: the *convolved feature* (the resulting 3×3 matrix) is generated by the convolution operation, flowing the *kernel filter* (the 3×3 matrix) on the *input image* (the 5×5 matrix):

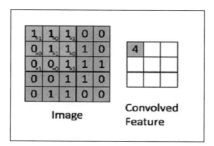

Figure 4: Input image (matrix 5×5 on the left), the kernel filter (matrix 3×3 on the input image), and convolved feature (matrix 3×3 on the right)

CNNs in action

Taking as an example the 5×5 input matrix shown earlier, a CNN is made up of an input layer consisting of 25 neurons (5×5) that has the task of acquiring the input value corresponding to each pixel and transferring it to the next layer.

In a multilayer network, the output from all of the neurons in the input layer would be connected to each neuron in the hidden layer (the fully connected layer). In CNN networks, however, the connection scheme that defines the **convolutional layer** that we are going to describe is significantly different. As you may be able to guess, this is the main type of layer: the use of one or more of these layers in a CNN is indispensable.

In a convolutional layer, each neuron is *connected* to a certain region of the input area called the **receptive field**. For example, using a 3×3 kernel filter, each neuron will have a bias and 9 weights (3×3) connected to a single receptive field. To effectively recognize an image, we need various different kernel filters to be applied to the same receptive field because each filter should recognize images from a different feature. The set of neurons that identifies the same feature defines a single **feature map**.

The following figure shows a CNN architecture *in action*: the 28×28 input image will be analyzed by a convolutional layer composed of a 28x28x32 feature map. The figure also shows a *receptive field* and a 3×3 *kernel filter*:

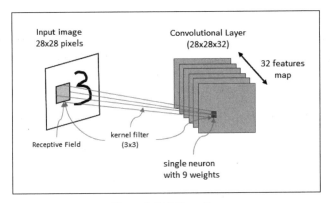

Figure 5: CNN in action

A CNN may consist of several convolution layers connected by cascade connections. The output of each convolutional layer is a set of feature maps (each generated by a single kernel filter). Each of these matrices defines a new input that *will be used* by the next layer.

Usually, in a CNN each neuron produces an output up to an *activation threshold*, which is proportional to the input and is not bounded.

CNNs also use *pooling layers* positioned immediately after the convolutional layers. A pooling layer divides a convolutional region into subregions. The pooling layer then selects a single representative value (max-pooling or average pooling) to reduce the computational time of subsequent layers and increase the robustness of the feature with respect to its spatial position. The last layer of a convolutional network is generally a *fully connected network* with a **softmax** activation function for the *output layer*. In the next few sections, the architectures of the most important CNNs will be analyzed in detail.

LeNet5

The LeNet5 CNN architecture was invented by *Yann LeCun* in 1998 and was the first CNN. It is a multilayered feed-forward network specifically designed to classify handwritten digits. It was used in LeCun's experiments and consists of seven layers containing trainable weights. The LeNet5 architecture looks like this:

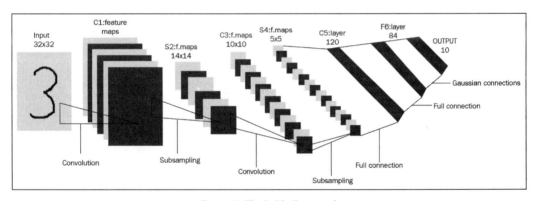

Figure 6: The LeNet5 network

The LeNet5 architecture consists of three convolutional layers and two alternating sequence pooling layers. The last two layers correspond to a traditional fully connected neural network, that is, a fully connected layer followed by an output layer. The main function of the output layer is to calculate the Euclidean distance between the input vector and the parameter vector. The output functions identify the difference between the measurements of the input pattern and our model. The output is kept minimal in order to achieve the best model. Therefore, the fully connected layer is configured so that the difference between the measurements of the input pattern and our model is minimized. Although it performs well on the MNIST dataset, the performance drops on datasets that have more images with higher resolution and more classes.

> See http://yann.lecun.com/exdb/lenet/index.html for basic references on LeNet family models.

Implementing a LeNet-5 step by step

In this section, we will learn how to build a *LeNet-5* architecture to classify images in the MNIST dataset. The next figure shows how the data flows in the first two convolutional layers: the input image is processed in the first convolutional layer using the filter weights. This results in 32 new images, one for each filter in the convolutional layer. The images are also *down-sampled* with the pooling operation, so the image resolution is decreased from 28×28 to 14×14. These 32 smaller images are then processed in the second convolutional layer. We need filter weights again for each of these 32 images and we need filter weights for each output channel of this layer. The images are again *down-sampled* with a pooling operation, so that the image resolution is decreased from 14×14 to 7×7. The total number of features for this convolutional layer is 64.

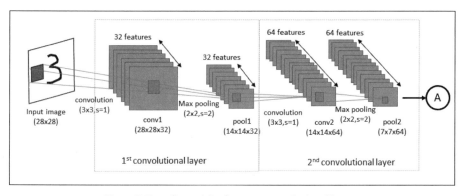

Figure 7: Data flow of the first two convolutional layers

The 64 resulting images are filtered again by a (3×3) third convolutional layer. No pooling operation is applied to this layer. The output of the third convolutional layer is 128 7×7-pixel images. These images are then *flattened* to become a single vector, of length 4×4×128 = 2048, which is used as input to a *fully connected layer*.

The *output layer* of the LeNet-5 consists of 625 neurons as input (that is, the output of the fully connected layer), and 10 neurons as output, which is used to determine the class of the image, which number is depicted in the image.

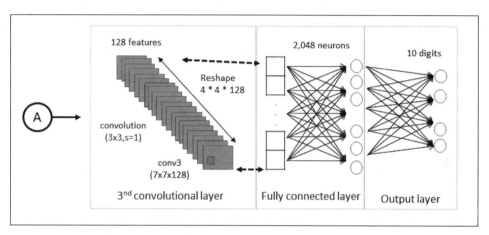

Figure 8: Data flow of the last three convolutional layers

The convolutional filters are initially *chosen at random*. The difference between the *predicted* and the *actual* class of the input image is referred to as the *cost function*, and this generalizes our network beyond the training data. The optimizer then automatically propagates this cost function back through the CNN and updates the filter weights to improve the classification error. This is done iteratively thousands of times until the classification error is sufficiently low.

Now let's see in detail how to code our first CNN. Let's start by importing the TensorFlow libraries we need for our implementation:

```
import tensorflow as tf
import numpy as np
from tensorflow.examples.tutorials.mnist import input_data
```

Set the following parameters. They indicate the number of samples to use in the training phase (`128`) and the test phase (`256`):

```
batch_size = 128
test_size = 256
```

When we define the following parameter, the value is 28 because a MNIST image is 28 pixels in height and width:

```
img_size = 28
```

For the *number of classes*, the value 10 means that we'll have one class for each 0 to 9 digits:

```
num_classes = 10
```

A placeholder variable, X, is defined for the input images. The data type of this tensor is set to float32, and the shape is set to [None, img_size, img_size, 1], where None means that the tensor may hold an arbitrary number of images:

```
X = tf.placeholder("float", [None, img_size, img_size, 1])
```

Then we set another placeholder variable, Y, for the labels that were associated correctly with input images in the placeholder variable, X. The shape of this placeholder variable is [None, num_classes], which means that it may hold an arbitrary number of labels. Each label is a vector of length num_classes, which is 10 in this case:

```
Y = tf.placeholder("float", [None, num_classes])
```

We collect the MNIST data, which will be copied into the data folder:

```
mnist = input_data.read_data_sets("MNIST-data", one_hot=True)
```

We build the datasets for *training* (trX, trY) and *testing* the network (teX, teY):

```
trX, trY, teX, teY = mnist.train.images, \
                     mnist.train.labels, \
                     mnist.test.images, \
                     mnist.test.labels
```

The trX and teX image sets must be reshaped to match the input shape:

```
trX = trX.reshape(-1, img_size, img_size, 1)
teX = teX.reshape(-1, img_size, img_size, 1)
```

We shall now proceed to defining the network's weights.

The init_weights function builds new *variables* in the shape provided and initializes the network's weights with *random values*:

```
def init_weights(shape):
    return tf.Variable(tf.random_normal(shape, stddev=0.01))
```

Each neuron of the *first convolutional layer is convoluted* to a small subset of the input tensor, with the dimensions 3×3×1. The value 32 is just the number of feature maps we are considering for this first layer. The weight, w, is then defined:

```
w = init_weights([3, 3, 1, 32])
```

The number of inputs is then increased to 32, which means that each neuron in the *second convolutional* layer is convoluted to **3x3x32** neurons of the first convolutional layer. The w2 weight is as follows:

```
w2 = init_weights([3, 3, 32, 64])
```

The value 64 represents the number of *output features* obtained. The *third convolutional* layer is convoluted to **3x3x64** neurons of the previous layer, while 128 are the resulting features.

```
w3 = init_weights([3, 3, 64, 128])
```

The *fourth layer* is **fully connected** and receives **128x4x4** inputs, while the output is equal to 625:

```
w4 = init_weights([128 * 4 * 4, 625])
```

The output layer receives 625 inputs, and the output is the number of classes:

```
w_o = init_weights([625, num_classes])
```

Note that these initializations are not actually done at this point. They are merely being defined in the TensorFlow graph.

```
p_keep_conv = tf.placeholder("float")
p_keep_hidden = tf.placeholder("float")
```

It's time to define the network model. Like the network's weights definition, it will be a *function*. It receives the X tensor, the *weights tensors*, and the *dropout parameters* as input for the convolutional and fully connected layer:

```
def model(X, w, w2, w3, w4, w_o, p_keep_conv, p_keep_hidden):
```

tf.nn.conv2d() executes the TensorFlow operation for convolution. Note that the strides for all dimensions *are set to 1*. In fact, the first and last stride must always be 1, because the first stride is for the image number and the last stride is for the input channel. The *padding parameter* is set to 'SAME', which means that the input image *is padded with zeroes*, so the size of the output is the same:

```
conv1 = tf.nn.conv2d(X, w,strides=[1, 1, 1, 1],\
                     padding='SAME')
```

Convolutional Neural Networks

Then we pass the `conv1` layer to the ReLU layer. It calculates the `max(x, 0)` function for each input pixel, x, adding some non-linearity to the formula, and allows us to learn more complicated functions:

```
conv1_a = tf.nn.relu(conv1)
```

The resulting layer is then pooled by the `tf.nn.max_pool` operator:

```
conv1 = tf.nn.max_pool(conv1_a, ksize=[1, 2, 2, 1]\
                ,strides=[1, 2, 2, 1],\
                padding='SAME')
```

It is a 2×2 max-pooling, which means that we are examining 2×2 windows and selecting the largest value in each window. Then we move *2 pixels* to the next window. We try to reduce overfitting via the `tf.nn.dropout()` function, to which we pass the `conv1` layer and the `p_keep_conv` probability value:

```
conv1 = tf.nn.dropout(conv1, p_keep_conv)
```

As you can see, the next two convolutional layers, `conv2` and `conv3`, are defined in the same way as `conv1`:

```
conv2 = tf.nn.conv2d(conv1, w2,\
                strides=[1, 1, 1, 1],\
                padding='SAME')
conv2_a = tf.nn.relu(conv2)
conv2 = tf.nn.max_pool(conv2, ksize=[1, 2, 2, 1],\
                strides=[1, 2, 2, 1],\
                padding='SAME')
conv2 = tf.nn.dropout(conv2, p_keep_conv)

conv3=tf.nn.conv2d(conv2, w3,\
                strides=[1, 1, 1, 1]\
                ,padding='SAME')

conv3 = tf.nn.relu(conv3)
```

The fully connected layers are added to the network. The input of the first `FC_layer` is the *convolutional layer* from the *previous convolution*:

```
FC_layer = tf.nn.max_pool(conv3, ksize=[1, 2, 2, 1],\
                strides=[1, 2, 2, 1],\
                padding='SAME')

FC_layer = tf.reshape(FC_layer,\
                [-1, w4.get_shape().as_list()[0]])
```

A dropout function is again used to reduce overfitting:

```
FC_layer = tf.nn.dropout(FC_layer, p_keep_conv)
```

The output layer receives `FC_layer` and the `w4` weight tensor as input. **ReLU** and **dropout** operators are applied:

```
output_layer = tf.nn.relu(tf.matmul(FC_layer, w4))
output_layer = tf.nn.dropout(output_layer, p_keep_hidden)
```

The **result** is a vector with a length of 10. This is used to determine which of the 10 input classes the image belongs to:

```
result = tf.matmul(output_layer, w_o)
return result
```

Cross-entropy is the performance measure we used in this classifier. Cross-entropy is a continuous function that is *always positive* and is *equal to zero* if the predicted output exactly matches the desired output. The goal of this optimization is therefore *to minimize* the cross-entropy, so it is as close to zero as possible, by changing the variables in the network layers. TensorFlow has a built-in function for calculating the cross-entropy. Note that the function calculates softmax internally, so we must use the output of `py_x` directly:

```
py_x = model(X, w, w2, w3, w4, w_o, p_keep_conv, p_keep_hidden)
Y_ = tf.nn.softmax_cross_entropy_with_logits_v2\
     (labels=Y,logits=py_x)
```

Now that we have defined the cross-entropy for each classified image, we have a measure of how well the model performs on each image. We need a *single scalar value* to use the cross-entropy to optimize the networks' variables, so we simply take the *average of the cross-entropy* for all of the classified images:

```
cost = tf.reduce_mean(Y_)
```

To minimize the evaluated `cost`, we must define an *optimizer*. In this case, we will use `RMSPropOptimizer`, which is an advanced form of GD. `RMSPropOptimizer` implements the *RMSProp algorithm,* which is an unpublished adaptive learning rate method that was proposed by *Geoff Hinton* in Lecture 6e of his Coursera Class. (http://www.cs.toronto.edu/~tijmen/csc321/slides/lecture_slides_lec6.pdf)

 You can find Geoff Hinton's course at https://www.coursera.org/learn/neural-networks.

`RMSPropOptimizer` also divides the learning rate by an exponentially decaying average of squared gradients. Hinton suggests setting the decay parameter to `0.9`, while a good default value for the learning rate is `0.001`:

```
optimizer = tf.train.RMSPropOptimizer(0.001, 0.9).minimize(cost)
```

Basically, the common SGD algorithm has a problem in that learning rates must scale with *1/T* (where *T* is the iteration number) to achieve convergence. *RMSProp* tries to get around this by automatically adjusting the step size so that the step is on the same scale as the gradients. As the average gradient gets smaller, the coefficient in the SGD update gets bigger to compensate.

An interesting reference about this algorithm can be found here:
http://www.cs.toronto.edu/%7Etijmen/csc321/slides/lecture_slides_lec6.pdf

Finally, we define `predict_op`, which is the *index with the largest value* across dimensions from the output of the mode:

```
predict_op = tf.argmax(py_x, 1)
```

Note that *optimization is not performed at this point*. Nothing is calculated at all because we'll just add the optimizer object to the TensorFlow graph for later execution.

We now come to defining the network's running session. There are 55,000 images in the training set, so it will take a long time to calculate the gradient of the model using all of these images. Therefore, we'll use a *small batch* of images in each iteration of the optimizer. If your computer crashes or becomes very slow because you run out of RAM, then you can reduce this number, but you may then need to perform more optimization iterations.

Now we can proceed to implementing a TensorFlow *session*:

```
with tf.Session() as sess:
    tf.global_variables_initializer().run()
    for i in range(100):
```

We get a batch of training examples, and the `training_batch` tensor now holds a subset of images and the corresponding labels:

```
training_batch = zip(range(0, len(trX), batch_size),\
                     range(batch_size, \
                     len(trX)+1, \
                     batch_size))
```

Put the batch into a `feed_dict` with appropriate names for the placeholder variables in the graph. We can now run the optimizer using this batch of training data. TensorFlow assigns the variables in feed to the placeholder variables and then runs the optimizer:

```
for start, end in training_batch:
    sess.run(optimizer, feed_dict={X: trX[start:end],\
                                   Y: trY[start:end],\
                                   p_keep_conv: 0.8,\
                                   p_keep_hidden: 0.5})
```

At the same time, we get a *shuffled* batch of test samples:

```
test_indices = np.arange(len(teX))
np.random.shuffle(test_indices)
test_indices = test_indices[0:test_size]
```

For each iteration, we display the evaluated accuracy of the batch:

```
print(i, np.mean(np.argmax(teY[test_indices], axis=1) ==\
      sess.run\
      (predict_op,\
       feed_dict={X: teX[test_indices],\
                  Y: teY[test_indices], \
                  p_keep_conv: 1.0,\
                  p_keep_hidden: 1.0})))
```

Training a network can take several hours, depending on the computational resources used. The results on my machine are as follows:

```
Successfully downloaded train-images-idx3-ubyte.gz 9912422 bytes.
Successfully extracted to train-images-idx3-ubyte.mnist 9912422 bytes.
Loading ata/train-images-idx3-ubyte.mnist
Successfully downloaded train-labels-idx1-ubyte.gz 28881 bytes.
Successfully extracted to train-labels-idx1-ubyte.mnist 28881 bytes.
Loading ata/train-labels-idx1-ubyte.mnist
Successfully downloaded t10k-images-idx3-ubyte.gz 1648877 bytes.
Successfully extracted to t10k-images-idx3-ubyte.mnist 1648877 bytes.
Loading ata/t10k-images-idx3-ubyte.mnist
Successfully downloaded t10k-labels-idx1-ubyte.gz 4542 bytes.
Successfully extracted to t10k-labels-idx1-ubyte.mnist 4542 bytes.
Loading ata/t10k-labels-idx1-ubyte.mnist
(0, 0.95703125)
```

```
(1, 0.98046875)
(2, 0.9921875)
(3, 0.99609375)
(4, 0.99609375)
(5, 0.98828125)
(6, 0.99609375)
(7, 0.99609375)
(8, 0.98828125)
(9, 0.98046875)
(10, 0.99609375)
.
.
.
..
.
(90, 1.0)
(91, 0.9921875)
(92, 0.9921875)
(93, 0.99609375)
(94, 1.0)
(95, 0.98828125)
(96, 0.98828125)
(97, 0.99609375)
(98, 1.0)
(99, 0.99609375)
```

After 10,000 iterations, the model has an accuracy of 99.60%, which is not bad!

AlexNet

The AlexNet neural network is one of the first CNNs to achieve tremendous success. The winner of the 2012 ILSVRC, this neural network was the first to get good results on a very complex dataset such as ImageNet using the standard construction of neural networks that the LeNet-5 network had defined earlier.

> The ImageNet project is a large visual database designed for use in visual object recognition software research. As of 2016, over ten million URLs of images have been hand-annotated by ImageNet to indicate the objects in the images. In at least one million of the images, bounding boxes are also provided. The database of the annotations of the third-party image URLs is freely available directly from ImageNet.

The architecture of an AlexNet is represented in the following figure:

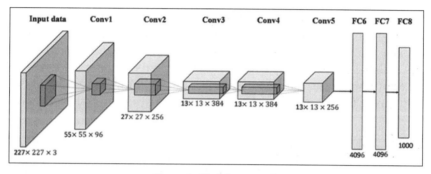

Figure 9: AlexNet network

In the AlexNet architecture, there are eight layers with trainable parameters: a series of five consecutive convolutional layers, followed by three fully connected layers. Each convolutional layer is followed by a ReLU layer, and optionally also by a max pooling layer, especially at the beginning of the network, in order to reduce the amount of space the network takes up.

All pooling layers have a 3x3 extension region and a step rate of 2: this means that you always use overlapping pooling. This is because this type of pooling provides slightly better network performance in comparison to normal pooling without overlapping. At the beginning of the network, between a pooling layer and the next convolutional layer, a couple of LRN standardization layers were always used: after some testing, it was seen that they tend to decrease the network error.

The first two fully connected layers possess 4,096 neurons, while the last one has 1,000 units, corresponding to the number of classes on the ImageNet dataset. Given the huge number of connections in the fully connected layers, a dropout layer with a ratio of 0.5 was added between each pair fully connected layers, that is, half of the neurons' activations are ignored each time. It has been noted in this case that the use of the dropout technique not only speeds up the processing of a single iteration, but also prevents overfitting quite well. Without the dropout layer, network makers claim that the original network had too much overfitting.

Transfer learning

Transfer learning consists of taking a network that has already been built and making appropriate changes to the parameters of the various layers so that it can adapt to another dataset. For example, you can use a pre-tested network on a large dataset, such as ImageNet, and train it again on a smaller dataset. Provided that our dataset is not drastically different in content to the original dataset, the pre-trained model will already have learned features that are relevant to our own classification problem.

If our dataset is not drastically different from the dataset that the pre-trained model was trained on, we can use the *fine-tuning* technique. Models that have been pre-trained on a large and diverse dataset may catch universal features such as curves and edges in its early layers that are relevant and useful in most classification problems. However, if our dataset is from a very specific domain, and no pre-trained networks in this domain can be found, we should consider training the network from scratch.

Pretrained AlexNet

We'll fine-tune a pre-trained AlexNet to distinguish between dogs and cats. AlexNet is pre-trained on the ImageNet dataset.

To execute this example, you also need to install scipy (see `https://www.scipy.org/install.html`) and **PIL** (**Pillow**), which is what scipy uses to read images: `pip install Pillow` or `pip3 install Pillow`.

Then you need to download the following files:

- `myalexnet_forward.py`: AlexNet implementation and testing code for 2017 versions of TensorFlow (Python 3.5)
- `bvlc_alexnet.npy`: The weights, which need to be in the working directory
- `caffe_classes.py`: The classes, in the same order as the outputs of the network
- `poodle.png, laska.png, dog.png, dog2.png, quail227.JPEG`: The test images (the images should be 227×227×3)

Download these files from the link: `http://www.cs.toronto.edu/~guerzhoy/tf_alexnet/` or from the code repository of this book.

First of all, we will test the network on the previously downloaded images. To do this, just run `myalexnet_forward.py` from the Python GUI.

As you can see by simply inspecting the source code (see the following snippet), the pre-trained network will be called to classify the following two images, `laska.png` and `poodle.png`, which were previously downloaded:

```
im1 = (imread("laska.png")[:,:,:3]).astype(float32)
im1 = im1 - mean(im1)
im1[:, :, 0], im1[:, :, 2] = im1[:, :, 2], im1[:, :, 0]

im2 = (imread("poodle.png")[:,:,:3]).astype(float32)
im2[:, :, 0], im2[:, :, 2] = im2[:, :, 2], im2[:, :, 0]
```

Figure 10: Images to classify

The weights and biases of the `bvlc_alexnet.npy` file are loaded by the following statement:

```
net_data = load(open("bvlc_alexnet.npy", "rb"), encoding="latin1").item()
```

The network is a set of convolutional and pooling layers followed by three fully connected states. The output of this model is a softmax function:

```
prob = tf.nn.softmax(fc8)
```

The outputs of the softmax function are classification ranks, because they indicate how strongly the network believes that the input image belongs to a class defined in the `caffe_classes.py` file.

If we run the code, we should get the following results:

```
Image 0
weasel 0.503177
black-footed ferret, ferret, Mustela nigripes 0.263265
polecat, fitch, foulmart, foumart, Mustela putorius 0.147746
mink 0.0649517
otter 0.00771955
Image 1
clumber, clumber spaniel 0.258953
komondor 0.165846
miniature poodle 0.149518
toy poodle 0.0984719
kuvasz 0.0848062
0.40007972717285156
>>>
```

In the preceding example, AlexNet gave a score of about 50% for weasel. This means the model is quite confident that the image shows a weasel and the remaining scores can be regarded as noise.

Dataset preparation

Our task is to build an image classifier that distinguishes between dogs and cats. We get some help from Kaggle, from which we can easily download the dataset: `https://www.kaggle.com/c/dogs-vs-cats/data`.

In this dataset, training set contains 20,000 labeled images, and the test and validation sets have 2,500 images.

To use the dataset, you must reshape each image to 227×227×3. In order to do this, you can use the Python code in `prep_images.py`. Otherwise, you can use the `trainDir.rar` and `testDir.rar` files from the repository of this book. They contain 6,000 reshaped images of dogs and cats for training, and 100 reshaped images for testing.

The following fine-tuning implementation, described in the section below, is implemented in `alexnet_finetune.py`, which is downloadable in the code repository of the book.

Fine-tuning implementation

Our classification task contains two categories, so the new softmax layer of the network will consist of 2 categories instead of 1,000 categories. Here is the input tensor, which is a 227×227×3 image, and the output tensor of rank 2:

```
n_classes = 2
train_x = zeros((1, 227,227,3)).astype(float32)
train_y = zeros((1, n_classes))
```

Fine-tuning implementation consists *of truncating the last layer* (the softmax layer) of the pre-trained network and replacing it with a new softmax layer that is relevant to our problem.

For example, the pre-trained network on ImageNet comes with a softmax layer with 1,000 categories.

The following code snippet defines the new softmax layer, fc8:

```
fc8W = tf.Variable(tf.random_normal\
                ([4096, n_classes]),\
                trainable=True, name="fc8w")
fc8b = tf.Variable(tf.random_normal\
                ([n_classes]),\
                trainable=True, name="fc8b")
fc8 = tf.nn.xw_plus_b(fc7, fc8W, fc8b)
prob = tf.nn.softmax(fc8)
```

Loss is a performance measure used in classification. It is a continuous function that is always positive, and if the predicted output of the model exactly matches the desired output then the cross-entropy equals zero. The goal of optimization is therefore to minimize the cross-entropy, by changing the weights and biases of the model, so it is as close to zero as possible.

TensorFlow has a built-in function for calculating cross-entropy. In order to use cross-entropy to optimize the model's variables we need a single scalar value, so we simply take the average of the cross-entropy for all the image classifications:

```
loss = tf.reduce_mean\
        (tf.nn.softmax_cross_entropy_with_logits_v2\
        (logits =prob, labels=y))
opt_vars = [v for v in tf.trainable_variables()\
            if (v.name.startswith("fc8"))]
```

Now that we have a cost measure that must be minimized, we can then create an `optimizer`:

```
optimizer = tf.train.AdamOptimizer\
            (learning_rate=learning_rate).minimize\
            (loss, var_list = opt_vars)
correct_pred = tf.equal(tf.argmax(prob, 1), tf.argmax(y, 1))
accuracy = tf.reduce_mean(tf.cast(correct_pred, tf.float32))
```

In this case, we use the `AdamOptimizer` in which the step size is set to 0.5. Note that optimization is not performed at this point. In fact, nothing is calculated at all, we just add the optimizer object to the TensorFlow graph for later execution. Then we run backpropagation on the network to fine-tune the pre-trained weights:

```
batch_size = 100
training_iters = 6000
display_step = 1
dropout = 0.85 # Dropout, probability to keep units

init = tf.global_variables_initializer()
with tf.Session() as sess:
    sess.run(init)
    step = 1
```

Keep training until we reach the maximum number of iterations:

```
while step * batch_size < training_iters:
    batch_x, batch_y = \
            next(next_batch(batch_size)) #.next()
```

Run the optimization operation (backpropagation):

```
sess.run(optimizer, \
        feed_dict={x: batch_x, \
                   y: batch_y, \
                   keep_prob: dropout})

if step % display_step == 0:
```

Calculate the batch loss and accuracy:

```
cost, acc = sess.run([loss, accuracy],\
                    feed_dict={x: batch_x, \
                               y: batch_y, \
                               keep_prob: 1.})
print ("Iter " + str(step*batch_size) \
       + ", Minibatch Loss= " + \
```

```
                            "{:.6f}".format(cost) + \
                            ", Training Accuracy= " + \
                            "{:.5f}".format(acc))

            step += 1
    print ("Optimization Finished!")
```

The training of the network produces the following results:

```
Iter 100, Minibatch Loss= 0.555294, Training Accuracy= 0.76000
Iter 200, Minibatch Loss= 0.584999, Training Accuracy= 0.73000
Iter 300, Minibatch Loss= 0.582527, Training Accuracy= 0.73000
Iter 400, Minibatch Loss= 0.610702, Training Accuracy= 0.70000
Iter 500, Minibatch Loss= 0.583640, Training Accuracy= 0.73000
Iter 600, Minibatch Loss= 0.583523, Training Accuracy= 0.73000
................................................
................................................
Iter 5400, Minibatch Loss= 0.361158, Training Accuracy= 0.95000
Iter 5500, Minibatch Loss= 0.403371, Training Accuracy= 0.91000
Iter 5600, Minibatch Loss= 0.404287, Training Accuracy= 0.91000
Iter 5700, Minibatch Loss= 0.413305, Training Accuracy= 0.90000
Iter 5800, Minibatch Loss= 0.413816, Training Accuracy= 0.89000
Iter 5900, Minibatch Loss= 0.413476, Training Accuracy= 0.90000
Optimization Finished!
```

To test our model, we compare the forecasts with the label set (cat = 0, dog = 1):

```
            output = sess.run(prob, feed_dict = {x:imlist, keep_prob: 1.})
            result = np.argmax(output,1)
            testResult = [1,1,1,1,0,0,0,0,0,0,\
                          0,1,0,0,0,0,1,1,0,0,\
                          1,0,1,1,0,1,1,0,0,1,\
                          1,1,1,0,0,0,0,0,1,0,\
                          1,1,1,1,0,1,0,1,1,0,\
                          1,0,0,1,0,0,1,1,1,0,\
                          1,1,1,1,1,0,0,0,0,0,\
                          0,1,1,1,0,1,1,1,1,0,\
                          0,0,1,0,1,1,1,1,0,0,\
                          0,0,0,1,1,0,1,1,0,0]
            count = 0
            for i in range(0,99):
```

```
        if result[i] == testResult[i]:
            count=count+1

    print("Testing Accuracy = " + str(count) +"%")
```

Finally, we have the accuracy of our model:

```
Testing Accuracy = 82%
```

VGG

VGG is the name of a team of people who presented their neural networks during ILSVRC 2014. We are talking about networks, plural, since more than one version of the same network was created, each possessing a different number of layers. Depending on the number of layers, n, with weight that one of these networks has, each of them is usually called **VGG-n**. All of these networks are more deep than AlexNet. This means that they are made up of a number of layers with more workable parameters than AlexNet, in this case 11 to 19 total trained layers. Often, only the workable layers are considered, because they are the ones that affect the processing and size of the model, as seen in the previous paragraph. However, the overall structure remains very similar: there is always an initial series of convolutional layers and a final series of fully connected layers, the latter being exactly the same as in AlexNet. What changes is therefore the number of convolutional layers used and, of course, their parameters. The following table shows all the variants built by the VGG team.

Each column, starting from the left and going to the right, shows a certain VGG network, from the deepest to the shallowest. Bold terms show what has been added in each version compared to the previous version. The ReLU layer is not shown in the table, but in the network it exists after each convolutional layer. All convolutional layers use a stride of 1:

ConvNet Configuration					
A	A-LRN	B	C	D	E
11 weight layers	11 weight layers	13 weight layers	16 weight layers	16 weight layers	19 weight layers
input (224×224 RGB image)					
conv3-64	conv3-64 LRN	conv3-64 conv3-64	conv3-64 conv3-64	conv3-64 conv3-64	conv3-64 conv3-64
maxpool					
conv3-128	conv3-128	conv3-128 conv3-128	conv3-128 conv3-128	conv3-128 conv3-128	conv3-128 conv3-128
maxpool					
conv3-256 conv3-256	conv3-256 conv3-256	conv3-256 conv3-256	conv3-256 conv3-256 conv1-256	conv3-256 conv3-256 conv3-256	conv3-256 conv3-256 conv3-256 conv3-256
maxpool					
conv3-512 conv3-512	conv3-512 conv3-512	conv3-512 conv3-512	conv3-512 conv3-512 conv1-512	conv3-512 conv3-512 conv3-512	conv3-512 conv3-512 conv3-512 conv3-512
maxpool					
conv3-512 conv3-512	conv3-512 conv3-512	conv3-512 conv3-512	conv3-512 conv3-512 conv1-512	conv3-512 conv3-512 conv3-512	conv3-512 conv3-512 conv3-512 conv3-512
maxpool					
FC-4096					
FC-4096					
FC-1000					
soft-max					

Table: VGGs network architectures

Note that AlexNet does not have convolutional layers with a fairly large receptive field: here, all receptive fields are 3×3, except for a couple of convolutional layers in VGG-16 that have a 1×1 receptive field. Recall that a convex layer with a 1-step gradient does not change the input space size while modifying the depth value that becomes the same as the number of kernels used. Consequently, the VGG convolutional layers do not ever affect the width and height of the input volumes; only the pooling layers do that. The idea of using a series of convolutional layers with a smaller receptive field, which in the end overall simulates a single convolutional layer with a larger receptive field, is motivated by the fact that in this way multiple ReLU layers are used instead of one alone, thereby increasing the nonlinearity of the activation function and thus making it more discriminating. It also serves to reduce the number of parameters used. These networks are considered an evolution of AlexNet because, overall, and with the same dataset, they perform better than AlexNet. The main concept demonstrated with VGG networks is that more a congestion neural network is profound and more its performance increases. However, it is necessary to have more and more powerful hardware, otherwise network training would become problematic.

For the VGGs, four NVIDIA Titan Blacks were used with 6 GB of memory each. VGGs therefore have better performance but need a lot of hardware for training and also use a very large number of parameters: the VGG-19 model, for example, is about 550 MB (twice as much as AlexNet). Smaller VGG networks still have a model of about 507 MB.

Artistic style learning with VGG-19

In this project, we'll use a pretrained VGG-19 to learn the style and patterns created by an artist and transfer them to an image (the project file is style_transfer.py in the GitHub repository of this book). This technique is called artistic style learning (see the paper A Neural Algorithm of Artistic Style (https://arxiv.org/pdf/1508.06576.pdf) by *Gatys and others*). According to the academic literature, artistic style learning is defined as follows: given two images as input, synthesize a third image that has the semantic content of the first image and the texture/style of the second.

For this to work properly, we need to train a deep convolutional neural network to build the following:

- A *content extractor* to determine the content of **image A**
- A *style extractor* to determine the style of **image B**
- A *merger* to merge some arbitrary content with another arbitrary style to obtain the **final result**

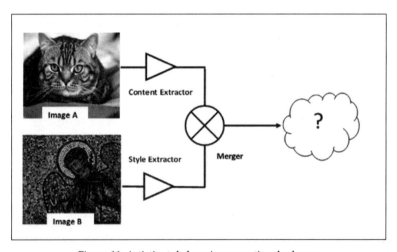

Figure 11: Artistic style learning operational schema

Input images

The input images, each of which is 478×478 pixels, are the following images (`cat.jpg`, and `mosaic.jpg`) that you will also find in the code repository for this book:

Figure 12: Input images in Artistic Style Learning

In order to be analyzed by the VGG model, these images need to be *preprocessed*:

1. Adding an extra dimension
2. Subtracting `MEAN_VALUES` from the input image:

   ```
   MEAN_VALUES = np.array([123.68, 116.779, 103.939]).
   reshape((1,1,1,3))
   content_image = preprocess('cat.jpg')
   style_image = preprocess('mosaic.jpg')

   def preprocess(path):
       image = plt.imread(path)
       image = image[np.newaxis]
       image = image - MEAN_VALUES
       return image
   ```

Content extractor and loss

To isolate the semantic content of an image, we use a **pre-trained** VGG-19 neural network, made some slight tweaks in the weights to adapt to this problem, and then used the output of one of the hidden layers as a content extractor. The following figure shows the CNN used for this problem:

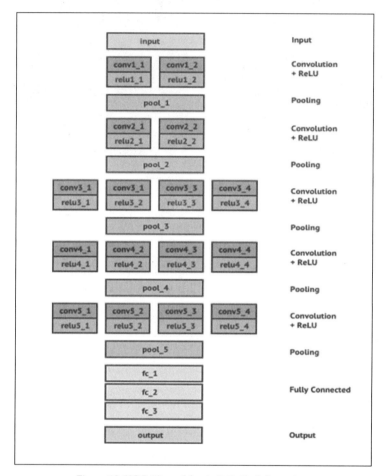

Figure 13: VGG-19 used for Artistic Style Learning

The pre-trained VGG is loaded using the following code:

```
import scipy.io
vgg = scipy.io.loadmat('imagenet-vgg-verydeep-19.mat')
```

The `imagenet-vgg-verydeep-19.mat` model should be downloaded from http://www.vlfeat.org/matconvnet/models/imagenet-vgg-verydeep-19.mat.

This model has 43 layers, 19 of which are convolutional layers. The rest are max pooling/activation/fully connected layers.

We can check the shape of each convolutional layer:

```
[print (vgg_layers[0][i][0][0][2][0][0].shape,\
       vgg_layers[0][i][0][0][0][0]) for i in range(43)
if 'conv' in vgg_layers[0][i][0][0][0][0] \
or 'fc' in vgg_layers[0][i][0][0][0][0]]
```

The result of the preceding code is as follows:

(3, 3, 3, 64) conv1_1
(3, 3, 64, 64) conv1_2
(3, 3, 64, 128) conv2_1
(3, 3, 128, 128) conv2_2
(3, 3, 128, 256) conv3_1
(3, 3, 256, 256) conv3_2
(3, 3, 256, 256) conv3_3
(3, 3, 256, 256) conv3_4
(3, 3, 256, 512) conv4_1
(3, 3, 512, 512) conv4_2
(3, 3, 512, 512) conv4_3
(3, 3, 512, 512) conv4_4
(3, 3, 512, 512) conv5_1
(3, 3, 512, 512) conv5_2
(3, 3, 512, 512) conv5_3
(3, 3, 512, 512) conv5_4
(7, 7, 512, 4096) fc6
(1, 1, 4096, 4096) fc7
(1, 1, 4096, 1000) fc8

Each shape is represented in the following way: [kernel height, kernel width, number of input channels, number of output channels].

The first layer has 3 input channels because the input is an RGB image, while the number of output channels goes from 64 to 512 for the convolutional layers, and all kernels are **3x3** matrices.

Convolutional Neural Networks

Then we apply the *transfer learning* technique in order to adapt the VGG-19 network to our problem:

1. Fully connected layers are not needed because they are used for object recognition.
2. Max pooling layers are substituted for average pool layers in order to achieve better results. Average layers work in the same way as the kernels in the convolutional layers.

```
IMAGE_WIDTH = 478
IMAGE_HEIGHT = 478
INPUT_CHANNELS = 3
model = {}
model['input'] = tf.Variable(np.zeros((1, IMAGE_HEIGHT,\
                                      IMAGE_WIDTH,\
                                      INPUT_CHANNELS)),\
                             dtype = 'float32')

model['conv1_1']  = conv2d_relu(model['input'], 0, 'conv1_1')
model['conv1_2']  = conv2d_relu(model['conv1_1'], 2, 'conv1_2')
model['avgpool1'] = avgpool(model['conv1_2'])

model['conv2_1']  = conv2d_relu(model['avgpool1'], 5, 'conv2_1')
model['conv2_2']  = conv2d_relu(model['conv2_1'], 7, 'conv2_2')
model['avgpool2'] = avgpool(model['conv2_2'])

model['conv3_1']  = conv2d_relu(model['avgpool2'], 10, 'conv3_1')
model['conv3_2']  = conv2d_relu(model['conv3_1'], 12, 'conv3_2')
model['conv3_3']  = conv2d_relu(model['conv3_2'], 14, 'conv3_3')
model['conv3_4']  = conv2d_relu(model['conv3_3'], 16, 'conv3_4')
model['avgpool3'] = avgpool(model['conv3_4'])

model['conv4_1']  = conv2d_relu(model['avgpool3'], 19, 'conv4_1')
model['conv4_2']  = conv2d_relu(model['conv4_1'], 21, 'conv4_2')
```

```
    model['conv4_3']   = conv2d_relu(model['conv4_2'], 23,
      'conv4_3')
    model['conv4_4']   = conv2d_relu(model['conv4_3'], 25,
      'conv4_4')
    model['avgpool4']  = avgpool(model['conv4_4'])

    model['conv5_1']   = conv2d_relu(model['avgpool4'], 28,
      'conv5_1')
    model['conv5_2']   = conv2d_relu(model['conv5_1'], 30,
      'conv5_2')
    model['conv5_3']   = conv2d_relu(model['conv5_2'], 32,
      'conv5_3')
    model['conv5_4']   = conv2d_relu(model['conv5_3'], 34,
      'conv5_4')
    model['avgpool5']  = avgpool(model['conv5_4'])
```

Here we defined the `contentloss` function that measures the *difference in content* between two images p and x:

```
def contentloss(p, x):
    size = np.prod(p.shape[1:])
    loss = (1./(2*size)) * tf.reduce_sum(tf.pow((x - p),2))
    return loss
```

This function tends to be 0 when the input images are very close to each other in terms of content and grows as their content *deviates*.

We'll use `contentloss` on the `conv5_4` layer. This is the *output layer* and its output would be prediction, hence we need to compare this prediction with actual one using the `contentloss` function:

```
content_loss = contentloss\
             (sess.run(model['conv5_4']), model['conv5_4'])
```

Minimizing the `content_loss` means that the mixed image has feature activation in the given layers that is *very similar* to the activation of the content image.

Style extractor and loss

Style extractor uses the **Gram** matrix of the filters for a given hidden layer. Simply speaking, using this matrix, we can *destroy* the semantic of the image *preserving* its basic components and making it a good *texture extractor*:

```
def gram_matrix(F, N, M):
    Ft = tf.reshape(F, (M, N))
    return tf.matmul(tf.transpose(Ft), Ft)
```

The `style_loss`, measures how *close in style* two images are to one another. This function is the sum of the squared difference of the elements of the Gram matrix produced by the style image and input `noise_image`:

```
noise_image = np.random.uniform\
              (-20, 20,\
               (1, IMAGE_HEIGHT, \
                IMAGE_WIDTH,\
                INPUT_CHANNELS)).astype('float32')

def style_loss(a, x):
    N = a.shape[3]
    M = a.shape[1] * a.shape[2]
    A = gram_matrix(a, N, M)
    G = gram_matrix(x, N, M)
    result = (1/(4 * N**2 * M**2))* tf.reduce_sum(tf.pow(G-A,2))
    return result
```

`style_loss` grows as its two input images (a and x) tend *to deviate* in style.

Merger and total loss

We can merge the content and style loss so that the input `noise_image` is trained to output (in the layers) a *similar style* as the style image, along with features that are similar to the content image:

```
alpha = 1
beta = 100
total_loss = alpha * content_loss + beta * styleloss
```

Training

Minimize the loss in the network so that the style loss (the loss between the output image's style and the style of the style image), content loss (loss between the content image and the output image), and the total variation loss are as low as possible:

```
train_step = tf.train.AdamOptimizer(1.5).minimize(total_loss)
```

Chapter 4

The output image generated from such a network should resemble the input image and have the stylist attributes of the style image.

Finally, we can prepare the network for training:

```
sess.run(tf.global_variables_initializer())
sess.run(model['input'].assign(input_noise))
for it in range(2001):
    sess.run(train_step)
    if it%100 == 0:
        mixed_image = sess.run(model['input'])
        print('iteration:',it,'cost: ', sess.run(total_loss))
        filename = 'out2/%d.png' % (it)
        deprocess(filename, mixed_image)
```

The training time could be very time-consuming, but the results could be very interesting:

```
iteration: 0 cost:    8.14037e+11
iteration: 100 cost:  1.65584e+10
iteration: 200 cost:  5.22747e+09
iteration: 300 cost:  2.72995e+09
iteration: 400 cost:  1.8309e+09
iteration: 500 cost:  1.36818e+09
iteration: 600 cost:  1.0804e+09
iteration: 700 cost:  8.83103e+08
iteration: 800 cost:  7.38783e+08
iteration: 900 cost:  6.28652e+08
iteration: 1000 cost: 5.41755e+08
```

After 1,000 iterations, we have created a new mosaic:

Figure 14: Output image in Artistic Style Learning

That's really amazing! You can finally train your neural network to paint like Picasso...have fun!

Inception-v3

The Inception micro-architecture was first introduced by *Szegedy and others* in their 2014 paper, *Going Deeper with Convolutions*:

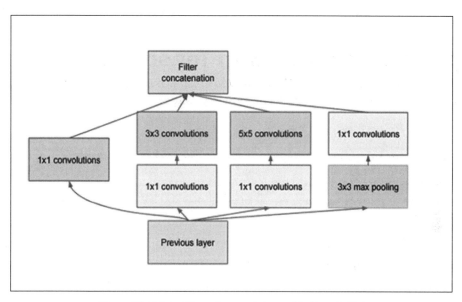

Figure 15: Original Inception module used in GoogLeNet

The goal of the inception module is to act as a multi-level feature extractor by computing **1×1**, **3×3**, and **5×5** convolutions within the same module of the network—the output of these filters is then stacked along the channel dimension before being fed into the next layer in the network. The original incarnation of this architecture was called **GoogLeNet**, but subsequent manifestations have simply been called **Inception vN**, where **N** refers to the version number put out by Google.

You might wonder why we are using different types of convolution on the same input. The answer is that it is not always possible to obtain enough useful features to perform an accurate classification with a single convolution, as far as its parameters have been carefully studied. In fact, with some input it works better with convolutions small kernels, while others get better results with other types of kernels. It is probably for this reason that the GoogLeNet team wants to consider some alternatives within their own network. As mentioned earlier, GoogLeNet uses three types of convolutional layer at the same network level (that is, they are in parallel) for this purpose: a 1×1 layer, a 3×3 layer, and a 5×5 layer.

The result of this 3-layer parallel local architecture is the combination of all their output values, chained into a single vector output, that will be the input of the next layer. This is done by using a layer concat. In addition to the three parallel convolutional layers, in the same local structure a pooling layer has been added, because pooling operations are essential to the success of a CNN.

Exploring Inception with TensorFlow

From the following link, `https://github.com/tensorflow/models`, you should be able to download the corresponding models repository.

Then type the following command:

`cd models/tutorials/image/imagenet python classify_image.py`

`classify_image.py` downloads the trained model from `tensorflow.org` when the program is run for the first time. You'll need about 200 MB of free space available on your hard disk.

The preceding command will classify the supplied image of a panda. If the model runs correctly, the script will produce the following output:

`giant panda, panda, panda bear, coon bear, Ailuropoda melanoleuca (score = 0.88493)`

`indri, indris, Indri indri, Indri brevicaudatus (score = 0.00878)`

`lesser panda, red panda, panda, bear cat, cat bear, Ailurus fulgens (score = 0.00317)`

`custard apple (score = 0.00149)`

`earthstar (score = 0.00127)`

If you wish to supply other JPEG images, you may do so by editing:

```
image_file argument:
python classify_image.py --image=image.jpg
```

You can have fun testing Inception by downloading images from the internet and seeing what results it produces.

For example, you can try the following image (we renamed it inception_image.jpg), taken from https://pixabay.com/it/:

Figure 16: Input image to classify with Inception-v3

The result is as follows:

```
python classify_image.py --image=inception_example.jpg
strawberry (score = 0.91541)
crayfish, crawfish, crawdad, crawdaddy (score = 0.01208)
chocolate sauce, chocolate syrup (score = 0.00628)
cockroach, roach (score = 0.00572)
grocery store, grocery, food market, market (score = 0.00264)
```

That sounds about right!

Emotion recognition with CNNs

One of the hardest problems to solve in deep learning has nothing to do with neural networks: it's the problem of getting the right data in the right format. However, the Kaggle platform (https://www.kaggle.com/) provides new problems, and new datasets to study.

Kaggle was founded in 2010 as a platform for predictive modeling and analytics competitions on which companies and researchers post their data and statisticians and data miners from all over the world compete to produce the best models. In this section, we show how to make a CNN for emotion detection from facial images. The train and test set of this example can be downloaded from https://inclass.kaggle.com/c/facial-keypoints-detector/data.

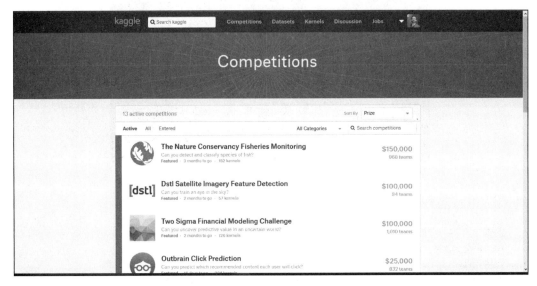

Figure 17: Kaggle competition page

The train set consists of 3,761 grayscale images that are 48×48 pixels in size and 3,761 labels, each with 7 elements.

Each element encodes an emotion: 0 = anger, 1 = disgust, 2 = fear, 3 = happiness, 4 = sadness, 5 = surprise, 6 = neutral.

In a classic Kaggle competition, the set of labels obtained from the test set must be evaluated by the platform. In this example, we will train a neural network from the train set, after which we will evaluate the model on a single image.

Before starting the CNN implementation, we'll take a look at the downloaded data by implementing a simple procedure (file download_and_display_images.py).

Import the libraries:

```
import tensorflow as tf
import numpy as np
from matplotlib import pyplot as plt
import EmotionUtils
```

The `read_data` function allows you to build all the datasets, starting with the downloaded data, which you can find in the `EmotionUtils` library in the code repository for this book:

```
FLAGS = tf.flags.FLAGS
tf.flags.DEFINE_string("data_dir",\
                       "EmotionDetector/",\
                       "Path to data files")
images = []
images = EmotionUtils.read_data(FLAGS.data_dir)

train_images = images[0]
train_labels = images[1]
valid_images = images[2]
valid_labels = images[3]
test_images  = images[4]
```

Then print the shape of the train and test images:

```
print ("train images shape = ",train_images.shape)
print ("test labels shape = ",test_images.shape)
```

Display the first image of the train set and its correct label:

```
image_0 = train_images[0]
label_0 = train_labels[0]
print ("image_0 shape = ",image_0.shape)
print ("label set = ",label_0)
image_0 = np.resize(image_0,(48,48))

plt.imshow(image_0, cmap='Greys_r')
plt.show()
```

There are 3,761 48×48-pixel grayscale images:

```
train images shape =  (3761, 48, 48, 1)
```

There are 3,761 class labels, with each class containing seven elements:

```
train labels shape =  (3761, 7)
```

The test set is formed of 1,312 48x48-pixel grayscale images:

```
test labels shape =  (1312, 48, 48, 1)
```

A single image has the following shape:

```
image_0 shape =  (48, 48, 1)
```

The label set for the first image is as follows:

```
label set = [ 0.  0.  0.  1.  0.  0.  0.]
```

This label corresponds to happy, and the image is visualized in the following matplot figure:

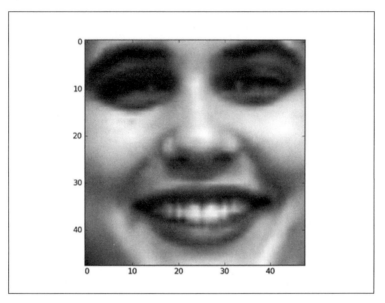

Figure 18: First image from the emotion detection face dataset

We shall now move on to the CNN architecture.

The following figure shows how the data flows in the CNN that will be implemented:

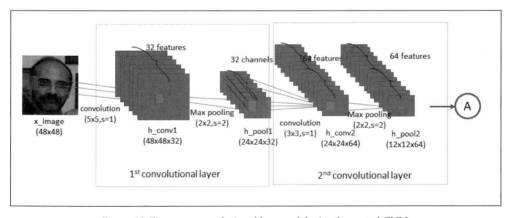

Figure 19: First two convolutional layers of the implemented CNN

Chapter 4

The network has *two convolutional layers, two fully-connected layers,* and finally a *softmax classification layer*. The input image (48×48 pixels) is processed in the first convolutional layer using a 5×5 convolutional kernel. This results in 32 images, one for each filter used. The images are also downsampled by a max pooling operation to decrease the images from 48×48 to 24×24 pixels. These 32 smaller images are then processed by a second convolutional layer; this results in 64 new images (see the preceding figure). The resulting images are downsampled again, to 12×12 pixels, by a second pooling operation.

The output of this second pooling layer is 64 12×12-pixel images. These are then flattened to a single vector of length 12 × 12 × 64 = 9,126, which is used as the input to a fully-connected layer with 256 neurons. This feeds into another fully-connected layer with 10 neurons, one for each of the classes, which is used to determine the class of the image, that is, which emotion is depicted in the image.

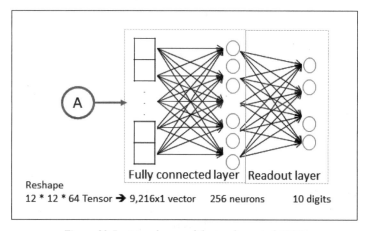

Figure 20: Last two layers of the implemented CNN

Let's move on to the *weights* and *bias* definitions. The following data structure represents the definition of the network's weights and summarizes what we have described so far:

```
weights = {
    'wc1': weight_variable([5, 5, 1, 32], name="W_conv1"),
    'wc2': weight_variable([3, 3, 32, 64],name="W_conv2"),
    'wf1': weight_variable([(IMAGE_SIZE // 4) * (IMAGE_SIZE // 4)
                            \* 64,256],name="W_fc1"),
    'wf2': weight_variable([256, NUM_LABELS], name="W_fc2")
}
```

Note that the convolutional filters *are randomly initialized*, so the classification is done randomly:

```
def weight_variable(shape, stddev=0.02, name=None):
    initial = tf.truncated_normal(shape, stddev=stddev)
    if name is None:
        return tf.Variable(initial)
    else:
        return tf.get_variable(name, initializer=initial)
```

In a similar way, we have defined the *bias*:

```
biases = {
    'bc1': bias_variable([32], name="b_conv1"),
    'bc2': bias_variable([64], name="b_conv2"),
    'bf1': bias_variable([256], name="b_fc1"),
    'bf2': bias_variable([NUM_LABELS], name="b_fc2")
}

def bias_variable(shape, name=None):
    initial = tf.constant(0.0, shape=shape)
    if name is None:
        return tf.Variable(initial)
    else:
        return tf.get_variable(name, initializer=initial)
```

An optimizer must propagate the error back through the CNN using the chain rule of differentiation and update the filter weights to improve the classification error. The difference between the predicted and true class of the input image is measured by the `loss` function. It takes as input the predicted output of the `pred` model and the desired output `label`:

```
def loss(pred, label):
    cross_entropy_loss =\
    tf.reduce_mean(tf.nn.softmax_cross_entropy_with_logits_v2\
                (logits=pred, labels=label))
    tf.summary.scalar('Entropy', cross_entropy_loss)
    reg_losses = tf.add_n(tf.get_collection("losses"))
    tf.summary.scalar('Reg_loss', reg_losses)
        return cross_entropy_loss + REGULARIZATION * reg_losses
```

The `tf.nn.softmax_cross_entropy_with_logits_v2(pred, label)` function computes `cross_entropy_loss` of the result after applying the softmax function (but it does it all together in a mathematically careful way). It's like the result of the following:

```
a = tf.nn.softmax(x)
b = cross_entropy(a)
```

We calculate `cross_entropy_loss` for each of the classified images, so we'll measure how well the model performs on each image individually.

We take the cross-entropy's average for the classified images:

```
cross_entropy_loss =    tf.reduce_mean(tf.nn.softmax_cross_entropy_
with_logits_v2 (logits=pred, labels=label))
```

To prevent overfitting, we will use L2 regularization, which consists of inserting an additional term to the `cross_entropy_loss`:

```
reg_losses = tf.add_n(tf.get_collection("losses"))
return cross_entropy_loss + REGULARIZATION * reg_losses
```

where:

```
def add_to_regularization_loss(W, b):
    tf.add_to_collection("losses", tf.nn.l2_loss(W))
    tf.add_to_collection("losses", tf.nn.l2_loss(b))
```

 See http://www.kdnuggets.com/2015/04/preventing-overfitting-neural-networks.html/2 for more information.

We have built the network's *weights* and *bias* and the optimization procedure. However, as with all the implemented networks we must start the implementation by importing all the necessary libraries:

```
import tensorflow as tf
import numpy as np
from datetime import datetime
import EmotionUtils
import os, sys, inspect
from tensorflow.python.framework import ops
import warnings

warnings.filterwarnings("ignore")
os.environ['TF_CPP_MIN_LOG_LEVEL'] = '3'
ops.reset_default_graph()
```

Convolutional Neural Networks

We then set the paths for storing the dataset on your computer, and the network parameters:

```
FLAGS = tf.flags.FLAGS
tf.flags.DEFINE_string("data_dir",\
                    "EmotionDetector/",\
                    "Path to data files")
tf.flags.DEFINE_string("logs_dir",\
                    "logs/EmotionDetector_logs/",\
                    "Path to where log files are to be saved")
tf.flags.DEFINE_string("mode",\
                    "train",\
                    "mode: train (Default)/ test")
BATCH_SIZE = 128
LEARNING_RATE = 1e-3
MAX_ITERATIONS = 1001
REGULARIZATION = 1e-2
IMAGE_SIZE = 48
NUM_LABELS = 7
VALIDATION_PERCENT = 0.1
```

The `emotion_cnn` function implements our model:

```
def emotion_cnn(dataset):
    with tf.name_scope("conv1") as scope:
        tf.summary.histogram("W_conv1", weights['wc1'])
        tf.summary.histogram("b_conv1", biases['bc1'])
        conv_1 = tf.nn.conv2d(dataset, weights['wc1'],\
                        strides=[1, 1, 1, 1],\
                        padding="SAME")

        h_conv1 = tf.nn.bias_add(conv_1, biases['bc1'])
        h_1 = tf.nn.relu(h_conv1)
        h_pool1 = max_pool_2x2(h_1)
        add_to_regularization_loss(weights['wc1'], biases['bc1'])

    with tf.name_scope("conv2") as scope:
        tf.summary.histogram("W_conv2", weights['wc2'])
        tf.summary.histogram("b_conv2", biases['bc2'])
        conv_2 = tf.nn.conv2d(h_pool1, weights['wc2'],\
                        strides=[1, 1, 1, 1], \
                        padding="SAME")
        h_conv2 = tf.nn.bias_add(conv_2, biases['bc2'])
```

```
            h_2 = tf.nn.relu(h_conv2)
            h_pool2 = max_pool_2x2(h_2)
            add_to_regularization_loss(weights['wc2'], biases['bc2'])

        with tf.name_scope("fc_1") as scope:
            prob=0.5
            image_size = IMAGE_SIZE // 4
            h_flat = tf.reshape(h_pool2,[-1,image_size*image_size*64])
            tf.summary.histogram("W_fc1", weights['wf1'])
            tf.summary.histogram("b_fc1", biases['bf1'])
            h_fc1 = tf.nn.relu(tf.matmul\
                    (h_flat, weights['wf1']) + biases['bf1'])
            h_fc1_dropout = tf.nn.dropout(h_fc1, prob)

        with tf.name_scope("fc_2") as scope:
            tf.summary.histogram("W_fc2", weights['wf2'])
            tf.summary.histogram("b_fc2", biases['bf2'])
            pred = tf.matmul(h_fc1_dropout, weights['wf2']) +\
                   biases['bf2']
        return pred
```

Then we define a `main` function, in which we'll define the dataset, the input and output placeholder variables, and the main session, in order to start the training procedure:

```
def main(argv=None):
```

The first operation in this function is to load the dataset for training and validation. We'll use the training set to teach the classifier to recognize the to-be-predicted labels, and the we'll use the validation set to evaluate the classifier's performance:

```
    train_images,\
    train_labels,\
    valid_images,\
    valid_labels,\ test_images=EmotionUtils.read_data(FLAGS.data_dir)
    print("Train size: %s" % train_images.shape[0])
    print('Validation size: %s' % valid_images.shape[0])
    print("Test size: %s" % test_images.shape[0])
```

We define the placeholder variable for the input images. This allows us to change the images that are input to the TensorFlow graph. The datatype is set to `float32`, the shape is set to `[None, img_size, img_size, 1]` (where `None` means that the tensor may hold an arbitrary number of images with each image being `img_size` pixels high and `img_size` pixels wide), and `1` is the number of color channels:

```
input_dataset = tf.placeholder(tf.float32, \
                               [None, \
                               IMAGE_SIZE, \
                               IMAGE_SIZE, 1],name="input")
```

Next, we have the placeholder variable for the labels correctly associated with the images that were input in the placeholder variable, `input_dataset`. The shape of this placeholder variable is `[None, NUM_LABELS]`, which means it may hold an arbitrary number of labels, and each label is a vector of length `NUM_LABELS`, which is 7 in this case:

```
input_labels = tf.placeholder(tf.float32,\
                              [None, NUM_LABELS])
```

`global_step` keeps track of the number of optimization iterations performed so far. We want to save this variable with all the other TensorFlow variables in the checkpoints. Note that `trainable=False`, which means that TensorFlow will not try to optimize this variable:

```
global_step = tf.Variable(0, trainable=False)
```

The following variable, `dropout_prob`, is for dropout optimization:

```
dropout_prob = tf.placeholder(tf.float32)
```

Now create the neural network for the test phase. The `emotion_cnn()` function returns the predicted class labels `pred` for the `input_dataset`:

```
pred = emotion_cnn(input_dataset)
```

`output_pred` is the predictions for the test and validation, which we'll compute in the running session:

```
output_pred = tf.nn.softmax(pred,name="output")
```

`loss_val` contains the difference between the predicted class (`pred`) and the actual class of the input image (`input_labels`):

```
loss_val = loss(pred, input_labels)
```

`train_op` defines the optimizer used to minimize the cost function. In this case, we again use `AdamOptimizer`:

```
train_op = tf.train.AdamOptimizer\
            (LEARNING_RATE).minimize\
                    (loss_val, global_step)
```

`summary_op` is used for TensorBoard visualizations:

```
summary_op = tf.summary.merge_all()
```

Once the graph has been created, we have to create a TensorFlow session, which is used to execute the graph:

```
with tf.Session() as sess:
    sess.run(tf.global_variables_initializer())
    summary_writer = tf.summary.FileWriter(FLAGS.logs_dir,
    sess.graph)
```

We define a `saver` to restore the model:

```
saver = tf.train.Saver()
ckpt = tf.train.get_checkpoint_state(FLAGS.logs_dir)
if ckpt and ckpt.model_checkpoint_path:
    saver.restore(sess, ckpt.model_checkpoint_path)
    print ("Model Restored!")
```

We next need to get a batch of training examples. `batch_image` now holds a batch of images and `batch_label` contains the correct labels for those images:

```
for step in xrange(MAX_ITERATIONS):
        batch_image, batch_label =
    get_next_batch(train_images,\
                            train_labels,\
                            step)
```

We put the batch into a `dict` containing the proper names for the placeholder variables in the TensorFlow graph:

```
feed_dict = {input_dataset: batch_image, \
             input_labels: batch_label}
```

We run the optimizer using this batch of training data. TensorFlow assigns the variables in `feed_dict_train` to the placeholder variables and then runs the optimizer:

```
sess.run(train_op, feed_dict=feed_dict)
if step % 10 == 0:
    train_loss,\
            summary_str =\
                sess.run([loss_val,summary_op],\
                        feed_dict=feed_dict)
    summary_writer.add_summary(summary_str,\
                            global_step=step)
    print ("Training Loss: %f" % train_loss)
```

When the running step is a multiple of 100 we run the trained model on the validation set:

```
if step % 100 == 0:
    valid_loss = \
            sess.run(loss_val, \
                feed_dict={input_dataset:
  valid_images, input_labels: valid_labels})
```

Then we print out the loss value:

```
print ("%s Validation Loss: %f" \
    % (datetime.now(), valid_loss))
```

At the end of the training session, the model is saved:

```
saver.save(sess, FLAGS.logs_dir\
        + 'model.ckpt', \
        global_step=step)

if __name__ == "__main__":
    tf.app.run()
```

Here is the output. As you can see, the loss function decreases during the simulation:

```
Reading train.csv ...
(4178, 48, 48, 1)
(4178, 7)
Reading test.csv ...
Picking ...
Train size: 3761
Validation size: 417
```

```
Test size: 1312
2018-02-24 15:17:45.421344 Validation Loss: 1.962773
2018-02-24 15:19:09.568140 Validation Loss: 1.796418
2018-02-24 15:20:35.122450 Validation Loss: 1.328313
2018-02-24 15:21:58.200816 Validation Loss: 1.120482
2018-02-24 15:23:24.024985 Validation Loss: 1.066049
2018-02-24 15:24:38.838554 Validation Loss: 0.965881
2018-02-24 15:25:54.761599 Validation Loss: 0.953470
2018-02-24 15:27:15.592093 Validation Loss: 0.897236
2018-02-24 15:28:39.881676 Validation Loss: 0.838831
2018-02-24 15:29:53.012461 Validation Loss: 0.910777
2018-02-24 15:31:14.416664 Validation Loss: 0.888537
>>>
```

However, the model can be improved by acting on hyperparameters or changing the architecture.

In the next section, we will see how to effectively test the model on your own images.

Testing the model on your own image

The dataset we are using is standardized. All faces are pointing at the camera and the expressions are exaggerated and even comical in some situations. Let's see now what happens if we use a more natural image. Make sure that there is no text overlaid on the face, the emotion is recognizable, and the face is pointing mostly at the camera.

I started with this JPEG image (it's a color image that you can download from the book's code repository):

Figure 21: Input image

Using Matplotlib and other NumPy Python libraries, we convert the input color image into a *valid input* for the network, that is, a *grayscale* image:

```
img = mpimg.imread('author_image.jpg')
gray = rgb2gray(img)
```

The conversion function is as follows:

```
def rgb2gray(rgb):
    return np.dot(rgb[...,:3], [0.299, 0.587, 0.114])
```

The result is shown in the following figure:

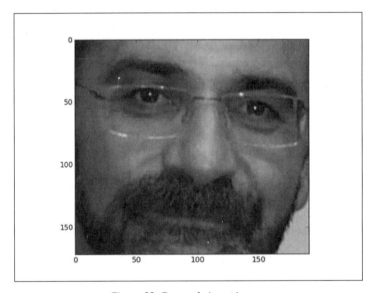

Figure 22: Grayscale input image

Finally, we can feed the network with this image, but first we must define a running TensorFlow session:

```
sess = tf.InteractiveSession()
```

Then we can recall the previously saved model:

```
new_saver = tf.train.\
import_meta_graph('logs/EmotionDetector_logs/model.ckpt-1000.meta')
new_saver.restore(sess,'logs/EmotionDetector_logs/model.ckpt-1000')
tf.get_default_graph().as_graph_def()
x = sess.graph.get_tensor_by_name("input:0")
y_conv = sess.graph.get_tensor_by_name("output:0")
```

To test an image, we must reshape it into a valid 48×48×1 format for the network:

```
image_test = np.resize(gray,(1,48,48,1))
```

We evaluate the same picture several times (`1000`) in order to get a range of possible emotions present in the input image:

```
tResult = testResult()
num_evaluations = 1000
for i in range(0,num_evaluations):
    result = sess.run(y_conv, feed_dict={x:image_test})
    label = sess.run(tf.argmax(result, 1))
    label = label[0]
    label = int(label)
    tResult.evaluate(label)

tResult.display_result(num_evaluations)
```

After few seconds, a result like this should appear:

```
>>>
anger = 0.1%
disgust = 0.1%
fear = 29.1%
happy = 50.3%
sad = 0.1%
surprise = 20.0%
neutral = 0.3%
>>>
```

The highest percentage confirms (`happy = 50.3%`) that we are on the right track. Of course, this doesn't mean that our model is accurate. Possible improvements can result from a greater and more diverse training set, changing the network's parameters, or modifying the network's architecture.

Source code

The second part of the implemented classifier is listed here:

```
from scipy import misc
import numpy as np
import matplotlib.cm as cm
import tensorflow as tf
from matplotlib import pyplot as plt
```

Convolutional Neural Networks

```python
import matplotlib.image as mpimg
import EmotionUtils
from EmotionUtils import testResult


def rgb2gray(rgb):
    return np.dot(rgb[...,:3], [0.299, 0.587, 0.114])

img = mpimg.imread('author_image.jpg')
gray = rgb2gray(img)
plt.imshow(gray, cmap = plt.get_cmap('gray'))
plt.show()

sess = tf.InteractiveSession()
new_saver = tf.train.import_meta_graph('logs/model.ckpt-1000.meta')
new_saver.restore(sess, 'logs/model.ckpt-1000')
tf.get_default_graph().as_graph_def()
x = sess.graph.get_tensor_by_name("input:0")
y_conv = sess.graph.get_tensor_by_name("output:0")

image_test = np.resize(gray,(1,48,48,1))
tResult = testResult()
num_evaluations = 1000
for i in range(0,num_evaluations):
    result = sess.run(y_conv, feed_dict={x:image_test})
    label = sess.run(tf.argmax(result, 1))
    label = label[0]
    label = int(label)
    tResult.evaluate(label)

tResult.display_result(num_evaluations)
```

We implement the `testResult` Python class to display the resulting percentages. It can be found in the `EmotionUtils` file.

Here is the implementation of this class:

```python
class testResult:

    def __init__(self):
        self.anger = 0
        self.disgust = 0
        self.fear = 0
```

```
            self.happy = 0
            self.sad = 0
            self.surprise = 0
            self.neutral = 0

        def evaluate(self,label):

            if (0 == label):
                self.anger = self.anger+1
            if (1 == label):
                self.disgust = self.disgust+1
            if (2 == label):
                self.fear = self.fear+1
            if (3 == label):
                self.happy = self.happy+1
            if (4 == label):
                self.sad = self.sad+1
            if (5 == label):
                self.surprise = self.surprise+1
            if (6 == label):
                self.neutral = self.neutral+1

        def display_result(self,evaluations):
            print("anger = "     +\
                str((self.anger/float(evaluations))*100)    + "%")
            print("disgust = "   +\
                str((self.disgust/float(evaluations))*100)  + "%")
            print("fear = "      +\
                str((self.fear/float(evaluations))*100)     + "%")
            print("happy = "     +\
                str((self.happy/float(evaluations))*100)    + "%")
            print("sad = "       +\
                str((self.sad/float(evaluations))*100)      + "%")
            print("surprise = "  +\
                str((self.surprise/float(evaluations))*100) + "%")
            print("neutral = "   +\
                str((self.neutral/float(evaluations))*100)  + "%")
```

Summary

In this chapter, we introduced CNNs. We have seen that CNNs are suitable for image classification problems, making the training phase faster and the test phase more accurate.

The most common CNN architectures have been described: the **LeNet-5** model, designed for handwritten and machine-printed character recognition; **AlexNet**, which competed in the ILSVRC in 2012; the **VGG** model, which achieves a top-5 test accuracy of 92.7% in ImageNet (a dataset of over 14 million images belonging to 1,000 classes); and finally the **Inception-v3** model, which was responsible for setting the standard for classification and detection in the ILSVRC in 2014.

The description of each CNN architecture was followed by a code example. Also, the AlexNet network and VGG examples have helped to explain the concepts of the *transfer* and *style learning* techniques.

Finally, we built a CNN to classify emotions in a dataset of images; we tested the network on a single image and evaluated the limits and the quality of our model.

The next chapter describes *autoencoders*: these algorithms are useful for dimensionality reduction, classification, regression, collaborative filtering, feature learning, and topic modeling. We will carry out further data analysis using autoencoders and measure classification performance using image datasets.

5
Optimizing TensorFlow Autoencoders

In **Machine Learning** (**ML**), the so-called *curse of dimensionality* is a progressive decline in performance with an increase in the input space, often with hundreds or thousands of dimensions, which does not occur in low-dimensional settings such as three-dimensional space. This occurs because the number of samples needed to obtain a sufficient sampling of the input space increases exponentially with the number of dimensions. To overcome this problem, some *optimizing* networks have been developed.

The first one is autoencoder networks. These are designed and trained to transform an input pattern in itself so that in the presence of a degraded or incomplete version of an input pattern, it is possible to obtain the original pattern. An autoencoder is a **Neural Network** (**NN**). The network is trained to create output data like those presented in the entrance and the hidden layer stores the compressed data.

The second optimizing networks are **Boltzmann Machines** (see *Chapter 3, Feed-Forward Neural Networks with TensorFlow* for more details). This type of network consists of a visible input/output layer and one hidden layer. The connections between the visible layer and the hidden one are non-directional—data can travel in both directions, visible-hidden and hidden-visible, and the different neuronal units can be fully connected or partially connected.

Autoencoders can be compared with **Principal Component Analysis** (**PCA**) (refer to `https://en.wikipedia.org/wiki/Principal_component_analysis`), which is used to represent a given input using fewer dimensions than originally present. However, in this chapter, we'll focus only on autoencoders.

In a nutshell, the following topics will be covered in this chapter:

- How does an autoencoder work?
- How to implement an autoencoder
- Improving autoencoder robustness
- Building denoising autoencoders
- Convolutional autoencoders
- Fraud analytics using autoencoders

How does an autoencoder work?

Autoencoding is a data compression technique where the compression and decompression functions are data-specific, lossy, and *learned automatically from samples* rather than human-crafted manual features. Additionally, in almost all contexts where the term **autoencoder** is used, the compression and decompression functions are implemented with NNs.

An autoencoder is a network with three or more layers, where the input and the output layers have the *same number* of neurons, and those intermediate (hidden layers) have a *lower number* of neurons. The network is trained to reproduce output simply, for each piece of input data, the same pattern of activity in the input.

The remarkable aspect of autoencoders is that, due to the lower number of neurons in the hidden layer, if the network can learn from examples and generalize to an acceptable extent, it performs **data compression**: the status of the hidden neurons provides, for each example, a *compressed version* of the *input* and *output common states*.

In the first examples of such networks, in the mid-1980s, a *compression of simple images* was obtained in this way. Some authors, who have developed an effective strategy for improving the learning process in this type of network (they are usually very slow and are not always effective), have recently revived interest in autoencoders through a prelearning procedure, that provides a good initial condition of the weights for the learning procedure.

Useful applications of autoencoders are *data denoising* and *dimensionality reduction* for data visualization. The following diagram shows how an autoencoder typically works—it reconstructs the received input through two phases: an *encoding* phase, which corresponds to a *dimensional reduction* for the original input, and a *decoding* phase, capable of *reconstructing* the original input from the encoded (*compressed*) representation:

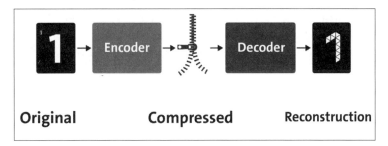

Figure 1: Encoder and decoder phases in an autoencoder

As mentioned earlier, an autoencoder is an NN, as well as an unsupervised learning (feature learning) algorithm. Less technically, it tries to learn an approximation of an identity function. However, we can impose constraints on the network such as *fewer* units in the hidden layer. In this way, an autoencoder represents original input from compressed, noisy, or corrupted data. The following diagram shows an autoencoder that consists of the narrow hidden layer between an encoder and a decoder:

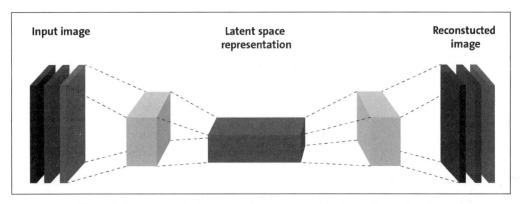

Figure 2: An unsupervised autoencoder as a network for latent feature learning

In the preceding diagram, the hidden layer or the intermediate layer is also called the latent space representation of the input data. Now, suppose we have a set of unlabeled training examples $\{x^{(1)}, x^{(2)}, x^{(3)}, ...\}$, where $x^{(i)} \in R^n$ and x is a vector, and $x^{(1)}$ refers to the first item in the vector.

An autoencoder NN is essentially an unsupervised learning algorithm that applies backpropagation, setting the target values to be equal to the inputs; it uses $y^{(i)} = x^{(i)}$.

The autoencoder tries to learn a function, $h_{W,b}(x) \approx x$. In other words, it is trying to learn an approximation to the identity function in order to output $x^\wedge$ that is similar to x. The identity function seems a particularly trivial function to be trying to learn, but by placing constraints on the network, such as by limiting the number of hidden units, we can discover interesting features of the data:

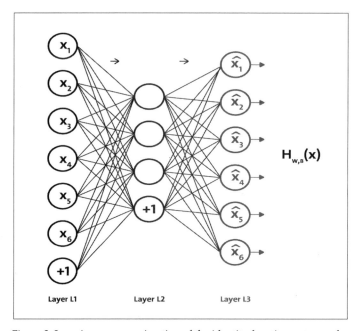

Figure 3: Learning an approximation of the identity function autoencoder

As a concrete example, suppose the inputs x are the pixel intensity values of a 10 × 10 image (100 pixels), so $n=100$, and there are $S_2 = 50$ hidden units in layer L_2 and $y \in R^{100}$. Since there are only 50 hidden units, the network is forced to learn a compressed representation of the input. It is only given the vector of hidden unit activations $a^{(2)} \in R^{50}$, so it must try to reconstruct the 100-pixel input, that is, $x_1, x_2, ..., x_{100}$ from the 50 hidden units. The preceding diagram shows only 6 inputs feeding into layer 1 and exactly 6 units feeding out from layer 3.

A neuron can be active (or firing) if its output value is close to 1, or inactive if its output value is close to 0. However, for simplicity, we assume that the neurons are inactive most of the time. This argument is true as long as we are talking about the sigmoid activation function. However, if you are using the tanh function as an activation function, then a neuron is inactive when it outputs values close to -1.

Implementing autoencoders with TensorFlow

Training an autoencoder is a simple process. It is an NN, where an *output is the same as its input*. There is an *input layer*, which is followed by a few hidden layers, and then after a certain depth, the hidden layers *follow the reverse architecture* until we reach a point where *the final layer is the same as the input layer*. We pass data into the network whose embedding we wish to learn.

In this example, we use images from the MNIST dataset as input. We begin our implementation by importing all the main libraries:

```
import tensorflow as tf
import numpy as np
import matplotlib.pyplot as plt
```

Then we prepare the MNIST dataset. We use the built-in `input_data` class from TensorFlow to load and set up the data. This class ensures that the data is downloaded and preprocessed to be consumed by the autoencoder. Therefore, basically, we don't need to do any feature engineering at all:

```
from tensorflow.examples.tutorials.mnist import input_data
mnist = input_data.read_data_sets("MNIST_data/",one_hot=True)
```

In the preceding code block, the `one_hot=True` parameter ensures that all the features are one hot encoded. One hot encoding is a technique by which categorical variables are converted into a form that could be fed into ML algorithms.

Next, we configure the network parameters:

```
learning_rate = 0.01
training_epochs = 20
batch_size = 256
display_step = 1
examples_to_show = 20
```

The size of input images is as follows:

```
n_input = 784
```

The sizes of the hidden features are as follows:

```
n_hidden_1 = 256
n_hidden_2 = 128
```

The final size corresponds to 28 × 28 = 784 pixels.

We need to define a placeholder variable for the input images. The data type for this tensor is set to `float` since the `mnist` values are in scale of [0, 1], and the shape is set to [None, n_input]. Defining the None parameter means that the tensor may hold an arbitrary number of images:

```
X = tf.placeholder("float", [None, n_input])
```

Then we can define the weights and biases of the network. The `weights` data structure contains the definition of the weights for the encoder and decoder. Notice that weights are chosen using `tf.random_normal`, which returns random values with a normal distribution:

```
weights = {
    'encoder_h1': tf.Variable\
    (tf.random_normal([n_input, n_hidden_1])),
    'encoder_h2': tf.Variable\
    (tf.random_normal([n_hidden_1, n_hidden_2])),
    'decoder_h1': tf.Variable\
    (tf.random_normal([n_hidden_2, n_hidden_1])),
    'decoder_h2': tf.Variable\
    (tf.random_normal([n_hidden_1, n_input])),
}
```

Similarly, we define the network's bias:

```
biases = {
    'encoder_b1': tf.Variable\
    (tf.random_normal([n_hidden_1])),
    'encoder_b2': tf.Variable\
    (tf.random_normal([n_hidden_2])),
    'decoder_b1': tf.Variable\
    (tf.random_normal([n_hidden_1])),
    'decoder_b2': tf.Variable\
    (tf.random_normal([n_input])),
}
```

We split the network modeling into two complementary *fully connected networks*: an *encoder* and a *decoder*. The encoder encodes the data; it takes as input an image, X, from the MNIST dataset, and performs the data encoding:

```
encoder_in = tf.nn.sigmoid(tf.add\
                    (tf.matmul(X, \
                          weights['encoder_h1']),\
                    biases['encoder_b1']))
```

The input data encoding is simply a matrix multiplication operation. The input data, x, of dimension 784 is reduced to a lower dimension, 256, using matrix multiplication:

$$(W * x + b) = encoder\_in$$

Here, *W* is the weight tensor, encoder_h1, and *b* is the bias tensor, encoder_b1. Through this operation, we have coded the initial image into a useful input for the autoencoder. The second step of the encoding procedure consists of data compression. The data represented by the input encoder_in tensor is reduced to a smaller size by means of a second matrix multiplication operation:

```
encoder_out = tf.nn.sigmoid(tf.add\
                    (tf.matmul(encoder_in,\
                            weights['encoder_h2']),\
                    biases['encoder_b2']))
```

The input data, encoder_in, of dimension 256 is then compressed to a lower tensor of size 128:

$$(W * encoder\_in + b) = encoder\_out$$

Here, *W* stands for the weight tensor, encoder_h2, while *b* stands for the bias tensor, encoder_b2. Notice that we used a sigmoid for the activation function for the encoder phase.

The decoder performs the inverse operation of the encoder. It decompresses the input to obtain an output of the same size of the network input. The first step of the procedure is to transform the encoder_out tensor of size 128 into a tensor of the intermediate representation of size 256:

```
decoder_in = tf.nn.sigmoid(tf.add\
                    (tf.matmul(encoder_out,\
                            weights['decoder_h1']),\
                    biases['decoder_b1']))
```

In formulas, it means this:

$$(W * encoder\_out + b) = decoder\_in$$

Here, W is the weight tensor, `decoder_h1`, of size 256 × 128, and b is the bias tensor, `decoder_b1`, of size 256. The final decoding operation is to decompress the data from its intermediate representation (of size 256) to a final representation (of dimension 784), which is the size of the original data:

```
decoder_out = tf.nn.sigmoid(tf.add\
                            (tf.matmul(decoder_in,\
                                       weights['decoder_h2']),\
                             biases['decoder_b2']))
```

The `y_pred` parameter is set equal to `decoder_out`:

```
y_pred = decoder_out
```

The network will learn whether the input data, X, is equal to the decoded data, so we define the following:

```
y_true = X
```

The point of the autoencoder is to create a reduction matrix that is good at reconstructing the original data. Thus, we want to minimize the `cost` function. Then we define the `cost` function as the mean squared error between `y_true` and `y_pred`:

```
cost = tf.reduce_mean(tf.pow(y_true - y_pred, 2))
```

To optimize the `cost` function, we use the following `RMSPropOptimizer` class:

```
optimizer = tf.train.RMSPropOptimizer(learning_rate).minimize(cost)
```

Then we prepare to launch the session:

```
init = tf.global_variables_initializer()
with tf.Session() as sess:
    sess.run(init)
```

We need to set the size of the batch images to train the network:

```
total_batch = int(mnist.train.num_examples/batch_size)
```

Start with the training cycle (the number of `training_epochs` is set to `10`):

```
for epoch in range(training_epochs):
```

While looping over all batches:

```
for i in range(total_batch):
    batch_xs, batch_ys =\
                 mnist.train.next_batch(batch_size)
```

Then we run the optimization procedure, feeding the execution graph with the batch set, batch_xs:

```
_, c = sess.run([optimizer, cost],\
                 feed_dict={X: batch_xs})
```

Next, we display the results for each epoch step:

```
if epoch % display_step == 0:
    print („Epoch:", ,%04d' % (epoch+1),
          „cost=", „{:.9f}".format(c))
print("Optimization Finished!")
```

Finally, we test the model, applying the encode or decode procedure. We feed the model a subset of images, where the value of example_to_show is set to 4:

```
encode_decode = sess.run(
    y_pred, feed_dict=\
    {X: mnist.test.images[:examples_to_show]})
```

We compare the original images with their reconstructions using Matplotlib:

```
f, a = plt.subplots(2, 10, figsize=(10, 2))
for i in range(examples_to_show):
    a[0][i].imshow(np.reshape(mnist.test.images[i], (28, 28)))
    a[1][i].imshow(np.reshape(encode_decode[i], (28, 28)))
f.show()
plt.draw()
plt.show()
```

When we run the session, we should have an output like this:

```
Extracting MNIST_data/train-images-idx3-ubyte.gz
Extracting MNIST_data/train-labels-idx1-ubyte.gz
Extracting MNIST_data/t10k-images-idx3-ubyte.gz
Extracting MNIST_data/t10k-labels-idx1-ubyte.gz
Epoch: 0001 cost= 0.208461761
Epoch: 0002 cost= 0.172908291
Epoch: 0003 cost= 0.153524384
Epoch: 0004 cost= 0.144243762
Epoch: 0005 cost= 0.137013704
Epoch: 0006 cost= 0.127291277
```

Optimizing TensorFlow Autoencoders

```
Epoch: 0007 cost= 0.125370100
Epoch: 0008 cost= 0.121299766
Epoch: 0009 cost= 0.111687921
Epoch: 0010 cost= 0.108801551
Epoch: 0011 cost= 0.105516203
Epoch: 0012 cost= 0.104304880
Epoch: 0013 cost= 0.103362709
Epoch: 0014 cost= 0.101118311
Epoch: 0015 cost= 0.098779991
Epoch: 0016 cost= 0.095374011
Epoch: 0017 cost= 0.095469855
Epoch: 0018 cost= 0.094381645
Epoch: 0019 cost= 0.090281256
Epoch: 0020 cost= 0.092290156
Optimization Finished!
```

Then we display the results. The first row is the original images, and the second row is the decoded images:

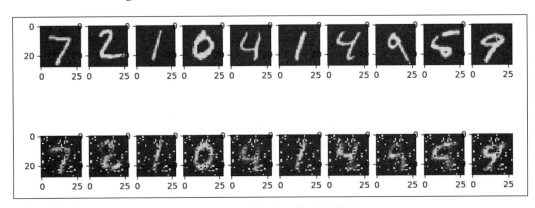

Figure 4: Original and the decoded MNIST images

As you can see, the number two differs from the original one (it still seems to be digit two like the number three). We can increase the number of epochs or change the network parameters to improve the result.

Improving autoencoder robustness

A successful strategy we can use to improve the model's robustness is to *introduce a noise* in the encoding phase. We call a **denoising autoencoder** a *stochastic version* of an autoencoder; in a denoising autoencoder, the input is stochastically corrupted, but the *uncorrupted version* of the same input is *used as the target* for the decoding phase.

Intuitively, a denoising autoencoder does two things: first, it tries to encode the input, preserving the relevant information; and then, it seeks to nullify the effect of the corruption process applied to the same input. In the next section, we'll show an implementation of a denoising autoencoder.

Implementing a denoising autoencoder

The network architecture is very simple. A 784-pixel input image is stochastically corrupted and then dimensionally reduced by an encoding network layer. The image size is reduced from 784 to 256 pixels.

In the decoding phase, we prepare the network for output, returning the image size to 784 pixels. As usual, we start loading all the necessary libraries into our implementation:

```
import numpy as np
import tensorflow as tf
import matplotlib.pyplot as plt
from tensorflow.examples.tutorials.mnist import input_data
```

Then we set the basic network parameters:

```
n_input    = 784
n_hidden_1 = 1024
n_hidden_2 = 2048
n_output   = 784
```

And we set the session's parameters:

```
epochs     = 100
batch_size = 100
disp_step  = 10
```

We build the training and testing sets. We again use the `input_data` feature imported from the `tensorflow.examples.tutorials.mnist`:

```
print ("PACKAGES LOADED")
mnist = input_data.read_data_sets('data/', one_hot=True)
trainimg   = mnist.train.images
trainlabel = mnist.train.labels
testimg    = mnist.test.images
testlabel  = mnist.test.labels
print ("MNIST LOADED")
```

Let's define a placeholder variable for the input images. The data type is set to `float` and the shape is set to `[None, n_input]`. The `None` parameter means that the tensor may hold an arbitrary number of images, and the size per image is n_input:

```
x = tf.placeholder("float", [None, n_input])
```

Next, we have a placeholder variable for the *true labels* associated with the images that were input in the placeholder variable, x. The shape of this placeholder variable is `[None, n_output]`, which means it may hold an arbitrary number of labels and that each label is a vector of length n_output, which is 10 in this case:

```
y = tf.placeholder("float", [None, n_output])
```

To reduce *overfitting*, we'll apply a dropout before the encoding and decoding procedure, so we must define a placeholder for the probability that a neuron's output is *kept during dropout*:

```
dropout_keep_prob = tf.placeholder("float")
```

On these definitions, we fix the weights and network biases:

```
weights = {
    'h1': tf.Variable(tf.random_normal([n_input, n_hidden_1])),
    'h2': tf.Variable(tf.random_normal([n_hidden_1, n_hidden_2])),
    'out': tf.Variable(tf.random_normal([n_hidden_2, n_output]))
}
biases = {
    'b1': tf.Variable(tf.random_normal([n_hidden_1])),
    'b2': tf.Variable(tf.random_normal([n_hidden_2])),
    'out': tf.Variable(tf.random_normal([n_output]))
}
```

The `weights` and `biases` values are chosen using `tf.random_normal`, which returns *random values* with a normal distribution. The encoding phase takes as input an image from the MNIST dataset, and then performs the data compression by applying a matrix multiplication operation:

```
encode_in = tf.nn.sigmoid\
            (tf.add(tf.matmul\
                (x, weights['h1']),\
                biases['b1']))
encode_out = tf.nn.dropout\
            (encode_in, dropout_keep_prob)
```

In the decoding phase we apply the same procedure:

```
decode_in = tf.nn.sigmoid\
            (tf.add(tf.matmul\
                 (encode_out, weights['h2']),\
                  biases['b2']))
```

The reduction in overfitting is performed by a dropout procedure:

```
decode_out = tf.nn.dropout(decode_in,\
                           dropout_keep_prob)
```

Finally, we are ready to build the *prediction* tensor, y_pred:

```
y_pred = tf.nn.sigmoid\
         (tf.matmul(decode_out,\
                    weights['out']) +\
          biases['out'])
```

We then define a *cost measure*, which is used to guide the variable optimization procedure:

```
cost = tf.reduce_mean(tf.pow(y_pred - y, 2))
```

We will minimize the cost function using the RMSPropOptimizer class:

```
optimizer = tf.train.RMSPropOptimizer(0.01).minimize(cost)
```

Finally, we can initialize the defined variables as follows:

```
init = tf.global_variables_initializer()
```

Then we set TensorFlow's running session:

```
with tf.Session() as sess:
    sess.run(init)
    print ("Start Training")
    for epoch in range(epochs):
        num_batch  = int(mnist.train.num_examples/batch_size)
        total_cost = 0.
        for i in range(num_batch):
```

For each *training epoch*, we select a smaller *batch set* from the *training dataset*:

```
            batch_xs, batch_ys = \
                   mnist.train.next_batch(batch_size)
```

Optimizing TensorFlow Autoencoders

Here is the focal point. We randomly corrupt the `batch_xs` set using the `randn` function from the NumPy package we imported earlier:

```
batch_xs_noisy = batch_xs + \
              0.3*np.random.randn(batch_size, 784)
```

We use these sets to feed the execution graph and then to run the session (`sess.run`):

```
feeds = {x: batch_xs_noisy, \
         y: batch_xs, \
         dropout_keep_prob: 0.8}
sess.run(optimizer, feed_dict=feeds)
total_cost += sess.run(cost, feed_dict=feeds)
```

Every ten epochs, the *average cost* value will be displayed:

```
if epoch % disp_step == 0:
    print("Epoch %02d/%02d average cost: %.6f"
           % (epoch, epochs, total_cost/num_batch))
```

Finally, we start to test the trained model:

```
print("Start Test")
```

To do this, we randomly select an image from the testing set:

```
randidx   = np.random.randint\
                  (testimg.shape[0], size=1)
orgvec    = testimg[randidx, :]
testvec   = testimg[randidx, :]
label     = np.argmax(testlabel[randidx, :], 1)
print("Test label is %d" % (label))
noisyvec = testvec + 0.3*np.random.randn(1, 784)
```

Then we run the *trained model* on the *selected image*:

```
outvec = sess.run(y_pred, \
              feed_dict={x: noisyvec, \
              dropout_keep_prob: 1})
```

As we'll see, the following `plotresult` function will display the *original image*, the *noisy image*, and the *predicted image*:

```
plotresult(orgvec,noisyvec,outvec)
print("restart Training")
```

When we run the session, we should see a result like this:

```
PACKAGES LOADED
Extracting data/train-images-idx3-ubyte.gz
Extracting data/train-labels-idx1-ubyte.gz
Extracting data/t10k-images-idx3-ubyte.gz
Extracting data/t10k-labels-idx1-ubyte.gz
MNIST LOADED
Start Training
```

For the sake of brevity, we have only reported the results after 100 epochs:

```
Epoch 100/100 average cost: 0.212313
Start Test
Test label is 6
```

These are the original and the noisy images (the number six, as you can see):

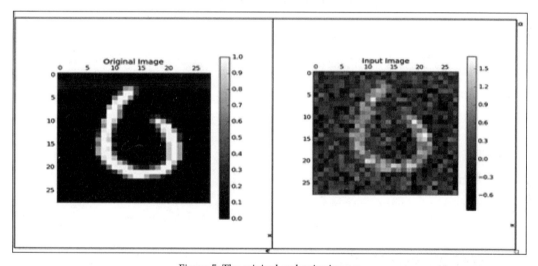

Figure 5: The original and noisy images

Optimizing TensorFlow Autoencoders

Here's a badly reconstructed image:

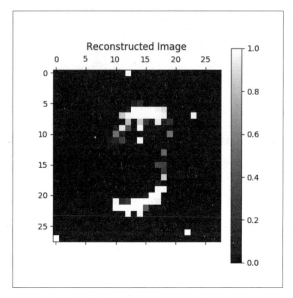

Figure 6: A badly reconstructed image

After 100 epochs, we have a better result:

```
Epoch 100/100 average cost: 0.018221
Start Test
Test label is 5
```

Here are the original and the noisy images:

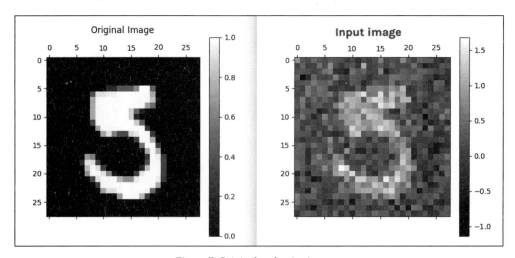

Figure 7: Original and noisy images

Here is a good reconstructed image:

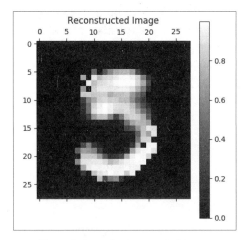

Figure 8: A good reconstructed image

Implementing a convolutional autoencoder

Until now, we have seen that autoencoder inputs are images. So, it makes sense to ask whether a *convolutional architecture* can work better on the autoencoder architectures that we showed earlier. We will analyze how encoders and decoders work in convolutional autoencoders.

Encoder

An encoder consists of three convolutional layers. The number of features changes from 1, the *input data*, to 16 for the first convolutional layer; then, from 16 to 32 for the second layer; and finally, from 32 to 64 for the last convolutional layer. While moving from a convolutional layer to another, the shape undergoes *image compression*:

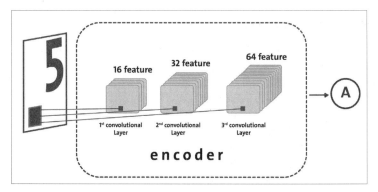

Figure 9: The data flow of the encoding phase

Decoder

The decoder consists of three deconvolutional layers arranged in sequence. For each deconvolution operation, we reduce the number of features to obtain an image *that must be* the same size as the original image. In addition to reducing the number of features, a deconvolution transforms the shape of the images:

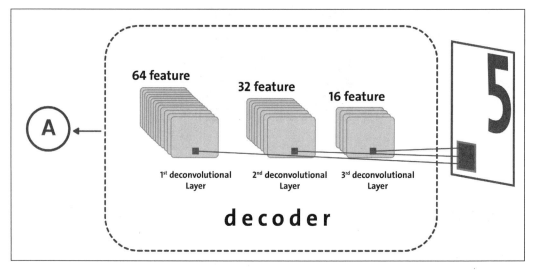

Figure 10: The data flow of decoding phase

We're ready to see how to implement a *convolutional autoencoder*; the first implementation step is loading the basic libraries:

```
import matplotlib.pyplot as plt
import numpy as np
import math
import tensorflow as tf
import tensorflow.examples.tutorials.mnist.input_data as input_data
```

Then we build the training and testing sets:

```
mnist = input_data.read_data_sets("data/", one_hot=True)
trainings   = mnist.train.images
trainlabels = mnist.train.labels
testings    = mnist.test.images
testlabels  = mnist.test.labels
ntrain      = trainings.shape[0]
ntest       = testings.shape[0]
dim         = trainings.shape[1]
nout        = trainlabels.shape[1]
```

Chapter 5

We need to define a placeholder variable for the input images:

```
x = tf.placeholder(tf.float32, [None, dim])
```

The data type is set to `float32` and the shape is set to `[None, dim]`, where `None` means that the tensor may hold an arbitrary number of images, with each image being a vector of length `dim`. Next, we have a placeholder variable for the output images. The shape of this variable is set to `[None, dim]` the same as the input shape:

```
y = tf.placeholder(tf.float32, [None, dim])
```

Then we define the `keepprob` variable, which is used to configure the dropout rate (https://www.tensorflow.org/tutorials/layers#dropout) used during the training of the network:

```
keepprob = tf.placeholder(tf.float32)
```

Also, we have to define the number of nodes in each of the network's layers:

```
n1 = 16
n2 = 32
n3 = 64
ksize = 5
```

The network contains a total number of *six layers*. The first three layers are *convolutional* and belong to the *encoding phase*, while the last three layers are *deconvolutional* and are part of the *decoding phase*:

```
weights = {
    'ce1': tf.Variable(tf.random_normal\
                      ([ksize, ksize, 1, n1],stddev=0.1)),
    'ce2': tf.Variable(tf.random_normal\
                      ([ksize, ksize, n1, n2],stddev=0.1)),
    'ce3': tf.Variable(tf.random_normal\
                      ([ksize, ksize, n2, n3],stddev=0.1)),
    'cd3': tf.Variable(tf.random_normal\
                      ([ksize, ksize, n2, n3],stddev=0.1)),
    'cd2': tf.Variable(tf.random_normal\
                      ([ksize, ksize, n1, n2],stddev=0.1)),
    'cd1': tf.Variable(tf.random_normal\
                      ([ksize, ksize, 1, n1],stddev=0.1))
}

biases = {
    'be1': tf.Variable\
    (tf.random_normal([n1], stddev=0.1)),
    'be2': tf.Variable\
```

```
            (tf.random_normal([n2], stddev=0.1)),
        'be3': tf.Variable\
            (tf.random_normal([n3], stddev=0.1)),
        'bd3': tf.Variable\
            (tf.random_normal([n2], stddev=0.1)),
        'bd2': tf.Variable\
            (tf.random_normal([n1], stddev=0.1)),
        'bd1': tf.Variable\
            (tf.random_normal([1], stddev=0.1))
    }
```

The following function, cae, builds the convolutional autoencoder: the inputs passed are the image, _X; the data structure *weights* and *bias*, _W and _b; and the _keepprob parameter:

```
    def cae(_X, _W, _b, _keepprob):
```

The initial 784-pixel image must be reshaped into a 28 × 28 matrix to be subsequently processed by the next convolutional layers:

```
        _input_r = tf.reshape(_X, shape=[-1, 28, 28, 1])
```

The *first convolutional layer* is _ce1. It has the _input_r tensor as input, relative to the input image:

```
        _ce1 = tf.nn.sigmoid\
               (tf.add(tf.nn.conv2d\
                    (_input_r, _W['ce1'],\
                     strides=[1, 2, 2, 1],\
                     padding='SAME'),\
                _b['be1']))
```

Before moving to the *second convolutional layer*, we apply the dropout operation:

```
        _ce1 = tf.nn.dropout(_ce1, _keepprob)
```

In the following two encoding layers, we apply the same convolution and dropout operations:

```
        _ce2 = tf.nn.sigmoid\
               (tf.add(tf.nn.conv2d\
                    (_ce1, _W['ce2'],\
                     strides=[1, 2, 2, 1],\
                     padding='SAME'),\
                _b['be2']))
        _ce2 = tf.nn.dropout(_ce2, _keepprob)
        _ce3 = tf.nn.sigmoid\
```

```
                  (tf.add(tf.nn.conv2d\
                         (_ce2, _W['ce3'],\
                          strides=[1, 2, 2, 1],\
                          padding='SAME'),\
                          _b['be3'])) 
   _ce3 = tf.nn.dropout(_ce2, _keepprob)
```

The number of features has increased from 1 (the input image) to 64, while the original shape image has been reduced to 28 × 28 to 7 × 7. In the decoding phase, the compressed (or encoded) and reshaped image must be as similar as possible to the original image. To achieve this, we used the conv2d_transpose TensorFlow function for the next three layers:

```
tf.nn.conv2d_transpose(value, filter, output_shape, strides,
padding='SAME')
```

This operation is sometimes called **deconvolution**; it is simply the gradient of conv2d. The arguments of this function are as follows:

- value: A 4D tensor of type float and shape [batch, height, width, in_channels].
- filter: A 4D tensor with the same type as value and shape [height, width, output_channels, in_channels]. The in_channels dimension must match that of value.
- output_shape: A 1D tensor representing the output shape of the deconvolution operation.
- strides: A list of *ints*. The stride of the sliding window for each dimension of the input tensor.
- padding: A string, either valid or SAME.

The conv2d_transpose function will return a tensor with the same type as the value argument. The first deconvolutional layer, _cd3, has the convolutional layer _ce3 as input. It returns the _cd3 tensor, whose shape is (1, 7, 7, 32):

```
    _cd3 = tf.nn.sigmoid\
           (tf.add(tf.nn.conv2d_transpose\
                   (_ce3, _W['cd3'],\
                    tf.stack([tf.shape(_X)[0], 7, 7, n2]),\
                    strides=[1, 2, 2, 1],\
                    padding='SAME'),\
                    _b['bd3']))
    _cd3 = tf.nn.dropout(_cd3, _keepprob)
```

Optimizing TensorFlow Autoencoders

To the *second deconvolutional layer*, _cd2, we pass as input the deconvolutional layer _cd3. It returns the _cd2 tensor, whose shape is (1, 14, 14, 16):

```
_cd2 = tf.nn.sigmoid\
       (tf.add(tf.nn.conv2d_transpose\
              (_cd3, _W['cd2'],\
               tf.stack([tf.shape(_X)[0], 14, 14, n1]),\
               strides=[1, 2, 2, 1],\
               padding='SAME'),\
               _b['bd2']))
_cd2 = tf.nn.dropout(_cd2, _keepprob)
```

The *third and final deconvolutional layer*, _cd1, has the _cd2 layer passed as input. It returns the resulting _out tensor, whose shape is (1, 28, 28, 1), the same as the input image:

```
_cd1 = tf.nn.sigmoid\
       (tf.add(tf.nn.conv2d_transpose\
              (_cd2, _W['cd1'],\
               tf.stack([tf.shape(_X)[0], 28, 28, 1]),\
               strides=[1, 2, 2, 1],\
               padding='SAME'),\
               _b['bd1']))
_cd1 = tf.nn.dropout(_cd1, _keepprob)
_out = _cd1
return _out
```

Then we define a *cost function as the mean squared error* between y and pred:

```
pred = cae(x, weights, biases, keepprob)
cost = tf.reduce_sum\
       (tf.square(cae(x, weights, biases, keepprob)\
                  - tf.reshape(y, shape=[-1, 28, 28, 1])))
learning_rate = 0.001
```

To optimize the cost, we'll use AdamOptimizer:

```
optm = tf.train.AdamOptimizer(learning_rate).minimize(cost)
```

In the next step, we configure the running session for our network:

```
init = tf.global_variables_initializer()
print ("Functions ready")
sess = tf.Session()
sess.run(init)
mean_img = np.zeros((784))
```

The size of the batch is set to 128:

```
batch_size = 128
```

The number of epochs is 50:

```
n_epochs   = 50
```

Then we start the loop session:

```
for epoch_i in range(n_epochs):
```

For each epoch, we get a batch set, trainbatch:

```
for batch_i in range(mnist.train.num_examples // batch_size):
    batch_xs, _ = mnist.train.next_batch(batch_size)
    trainbatch = np.array([img - mean_img for img in batch_xs])
```

We apply a random noise, just like with denoising autoencoders, to improve the learning:

```
trainbatch_noisy = trainbatch + 0.3*np.random.randn(\
    trainbatch.shape[0], 784)
sess.run(optm, feed_dict={x: trainbatch_noisy \
                , y: trainbatch, keepprob: 0.7})
print ("[%02d/%02d] cost: %.4f" % (epoch_i, n_epochs \
, sess.run(cost, feed_dict={x: trainbatch_noisy \
                , y: trainbatch, keepprob: 1.})))
```

For each training epoch, we randomly take five training examples:

```
if (epoch_i % 10) == 0:
    n_examples = 5
    test_xs, _ = mnist.test.next_batch(n_examples)
    test_xs_noisy = test_xs + 0.3*np.random.randn(
        test_xs.shape[0], 784)
```

Then we test the trained model on a little subset:

```
recon = sess.run(pred, feed_dict={x: test_xs_noisy,\
                                    keepprob: 1.})
fig, axs = plt.subplots(2, n_examples, figsize=(15, 4))
for example_i in range(n_examples):
    axs[0][example_i].matshow(np.reshape(
        test_xs_noisy[example_i, :], (28, 28))
        , cmap=plt.get_cmap('gray'))
```

Optimizing TensorFlow Autoencoders

Finally, we can display the inputs and the learned set using Matplotlib:

```
axs[1][example_i].matshow(np.reshape(
    np.reshape(recon[example_i, ...], (784,))
    + mean_img, (28, 28)), cmap=plt.get_cmap('gray'))
plt.show()
```

The execution will produce the following output:

```
>>>
Extracting data/train-images-idx3-ubyte.gz
Extracting data/train-labels-idx1-ubyte.gz
Extracting data/t10k-images-idx3-ubyte.gz
Extracting data/t10k-labels-idx1-ubyte.gz
Packages loaded
Network ready
Functions ready
Start training..
[00/05] cost: 8049.0332
[01/05] cost: 3706.8667
[02/05] cost: 2839.9155
[03/05] cost: 2462.7021
[04/05] cost: 2391.9460
>>>
```

Note that for each epoch, we'll visualize the input set and the corresponding learned set that are shown previously. As you can see in the first epoch, we have no idea which images have been learned:

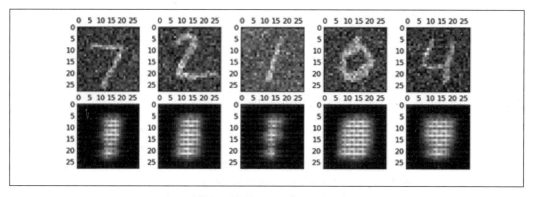

Figure 11: First epoch images

The idea becomes clearer in the second epoch:

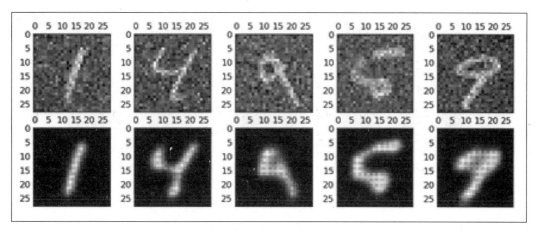

Figure 12: Second epoch images

This is the third epoch:

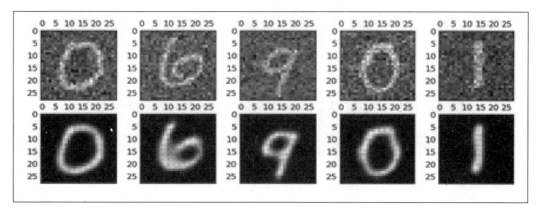

Figure 13: Third epoch images

It's better again in the fourth epoch:

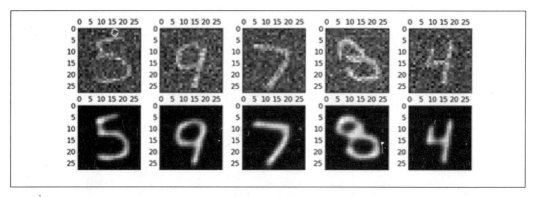

Figure 14: Fourth epoch images

We probably could have stopped in the previous epoch, but this is the fifth and final epoch:

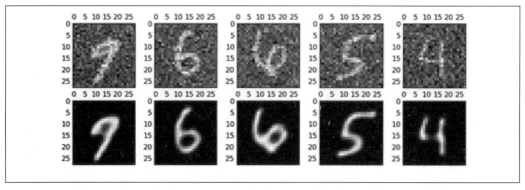

Figure 15: Fifth epoch images

So far, we have seen a different implementation of autoencoders and an improved version as well. However, applying this technique to the MNIST dataset does not tell its true power. Thus, it's time to see a more real-life problem in which we can apply the autoencoder technique.

Fraud analytics with autoencoders

Fraud detection and prevention in financial companies such as banks, insurance companies, and credit unions is an important task. So far, we have seen how, and where, to use **Deep Neural Networks (DNNs)** and **Convolutional Neural Network (CNNs)**.

Now it's time to use other unsupervised learning algorithm, such as autoencoders. In this section, we will be exploring a dataset of credit card transactions and trying to build an unsupervised machine-learning model that is able to tell whether a particular transaction is fraudulent or genuine.

More specifically, we will use autoencoders to pretrain a classification model and apply anomaly detection techniques to predict possible fraud. Before we start, we need to know the dataset.

Description of the dataset

For this example, we will be using the *Credit Card Fraud Detection* dataset from Kaggle. The dataset can be downloaded from https://www.kaggle.com/hunk3749/credit-card/data. Since I am using the dataset, it would a good idea to be transparent by citing the following publication:

Andrea Dal Pozzolo, Olivier Caelen, Reid A. Johnson and Gianluca Bontempi. Calibrating Probability with Undersampling for Unbalanced Classification. In Symposium on Computational Intelligence and Data Mining (CIDM), IEEE, 2015.

The dataset contains transactions made by European credit card holders in September 2013 over a span of two days. There are 285,299 transactions with only 492 frauds, meaning the dataset is highly unbalanced. The positive class (frauds) account for 0.172% of all transactions.

The dataset contains numerical input variables that are the result of a PCA transformation. Unfortunately, due to confidentiality issues, we cannot provide the original features and more background information about the data. There are 28 features, namely V1, V2, ...V27, which are principal components obtained using PCA, except for the Time and Amount features. The Class feature is the response variable and it takes the value 1 in a case of fraud and 0 otherwise.

There are two additional features, `Time` and `Amount`. The `Time` column signifies the time in seconds between each transaction and the first transaction, whereas the `Amount` column signifies how much money was transferred in this transaction. So, let's look at the input data (only V1, V2, V26, and V27 are shown, though) in *Figure 16*:

```
+----+-------------------+--------------------+--------------------+--------------------+------+-----+
|Time|                 V1|                  v2|                 V26|                 V27|Amount|Class|
+----+-------------------+--------------------+--------------------+--------------------+------+-----+
|   0|  -1.3598071336738|-0.0727811733098497|  -0.189114843888824|   0.133558376740387|149.62|    0|
|   0|   1.19185711131486|   0.26615071205963|   0.125894532368176| -0.00898309914322813|  2.69|    0|
|   1|  -1.35835406159823|   -1.34016307473609|  -0.139096571514147| -0.0553527940384261|378.66|    0|
|   1| -0.966271711572087| -0.185226008082898|  -0.221928844458407|   0.0627228487293033| 123.5|    0|
|   2|  -1.15823309349523|   0.877736754848451|   0.502292224181569|   0.219422229513348| 69.99|    0|
|   2| -0.425965884412454|   0.960523044882985|   0.105914779097957|   0.253844224739337|  3.67|    0|
|   4|   1.22965763450793|   0.141003507049326|  -0.257236845917139|   0.0345074297438413|  4.99|    0|
|   7| -0.644269442348146|   1.41796354547385| -0.0516342969262494|   -1.20692108094258|  40.8|    0|
|   7| -0.89428608220282|   0.286157196276544|  -0.384157307702294|   0.0117473564581996|  93.2|    0|
|   9| -0.33826175242575|   1.11959337641566|   0.0941988339514961|   0.246219304619926|  3.68|    0|
+----+-------------------+--------------------+--------------------+--------------------+------+-----+
only showing top 10 rows
```

Figure 16: A snapshot of the credit card fraud detection dataset

Problem description

For this example, we will use an autoencoder as an unsupervised feature-learning algorithm that learns and generalizes the common patterns shared by the training data. During the reconstruction phase, the RMSE will be much higher for those data points that have unusual patterns. Thus, those data points are outliers, or anomalies. Our assumption is that the anomalies are also equal to the fraudulent transactions we are after.

Now, during the evaluation step, we can select a threshold for RMSE based on validation data and flag all data with an RMSE above the threshold as fraudulent. Alternatively, if we believe 0.1% of all transactions are fraudulent, we can also rank the data based on the reconstruction error for each data point (that is, the RMSEs), then select the top 0.1% to be the fraudulent transactions.

Given the class imbalance ratio, measuring the accuracy using **Area Under the Precision-Recall Curve** (**AUPRC**) is recommended because the confusion matrix accuracy is not meaningful in unbalanced classification. In this case, using linear machine learning models, such as random forests, logistic regression, or support vector machines, by applying over-or-under sampling techniques, would be a better idea. Alternatively, we can try to find anomalies in the data since we assume that there are only a few fraud cases, that is, anomalies, within the dataset.

When dealing with such a severe imbalance of response labels, we also need to be careful when measuring model performance. There are only a handful of fraudulent instances, so a model that predicts everything as non-fraud will achieve an accuracy of more than 99%. However, despite their high accuracy, linear ML models (even tree ensembles) will not necessarily help us to find fraudulent cases.

For this example, we will build an unsupervised model: the model will be trained with both positive and negative data (frauds and non-frauds), but without providing the labels. Since we have many more normal transactions than fraudulent ones, we should expect the model to learn and memorize the patterns of normal transactions after training, and the model should be able to give a score for any transaction that is an outlier.

This unsupervised training will be quite useful for this purpose because we do not have enough labeled data. So, let's get started.

Exploratory data analysis

Before we implement our model, exploring the dataset would provide some insight. We start by importing the required packages and modules (including others that will be required for this example):

```
import pandas as pd
import numpy as np
import tensorflow as tf
import os
from datetime import datetime
from sklearn.metrics import roc_auc_score as auc
import seaborn as sns # for statistical data visualization
import matplotlib.pyplot as plt
import matplotlib.gridspec as gridspec
```

Installing seaborn

You can install `seaborn`, which is a Python module for statistical data visualization in a number of ways:

```
$ sudo pip install seaborn # for Python 2.7
$ sudo pip3 install seaborn # for Python 3.x
$ sudo conda install seaborn # using conda
# Directly from GitHub (use pip for Python 2.7)
$ pip3 install git+https://github.com/mwaskom/seaborn.git
```

Now, I am assuming that you have already downloaded the dataset from the aforementioned URL (that is `https://www.kaggle.com/hunk3749/credit-card/data`). The download comes with a CSV file called `creditcard.csv`.

So next, let's read the dataset and create a pandas DataFrame:

```
df = pd.read_csv('creditcard.csv')
print(df.shape)
>>>
(284807, 31)
```

So, the dataset has about 300,000 transactions, 30 features, and two binary labels (that is, 0/1). Now let's see the column names and their data types:

```
print(df.columns)
>>>
Index(['Time', 'V1', 'V2', 'V3', 'V4', 'V5', 'V6', 'V7', 'V8', 'V9',
'V10', 'V11', 'V12', 'V13', 'V14', 'V15', 'V16', 'V17', 'V18',
'V19', 'V20', 'V21', 'V22', 'V23', 'V24', 'V25', 'V26', 'V27', 'V28',
'Amount', 'Class'],
  dtype='object')

print(df.dtypes)
>>>
Time       float64
V1         float64
V2         float64
V3         float64
...
V25        float64
V26        float64
V27        float64
V28        float64
Amount     float64
   Class    int64
```

Now let's take a look at the dataset:

```
print(df.head())
>>>
```

	Time	V1	V2	V3	V4	V5	V6	V7	V8	V9	...	V21
0	0.0	-1.359807	-0.072781	2.536347	1.378155	-0.338321	0.462388	0.239599	0.098698	0.363787	...	-0.018307
1	0.0	1.191857	0.266151	0.166480	0.448154	0.060018	-0.082361	-0.078803	0.085102	-0.255425	...	-0.225775
2	1.0	-1.358354	-1.340163	1.773209	0.379780	-0.503198	1.800499	0.791461	0.247676	-1.514654	...	0.247998
3	1.0	-0.966272	-0.185226	1.792993	-0.863291	-0.010309	1.247203	0.237609	0.377436	-1.387024	...	-0.108300
4	2.0	-1.158233	0.877737	1.548718	0.403034	-0.407193	0.095921	0.592941	-0.270533	0.817739	...	-0.009431

Figure 17: A snapshot of the dataset

Now let's see the timespan for all the transactions:

```
print("Total time spanning: {:.1f} days".format(df['Time'].max() / (3600 * 24.0)))
>>>
Total time spanning: 2.0 days
```

Now let's have a look at the statistics for the classes:

```
print("{:.3f} % of all transactions are fraud. ".format(np.sum(df['Class']) / df.shape[0] * 100))
>>>
0.173 % of all transactions are fraud.
```

Therefore, we have only a few fraudulent transactions. This is also called a rare event detection in literature, and means that the dataset is highly unbalanced. Now, let's draw the histogram for first five features:

```
plt.figure(figsize=(12,5*4))
gs = gridspec.GridSpec(5, 1)
for i, cn in enumerate(df.columns[:5]):
    ax = plt.subplot(gs[i])
    sns.distplot(df[cn][df.Class == 1], bins=50)
    sns.distplot(df[cn][df.Class == 0], bins=50)
    ax.set_xlabel('')
    ax.set_title('histogram of feature: ' + str(cn))
plt.show()
>>>
```

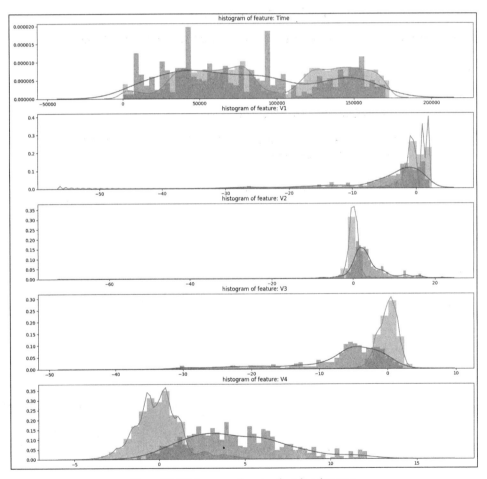

Figure 18: Histograms showing first five features

In the preceding screenshot, it can be seen that all of the features are either positively or negatively skewed. In addition, the dataset does not have many features, so trimming the tails would lose important information. So, for the time being, let's try not to do that, and use all the features.

Training, validation, and testing set preparation

Let's start the training by splitting the data into training, development (also known as *validation*), and test set. We are using first 80% of the data as the training and validation set. The remaining 20% will be used as the testing set:

```
TEST_RATIO = 0.20
df.sort_values('Time', inplace = True)
TRA_INDEX = int((1-TEST_RATIO) * df.shape[0])
train_x = df.iloc[:TRA_INDEX, 1:-2].values
train_y = df.iloc[:TRA_INDEX, -1].values
test_x = df.iloc[TRA_INDEX:, 1:-2].values
test_y = df.iloc[TRA_INDEX:, -1].values
```

Now, let's the statistics of the preceding split:

```
print("Total train examples: {}, total fraud cases: {}, equal to
{:.5f} % of total cases. ".format(train_x.shape[0], np.sum(train_y),
(np.sum(train_y)/train_x.shape[0])*100))

print("Total test examples: {}, total fraud cases: {}, equal to {:.5f}
% of total cases. ".format(test_x.shape[0], np.sum(test_y), (np.
sum(test_y)/test_y.shape[0])*100))

>>>
Total train examples: 227845, total fraud cases: 417, equal to 0.18302
% of total cases.
Total test examples: 56962, total fraud cases: 75, equal to 0.13167 %
of total cases.
```

Normalization

For better predictive accuracy, we can consider two types of standardization: z-score and min-max scaling:

- **Z-score**: This normalizes each column towards a mean of zero and standardization of ones. This is particularly suitable for activation functions such as tanh that output values on both sides of zero. Secondly, this will leave in extreme values, so there will be some extremeness left after normalization. This might be useful for detecting outliers in this case.
- **Min-max scaling**: This ensures all the values are between 0 and 1, that is, positive. This is the default approach if we are using sigmoid as our output activation.

We used a validation set to decide the data standardization approach and activation functions. Based on experiments, we've found that when used together with z-score normalization, tanh performs slightly better than sigmoid. Therefore, we chose tanh followed by z-score:

```
cols_mean = []
cols_std = []

for c in range(train_x.shape[1]):
    cols_mean.append(train_x[:,c].mean())
    cols_std.append(train_x[:,c].std())
    train_x[:, c] = (train_x[:, c] - cols_mean[-1]) / cols_std[-1]
    test_x[:, c] =  (test_x[:, c] - cols_mean[-1]) / cols_std[-1]
```

Autoencoder as an unsupervised feature learning algorithm

In this subsection, we will see how to use an autoencoder as an unsupervised feature learning algorithm. First, let's initialize the network hyperparameters:

```
learning_rate = 0.001
training_epochs = 1000
batch_size = 256
display_step = 10
n_hidden_1 = 15 # number of neurons is the num features
n_input = train_x.shape[1]
```

Since first and second layers contain 15 and 5 neurons respectively, we are building a network of such architecture: *28(input) -> 15 -> 5 -> 15 -> 28(output)*. So let's construct our autoencoder network.

Let's create a TensorFlow placeholder to hold the input:

```
X = tf.placeholder("float", [None, n_input])
```

Now we have to create the bias and the weight vectors with random initialization:

```
weights = {
    'encoder_h1': tf.Variable\
                (tf.random_normal([n_input, n_hidden_1])),
    'decoder_h1': tf.Variable\
                (tf.random_normal([n_hidden_1, n_input])),
}
biases = {
    'encoder_b1': tf.Variable(tf.random_normal([n_hidden_1])),
    'decoder_b1': tf.Variable(tf.random_normal([n_input])),
}
```

Now, we build a simple autoencoder. Here we have the `encoder()` function, which constructs the encoder. We encode the hidden layer with the `tanh` function as follows:

```
def encoder(x):
    layer_1 = tf.nn.tanh(tf.add\
                (tf.matmul(x, weights['encoder_h1']),\
                biases['encoder_b1']))
    return layer_1
```

Here is the `decoder()` function, which constructs the decoder. We decode the hidden layer with the `tanh` function as follows:

```
def decoder(x):
    layer_1 = tf.nn.tanh(tf.add\
                (tf.matmul(x, weights['decoder_h1']),\
                biases['decoder_b1']))
    return layer_1
```

After that, we construct the model by passing the TensorFlow placeholder for our input data. *Weights* and *biases* (the *W*s and *b*s of NNs) contain all parameters of the network that we will learn to optimize, as follows:

```
encoder_op = encoder(X)
decoder_op = decoder(encoder_op)
```

Once we have constructed the autoencoder network, it's time to make the prediction, where the targets are the input data:

```
y_pred = decoder_op
y_true = X
```

Now that we have made the prediction, it's time to define `batch_mse` to evaluate the performance:

```
batch_mse = tf.reduce_mean(tf.pow(y_true - y_pred, 2), 1)
```

> The **Mean Squared Error** (**MSE**) of an unobserved quantity measures the average (https://en.wikipedia.org/wiki/Expected_value) of the squares of the errors (https://en.wikipedia.org/wiki/Errors_and_residuals) or deviations (https://en.wikipedia.org/wiki/Deviation_(statistics)). From a statistical point of view, this is a measure of the quality of an estimator (it is always non-negative, and values closer to zero are better).

If $\hat{Y}$ is a vector of n predictions, and Y is the vector of observed values of the variable being predicted, then the within-sample MSE of the predictor is computed as follows:

$$\text{MSE} = \frac{1}{n}\sum_{i=1}^{n}\left(Y_i - \hat{Y}_i\right)^2$$

Therefore, an MSE is the mean $\left(\frac{1}{n}\sum_{i=1}^{n}\right)$ of the squares of the errors $\left(Y_i - \hat{Y}_i\right)^2$.

We have another `batch_mse` here that will return RMSEs for all the input data in a batch, which is a vector whose length equals the number of rows in the input data. These will be the predicted values—or fraud scores if you want—for the input (whether it is training, validation or testing data), which we can extract out after prediction. Then we define the loss and optimizer, and minimize the squared error:

```
cost_op = tf.reduce_mean(tf.pow(y_true - y_pred, 2))
optimizer = tf.train.RMSPropOptimizer(learning_rate).minimize(cost_op)
```

The activation functions for each layer used is `tanh`. The objective function here, or the cost, measures the total RMSE of our predicted and input arrays in one batch, which means it's a scalar. We then run the optimizer every time we want to do a batch update.

Fantastic! We're ready to start the training. However, before that, let's define the path where we will be saving our trained model:

```
save_model = os.path.join(data_dir, 'autoencoder_model.ckpt')
saver = tf.train.Saver()
```

Up to this point, we have defined many variables as well as hyperparameters, so we have to initialize the variables:

```
init_op = tf.global_variables_initializer()
```

Finally, we start the training. We loop over all batches in the training cycle. Then we run the optimization operation and cost operation to get the loss value. Then we display the logs per epoch step. Finally, we save the trained model:

```
epoch_list = []
loss_list = []
train_auc_list = []
data_dir = 'Training_logs/'
with tf.Session() as sess:
    now = datetime.now()
    sess.run(init_op)
    total_batch = int(train_x.shape[0]/batch_size)

    # Training cycle
    for epoch in range(training_epochs):
        # Loop over all batches
        for i in range(total_batch):
            batch_idx = np.random.choice(train_x.shape[0],\
                    batch_size)
            batch_xs = train_x[batch_idx]

            # Run optimization op (backprop) and
            # cost op (to get loss value)
            _, c = sess.run([optimizer, cost_op],\
                feed_dict={X: batch_xs})

        # Display logs per epoch step
        if epoch % display_step == 0:
            train_batch_mse = sess.run(batch_mse,\
                feed_dict={X: train_x})
            epoch_list.append(epoch+1)
            loss_list.append(c)
            train_auc_list.append(auc(train_y, train_batch_mse))
            print("Epoch:", '%04d,' % (epoch+1),
                "cost=", "{:.9f},".format(c),
                "Train auc=", "{:.6f},".format(auc(train_y, \
                train_batch_mse)),
    print("Optimization Finished!")
```

```
            save_path = saver.save(sess, save_model)
            print("Model saved in: %s" % save_path)

    save_model = os.path.join(data_dir, autoencoder_model_1L.ckpt')
    saver = tf.train.Saver()
```

The preceding code segment is straightforward. Each time, we randomly sample a mini-batch of size 256 from `train_x`, feed it into the model as the input of x, and run the *optimizer* to update the parameters through **Stochastic Gradient Descent (SGD)**:

```
>>>
Epoch: 0001, cost= 0.938937187, Train auc= 0.951383
Epoch: 0011, cost= 0.491790086, Train auc= 0.954131
...
Epoch: 0981, cost= 0.323749095, Train auc= 0.953185
Epoch: 0991, cost= 0.255667418, Train auc= 0.953107
Optimization Finished!
Model saved in: Training_logs/autoencoder_model.ckpt
Test auc score: 0.947296
```

The AUC score we have obtained for the valuation on `train_x` is around 0.95. Nevertheless, from the preceding logs, it's really difficult to understand how the training went:

```
# Plot Training AUC over time
plt.plot(epoch_list, train_auc_list, 'k--', label='Training AUC',
linewidth=1.0)
plt.title('Traing AUC per iteration')
plt.xlabel('Iteration')
plt.ylabel('Traing AUC')
plt.legend(loc='upper right')
plt.grid(True)

# Plot train loss over time
plt.plot(epoch_list, loss_list, 'r--', label='Training loss',
linewidth=1.0)
plt.title('Training loss')
plt.xlabel('Iteration')
plt.ylabel('Loss')
plt.legend(loc='upper right')
plt.grid(True)
plt.show()
>>>
```

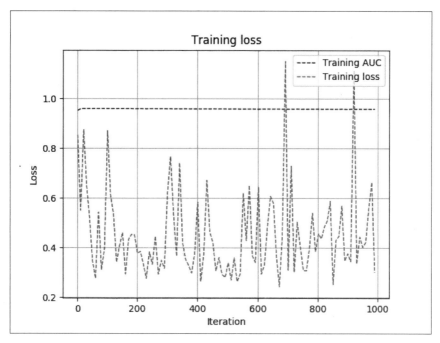

Figure 19: Training loss and AUC per iteration

In the preceding graph, we can see that the training error was a bit bumpy but the training AUC remains almost steady, around 95%. This might sound suspicious. You can also see that we used the same data for training as well as for validation. This might sound confusing too, but wait!

Since we are doing unsupervised training and the model never sees the labels during training, this will not lead to overfitting. This additional validation is used for monitoring early stopping as well as hyperparameter tuning.

Evaluating the model

After we have finished training our autoencoder model and hyperparameters, we can evaluate its performance on the 20% test dataset, which is shown here:

```
save_model = os.path.join(data_dir, autoencoder_model.ckpt')
saver = tf.train.Saver()

# Initializing the variables
init = tf.global_variables_initializer()

with tf.Session() as sess:
    now = datetime.now()
```

```
saver.restore(sess, save_model)
test_batch_mse = sess.run(batch_mse, feed_dict={X: test_x})

print("Test auc score: {:.6f}".format(auc(test_y, \
test_batch_mse)))
```

In this code, we have reused the trained model we made earlier. `test_batch_mse` is our fraud scores for test data:

```
>>>
Test auc score: 0.948843
```

Fantastic! Our trained model turned out to be a highly accurate model, showing an AUC of about 95%. Now that we have seen the evaluation, some visual analytics would be great. What do you think, guys? Let's plot the fraud score (MSE) distribution for non-fraud cases. The following code snippet does this:

```
plt.hist(test_batch_mse[test_y == 0.0], bins = 100)
plt.title("Fraud score (mse) distribution for non-fraud cases")
plt.xlabel("Fraud score (mse)")
plt.show()
>>>
```

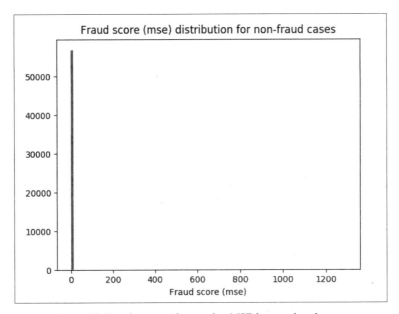

Figure 20: Fraud score with regard to MSE for non-fraud cases

The preceding screenshot is not understandable, so let's zoom it into the (0, 30) range and plot the graph again:

```
# Zoom into (0, 30) range
plt.hist(test_batch_mse[(test_y == 0.0) & (test_batch_mse < 30)], bins
= 100)
plt.title("Fraud score (mse) distribution for non-fraud cases")
plt.xlabel("Fraud score (mse)")
plt.show()
>>>
```

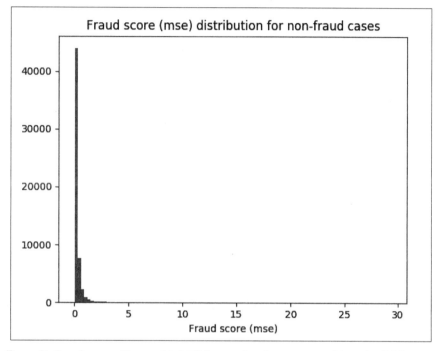

Figure 21: Fraud score with regard to MSE for non-fraud cases, zoomed in to the (0, 30) range

Now let's display only the fraud classes:

```
# Display only fraud classes
plt.hist(test_batch_mse[test_y == 1.0], bins = 100)plt.title("Fraud score (mse) distribution for fraud cases")
plt.xlabel("Fraud score (mse)")
plt.show()
>>>
```

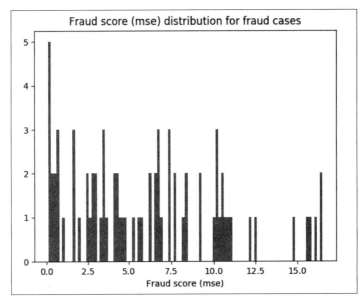

Figure 22: Fraud score with regard to MSE for fraud cases

Finally, let's show some related statistics. For example, we use 10 as our detection threshold. Now we can compute the number of detected cases above the threshold, the number of positive cases above the threshold, the percentage of accuracy above the threshold (that is, precision), and compare it to the average percentage of fraud in the testing set:

```
threshold = 10
print("Number of detected cases above threshold: {}, \n\
Number of pos cases only above threshold: {}, \n\
The percentage of accuracy above threshold (Precision): {:0.2f}%. \n\
Compared to the average percentage of fraud in test set: 0.132%".format( \
np.sum(test_batch_mse > threshold), \
np.sum(test_y[test_batch_mse > threshold]), \
np.sum(test_y[test_batch_mse > threshold]) / np.sum(test_batch_mse > threshold) * 100))
>>>
```

```
Number of detected cases above threshold: 198,
Number of positive cases only above threshold: 18,
The percentage of accuracy above threshold (Precision): 9.09%.
Compared to the average percentage of fraud in test set: 0.132%
```

To conclude, an autoencoder with just one hidden layer turned out to be enough (for training, at least) for our case. However, you could still try to adopt other variants, such as deconvolutional autoencoder and denoising autoencoder, to solve the same problem.

Summary

In this chapter, we implemented some optimizing networks called autoencoders. An autoencoder is basically a *data compression* network model. It is used *to encode* a given input into a representation of smaller dimensions, and then a decoder can be used to reconstruct the input back from the encoded version. All the autoencoders we implemented contain an encoding and a decoding part.

We also saw how to improve the autoencoders' performance by introducing noise during the network training and building a *denoising autoencoder*. Finally, we applied the concepts of CNNs introduced in *Chapter 4, Convolutional Neural Networks* with the implementation of *convolutional autoencoders*.

Even when the number of hidden units is large, we can still discover the interesting and hidden structure of the dataset using autoencoders by imposing other constraints on the network. In other words, if we impose a sparsity constraint on the hidden units, then the autoencoder will still discover interesting structure in the data, even if the number of hidden units is large. To prove this argument, we saw a real-life example, credit card fraud analytics, in which we successfully applied an autoencoder.

A **Recurrent Neural Network (RNN)** is a class of artificial neural network where the connections between units form a directed cycle. RNNs make use of information from the past such as time series forecasting. That way, they can make predictions about data with high temporal dependencies. This creates an internal state of the network, which allows it to exhibit dynamic temporal behavior.

In the next chapter, we'll examine RNNs. We will start by describing the basic principles of these networks, and then we'll implement some interesting examples of these architectures.

6
Recurrent Neural Networks

A **RNN** is a class of **ANN** where connections between units form a directed cycle. RNNs make use of information from the past. That way, they can make predictions in data with high temporal dependencies. This creates an internal state of the network, which allows it to exhibit dynamic temporal behavior. In this chapter, we will develop several real-life predictive models, using RNNs and their architectural variants, to make predictive analytics easier.

First, we will provide some theoretical background of RNNs. Then we will look at a few examples that will show a systematic way of implementing predictive models for image classification, sentiment analysis of movies, and spam predictions for **Natural Language Processing** (**NLP**).

Then we will show how to develop predictive models for time series data. Finally, we will see a how to develop a **LSTM** network for solving more advanced problems, such as human activity recognition.

Concisely, the following topics will be covered throughout this chapter:

- Working principles of RNNS
- RNNs and the gradient vanishing-exploding problem
- LSTM networks
- Implementing an RNN for spam prediction
- Developing an LSTM predictive model for time series data
- An LSTM predictive model for sentiment analysis
- Human activity recognition using an LSTM network

Working principles of RNNs

In this section, we will first provide some contextual information about RNNs. Then we will see some potential drawbacks of the classical RNN. Finally, we will see an improved variation of RNNs called LSTM to address the drawbacks.

Human beings do not start thinking from scratch. The human mind has so-called persistence of memory, the ability to associate the past with recent information. Traditional neural networks instead ignore past events. Take the movie scenes classifier as an example; it is not possible for a neural network to use past scenes to classify current ones. RNNs were developed to try to solve this problem.

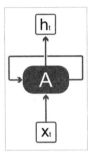

Figure 1: RNNs have loops

In contrast to conventional neural networks, RNNs are networks with a loop that allows the information to be persistent in a neural network. In the preceding diagram, with the network **A**, at some time **t**, it receives the input x_t and outputs a value of h_t. So, in the preceding figure, we think of an RNN as multiple copies of the same network, each passing a message to a successor.

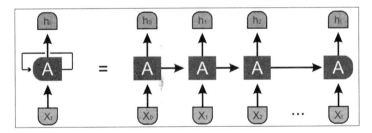

Now, if we unroll this network, what will we receive? Let's see a simple but real-life example. Suppose X_0=Monday, X_1=Tuesday, X_2=Wednesday, and so on. If h stores what you eat, then yesterday's meal decision would affect what you will eat tomorrow. This can be explained using the following figure:

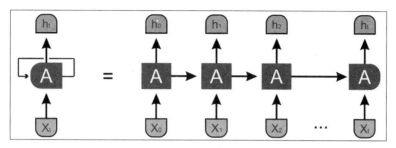

Figure 2: An unrolled representation of the same RNN represented in figure 1

However, the unrolled network does not provide detailed information about RNNs. An RNN is different from a traditional neural network in that it introduces a transition weight W to transfer information between times. RNNs process sequential input one piece at a time, updating a kind of vector state that contains information about all past elements of the sequence. The following diagram shows a neural network that takes a value of **X (t)** as input, and then outputs a value **Y (t)**:

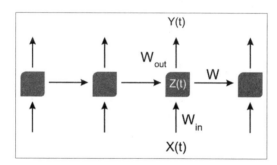

Figure 3: An RNN architecture can use the previous states of the network to its advantage

As shown in figure 1, the first half of the neural network is characterized by the $Z(t) = X(t) * W_{in}$ function, and the second half of the neural network takes the form $Y(t) = Z(t) * W_{out}$. If you prefer, the complete neural network is $Y(t) = (X(t) * W_{in}) * W_{out}$.

At each time t, when calling the learned model, this architecture does not take into account any knowledge about the previous runs. It is like predicting stock market trends by only looking at data from the current day. A better idea would be to exploit overarching patterns from a week's worth or a month's worth of data.

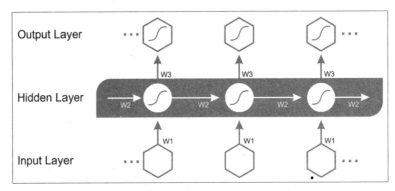

Figure 4: An RNN architecture where all the weights in all the layers have to be learned with time

A more explicit architecture can be found in figure 4, where the temporally shared weights **W2** (for the hidden layer) must be learned, in addition to **W1** (for the input layer) and **W3** (for the output layer).

From a computational point of view, an RNN takes many input vectors to process and generate output vectors. Imagine that each rectangle in the following diagram has a vectorial depth and other special hidden quirks. It can have many forms, such as one-to-one, one-to-many, and many-to-many.

As it can be seen that a one-to-one architecture would be a standard feedforward neural network. A many-to-one architecture accepts time series of feature vectors (one vector per time step) and converts them to a probability vector at the output for classification:

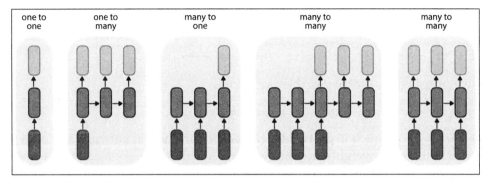

Figure 5: RNNs can have many forms

Therefore, RNNs can have many forms. In the preceding diagram, each rectangle is a vector and arrows represent functions (for example, matrix multiply). Input vectors are in green, output vectors are in yellow, and blue vectors hold the RNN's state (more on this soon).

Implementing basic RNNs in TensorFlow

TensorFlow has `tf.contrib.rnn.BasicRNNCell` and `tf.nn.rnn_cell.BasicRNNCell`, which provide the basic building blocks of RNNs. However, first let's implement a very simple RNN model, without using either of these. The idea is to have a better understanding of what goes on under the hood.

We will create an RNN composed of a layer of five recurrent neurons using the ReLU activation function. We will assume that the RNN runs over only two-time steps, taking input vectors of size 3 at each time step. The following code builds this RNN, unrolled through two-time steps:

```
n_inputs = 3
n_neurons = 5
X1 = tf.placeholder(tf.float32, [None, n_inputs])
X2 = tf.placeholder(tf.float32, [None, n_inputs])

Wx = tf.get_variable("Wx", shape=[n_inputs,n_neurons], dtype=tf.
float32, initializer=None, regularizer=None, trainable=True,
collections=None)
```

Recurrent Neural Networks

```
Wy = tf.get_variable("Wy", shape=[n_neurons,n_neurons], dtype=tf.
float32, initializer=None, regularizer=None, trainable=True,
collections=None)

b = tf.get_variable("b", shape=[1,n_neurons], dtype=tf.float32,
initializer=None, regularizer=None, trainable=True, collections=None)

Y1 = tf.nn.relu(tf.matmul(X1, Wx) + b)
Y2 = tf.nn.relu(tf.matmul(Y1, Wy) + tf.matmul(X2, Wx) + b)
```

Then we initialize the global variables as follows:

```
init_op = tf.global_variables_initializer()
```

This network looks much like a two-layer feedforward neural network, but both layers share the same weights and bias vectors. Additionally, we feed inputs at each layer and receive outputs from each layer.

```
X1_batch = np.array([[0, 2, 3], [2, 8, 9], [5, 3, 8], [3, 2, 9]]) # t
= 0
X2_batch = np.array([[5, 6, 8], [1, 0, 0], [8, 2, 0], [2, 3, 6]]) # t
= 1
```

These mini-batches contain four instances, each with an input sequence composed of exactly two inputs. At the end, `Y1_val` and `Y2_val` contain the outputs of the network at both time steps for all neurons and all instances in the mini-batch. Then we create a TensorFlow session and execute the computational graph as follows:

```
with tf.Session() as sess:
        init_op.run()
        Y1_val, Y2_val = sess.run([Y1, Y2], feed_dict={X1:
        X1_batch, X2: X2_batch})
```

Finally, we print the result:

```
print(Y1_val) # output at t = 0
print(Y2_val) # output at t = 1
```

The following is the output:

```
>>>
[[ 0.           0.           0.           2.56200171   1.20286    ]
 [ 0.           0.           0.          12.39334488   2.7824254  ]
 [ 0.           0.           0.          13.58520699   5.16213894]
 [ 0.           0.           0.           9.95982838   6.20652485]]

[[ 0.           0.           0.          14.86255169   6.98305273]
```

```
[  0.          0.         26.35326385  0.66462421 18.31009483]
[  5.12617588  4.76199865 20.55905533 11.71787453 18.92538261]
[  0.          0.         19.75175095  3.38827515 15.98449326]]
```

The network we created is simple, but if you run it over 100 time steps, for example, the graph is going to be very big. Now, let's look at how to create the same RNN using TensorFlow's `contrib` package. The `static_rnn()` function creates an unrolled RNN by chaining cells as follows:

```
basic_cell = tf.nn.rnn_cell.BasicRNNCell(num_units=n_neurons)
output_seqs, states = tf.contrib.rnn.static_rnn(basic_cell, [X1, X2], dtype=tf.float32)
Y1, Y2 = output_seqs
init_op = tf.global_variables_initializer()
X1_batch = np.array([[0, 2, 3], [2, 8, 9], [5, 3, 8], [3, 2, 9]]) # t = 0
X2_batch = np.array([[5, 6, 8], [1, 0, 0], [8, 2, 0], [2, 3, 6]]) # t = 1
with tf.Session() as sess:    init_op.run()
    Y1_val, Y2_val = sess.run([Y1, Y2], feed_dict={X1: X1_batch, X2: X2_batch})
print(Y1_val) # output at t = 0
print(Y2_val) # output at t = 1
```

The output is as follows:

```
>>>
[[-0.95058489  0.85824621  0.68384844 -0.55920446 -0.87788445]
 [-0.99997741  0.99928695  0.99601227 -0.98470896 -0.99964565]
 [-0.99321234  0.99998873  0.99999011 -0.83302033 -0.98657602]
 [-0.99771607  0.99999255  0.99997681 -0.74148595 -0.99279612]]

[[-0.99982268  0.99888307  0.999865   -0.98920071 -0.99867421]
 [-0.64704001 -0.87286478  0.34580848 -0.66372067 -0.52697307]
 [ 0.3253428   0.62628752  0.99945754 -0.887465   -0.17882833]
 [-0.99901992  0.9688856   0.99529684 -0.9469955  -0.99445421]]
```

However, if we use the `static_rnn()` function we can still build a computational graph containing one cell per time step. Now, imagine that there were 100 time steps; the graph would look very big and would be difficult to make sense of. To get rid of this problem, the `dynamic_rnn()` function provides a dynamic unrolling functionality over time:

```python
n_inputs = 3
n_neurons = 5
n_steps = 2

X = tf.placeholder(tf.float32, [None, n_steps, n_inputs])
seq_length = tf.placeholder(tf.int32, [None])

basic_cell = tf.nn.rnn_cell.BasicRNNCell(num_units=n_neurons)
output_seqs, states = tf.nn.dynamic_rnn(basic_cell, X,
dtype=tf.float32)

X_batch = np.array([
                [[0, 2, 3], [2, 8, 9]], # instance 0
                [[5, 6, 8], [0, 0, 0]], # instance 1 (padded
with a zero vector)
                [[6, 7, 8], [6, 5, 4]], # instance 2
                [[8, 2, 0], [2, 3, 6]], # instance 3
                ])
seq_length_batch = np.array([3, 4, 3, 5])
init_op = tf.global_variables_initializer()

with tf.Session() as sess:
        init_op.run()
        outputs_val, states_val = sess.run([output_seqs, states],
feed_dict={X: X_batch, seq_length: seq_length_batch})

print(outputs_val)
```

The following is the output of the preceding code:

```
>>>
[[[ 0.03708282  0.24521144 -0.65296066 -0.42676723  0.67448044]
  [ 0.50789726  0.98529315 -0.99976575 -0.84865189  0.96734977]]
 [[ 0.99343652  0.96998596 -0.99997932  0.59788793  0.00364922]
  [-0.51829755  0.56738734  0.78150493  0.16428296 -0.33302191]]
 [[ 0.99764818  0.99349713 -0.99999821  0.60863507 -0.02698862]
  [ 0.99159312  0.99838346 -0.99994278  0.83168954 -0.81424212]]
```

```
print(states_val)
>>>
[[ 0.99968255  0.99266654 -0.99999398  0.99020076 -0.99553883]
 [ 0.85630441  0.72372746 -0.90620565  0.60570842  0.1554002 ]]
[[ 0.50789726  0.98529315 -0.99976575 -0.84865189  0.96734977]
 [-0.51829755  0.56738734  0.78150493  0.16428296 -0.33302191]
 [ 0.99159312  0.99838346 -0.99994278  0.83168954 -0.81424212]
 [ 0.85630441  0.72372746 -0.90620565  0.60570842  0.1554002 ]]
```

Now, what happens under the hood? Well, during the backpropagation, the `dynamic_rnn()` function uses a `while_loop()` operation to run over the cell an appropriate number of times. Then, the tensor values for each iteration in the forward pass are stored so that the gradients during the reverse pass can be computed.

As we discussed in earlier chapters, overfitting is a major issue with RNNs. A dropout layer can help us avoid overfitting. We will see a user-friendly example of this later in the chapter.

RNN and the long-term dependency problem

RNNs are very powerful and popular too. However, we often only need to look at recent information to perform the present task rather than stored information or information that happened a long time ago. This happens frequently in NLP for language modeling. Let's see a common example:

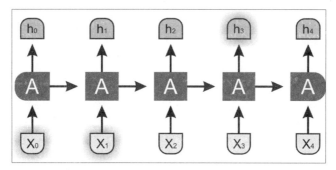

Figure 6: If the gap between the relevant information and the place that it is needed is small, RNNs can learn to use the past information

Recurrent Neural Networks

Suppose that a language model is trying to predict the next word based on the previous words. As a human being, if we try to predict the last word in "the sky is...," without further context it's most likely the next word that we will predict will be "blue." In such cases, the gap between the relevant information and the position is small. Thus, RNNs can learn to use past information easily.

Nevertheless, consider a longer text: "Reza grew up in Bangladesh. He studied in Korea. He speaks fluent...", we need more context. In this sentence, the most recent information advises that the next word would probably be the name of a language: however, if we want to narrow down which language it is, we need the context of Bangladesh, from the previous words.

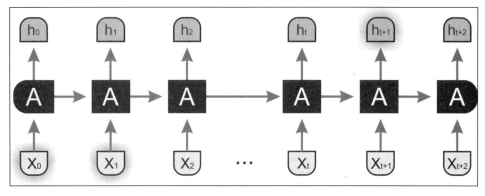

Figure 7: If the gap between the relevant information and the place that it's needed is bigger, RNNs can't learn to use past information

Here, the gap is larger than the previous example, so an RNN is unable to learn to connect the information. This is a serious drawback of RNNs. Here comes the LSTM as the savior. Let's see some commonly used architectures of RNNs, such as LSTM, bi-directional RNNs, and GRU, in the next subsection.

Bi-directional RNNs

Bi-directional RNNs are based on the idea that the output at time t may depend on the previous and future elements in the sequence. To deal with this, the output of two RNNs must be mixed: one executes the process in one direction, and the second runs the process in the opposite direction. The following diagram shows the basic difference between a regular RNN and a **Bi-directional RNN (BRNN)**.

A more explicit architecture of a BRNN can be found in the following diagram, in which the temporally shared weights **w2**, **w3**, **w4**, and **w5** (for the forward and the backward layer) must be learned in addition to input layer and the output layer:

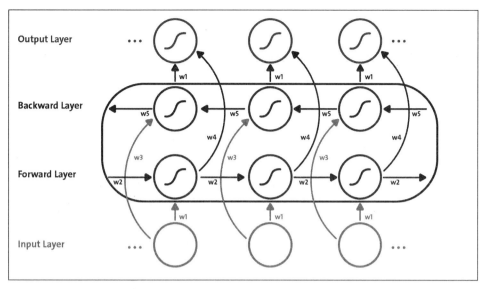

Figure 8: A BRNN architecture where all the weights in all the layers have to be learned over time

The unrolled architecture is also a very common implementation of BRNN. The unrolled architecture of B-RNN is depicted in the following diagram. The network splits the neurons of a regular RNN into two directions, one for the positive time direction (forward states), and another for the negative time direction (backward states). With this structure, the output layer can get information from the past and future states, as shown in figure 9:

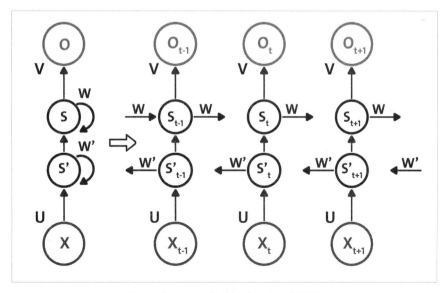

Figure 9: An unrolled bi-directional RNN

RNN and the gradient vanishing-exploding problem

Gradients for deeper layers are calculated as products of many gradients of activation functions in the multi-layer network. When those gradients are small or zero, it will easily vanish. On the other hand, when they are bigger than 1, it will possibly explode. So, it becomes very hard to calculate and update.

Let's explain them in more detail:

- If the weights are small, it can lead to a situation called vanishing gradients, where the gradient signal gets so small that learning either becomes very slow or stops working altogether. This is often referred to as *vanishing gradients*.
- If the weights in this matrix are large, it can lead to a situation where the gradient signal is so large that it can cause learning to diverge. This is often referred to as *exploding gradients*.

Thus, one of the major issues of RNN is the *vanishing-exploding gradient* problem, which directly affects performance. In fact, the backpropagation time rolls out the RNN, creating *a very deep* feed-forward neural network. The impossibility of getting a long-term context from the RNN is due precisely to this phenomenon: if the gradient vanishes or explodes within a few layers, the network will not be able to learn high temporal distance relationships between the data.

The next diagram shows schematically what happens: the *computed* and *backpropagated gradient* tends to decrease (or increase) at each *instant of time* and then, after a certain number of instants of time, the cost function tends to converge to zero (or explode to infinity).

We can get the exploding gradients by two ways. Since the purpose of activation function is to control the big changes in the network by squashing them, the weights we set must be non-negative and large. When these weights are multiplied along the layers, they cause a large change in the cost. When our neural network model is learning, the ultimate goal is to minimize the cost function and change the weights to reach the optimum cost.

For example, the cost function is the mean squared error. It is a pure convex function and the aim is to find the underlying cause of that convex. If your weights increase to a certain big amount, then the downward moment will increase and we will overshoot the optimum repeatedly and the model will never learn!

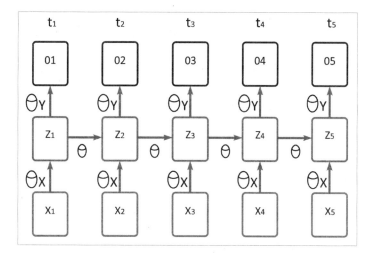

In the preceding figure, we have the following parameters:

- θ denotes the parameters of the hidden recurrent layer
- $θ_x$ denotes the parameter of input to the hidden layer
- $θ_y$ denotes the parameter of the output layer
- σ denotes the activation function of the hidden layer
- Input is represented as X_t
- The output from the hidden layer as h_t
- The final output as O_t for t (timestep)

Note that the preceding diagram denotes the time lapse of the recurrent neural network model given below. Now if you recall figure 1, the output can be formulated as follows:

$$h_t = θσ(h_{t-1}) + θx\, X_t \text{ and } O_t = θσ(h_t)$$

Now let E represent the loss at the output layer: $E = f(O_t)$. Then the above three equations tell us that the E depends upon the output O_t. Output O_t changes with respect to the change in the hidden state of the layer (h_t). The hidden state of the current timestep (h_t) depends upon the state of the neuron at the previous timestep (h_{t-1}). Now the following equation will clear the concept.

Recurrent Neural Networks

The rate of change of loss with respect to parameters chosen for the hidden layer = $\partial E / \partial \theta$, which is a chain rule that can be formulated as follows:

$$\partial E / \partial \theta = (k = 1 \text{ to } k = t) \sum (\partial E / \partial O_t) * (\partial O_t / \partial h_t) * (\partial h_t / \partial h_k) * (\partial h_k / \partial \theta) \quad \text{(I)}$$

In the preceding equation, the term $\partial ht / \partial hk$ is not only interesting but also useful.

$$\partial h_t / \partial h_k = (z = k+1 \text{ to } z = t) \prod \partial h_z / \partial h_{z-1} \quad \text{(II)}$$

Now let's consider $t = 5$ and $k = 1$ then

$$\partial h_5 / \partial h_1 = (\partial h_2 / \partial h_1) * (\partial h_3 / \partial h_2) * (\partial h_4 / \partial h_3) * (\partial h_5 / \partial h_4) \quad \text{(III)}$$

Differentiating equation (II) with respect to (h_{t-1}) gives us:

$$\partial h_t / \partial h_{t-1} = \theta \sigma'(h_{t-1}) \quad \text{(IV)}$$

Now if we combine equation (III) and (IV), we can have the following result:

$$\partial h_5 / \partial h_1 = (\theta \sigma'(h_1)) * (\theta \sigma'(h_2)) * (\theta \sigma'(h_3)) * (\theta \sigma'(h_4))$$

In these cases, θ also changes with timestep. The above equation shows the dependency of the current state with respect to the previous states. Now let's explain the anatomy of those two equations. Say you are at a timestep five ($t = 5$), then k will range from one to five ($k=1$ to 5) this means you have to calculate (k) for the following:

$$\partial h_5 / \partial h_1 \mid \partial h_5 / \partial h_2 \mid \partial h_5 / \partial h_3 \mid \partial h_5 / \partial h_4 \mid \partial h_5 / \partial h_5$$

Now come at equation (II), each of the above $\partial h_t / \partial h_k = (z = k+1 \text{ to } z = t) \prod \partial h_z / \partial h_{z-1}$. Moreover, it is dependent on the parameter of the recurrent layer θ. If your weights get large during training, which they will due to multiplications in equation (II) for each timestep (I). The problem of gradient exploding will occur.

To overcome the *vanishing-exploding problem*, various extensions of the basic RNN model have been proposed. LSTM networks, which will be introduced in the next section, are one of these.

LSTM networks

One type of RNN model is LSTM. The precise implementation details of LSTM are not in the scope of this book. An LSTM is a special RNN architecture that was originally conceived by Hochreiter and Schmidhuber in 1997.

This type of neural network has been recently rediscovered in the context of deep learning because it is free from the problem of vanishing gradients and offers excellent results and performance. LSTM-based networks are ideal for the prediction and classification of temporal sequences and are replacing many traditional approaches to deep learning.

The name signifies that short-term patterns are not forgotten in the long term. An LSTM network is composed of cells (LSTM blocks) linked to each other. Each LSTM block contains three types of the gate: an input gate, an output gate, and a forget gate, which implements the functions of writing, reading, and reset on the cell memory, respectively. These gates are not binary, but analog (generally managed by a sigmoidal activation function mapped in the range [0, 1], where 0 indicates total inhibition, and 1 shows total activation).

If you consider an LSTM cell as a black box, it can be used very much like a basic cell, except it will perform much better; training will converge more quickly and it will detect long-term dependencies in the data. In TensorFlow, you can simply use `BasicLSTMCell` instead of `BasicRNNCell`:

```
lstm_cell = tf.nn.rnn_cell.BasicLSTMCell(num_units=n_neurons)
```

LSTM cells manage two state vectors, and for performance reasons, they are kept separate by default. You can change this default behavior by setting state_is_tuple=False when creating BasicLSTMCell. So, how does an LSTM cell work? The architecture of a basic LSTM cell is shown in the following diagram:

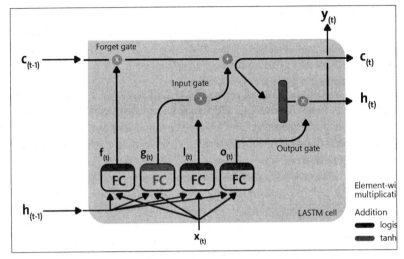

Figure 11: Block diagram of an LSTM cell

Now, let's see the mathematical notation behind this architecture. If we don't look at what's inside the LSTM box, the LSTM cell itself looks exactly like a regular memory cell, except that its state is split into two vectors, $h(t)$ and $c(t)$:

- c is a
- $h_{(t)}$ as the short-term s
- $c_{(t)}$ as the long-term state

Now, let's open the box! The key idea is that the network can learn the following:

- What to store in the long-term state
- What to throw away
- What to read it

As the long-term $c_{(t-1)}$ traverses the network from left to right, you can see that it first goes through a forget gate, dropping some memory, and then it adds some new memory via the addition operation (which adds the memory that was selected by an input gate). The result $c_{(t)}$ is sent straight out, without any further transformation.

Therefore, at each time step, some memory is dropped and some memory is added. Moreover, after the addition operation, the long-term state is then copied and passed through the tanh function, which produces outputs in the scale of [-1, +1].

Then the output gate filters the result. This produces the short-term s $h_{(t)}$ (which is equal to the cell's output for this time step $y_{(t)}$). Now, let's look at where new memories come from and how the gates work. First, the current input v $x_{(t)}$ and the previous short-term $h_{(t-1)}$ are fed to four different fully connected layers.

The presence of these gates allows LSTM cells *to remember* information for an indefinite period: in fact, if the *input gate* is below the activation threshold, the cell *will retain* the previous state, and if the current state is enabled, it will be combined with the input value. As the name suggests, the **forget gate** resets the current state of the cell (when its value is cleared to 0), and the **output gate** decides whether the value of the cell *must be carried out or not*.

The following equations are used to do the LSTM computations of a cell's long-term state, its short-term state, and its output at each time step for a single instance:

$$i_{(t)} = \sigma\left(W_{xi}^T . x_{(t)} + W_{hi}^T . h_{(t-1)} + b_i\right)$$

$$f_{(t)} = \sigma\left(W_{xf}^T . x_{(t)} + W_{hf}^T . h_{(t-1)} + b_f\right)$$

$$o_{(t)} = \sigma\left(W_{xo}^T . x_{(t)} + W_{ho}^T . h_{(t-1)} + b_o\right)$$

$$g_{(t)} = \tanh\left(W_{xg}^T . x_{(t)} + W_{hg}^T . h_{(t-1)} + b_g\right)$$

$$c_{(t)} = f_{(t)} \otimes c_{(t-1)} + i_{(t)} \otimes g_{(t)}$$

$$y_{(t)} = h_{(t)} = o_{(t)} \otimes \tanh\left(c_{(t)}\right)$$

In the preceding equation, W_{xi}, W_{xf}, W_{xo}, and W_{xg} are the weight matrices of each of the four layers for their connection to the input vector $x_{(t)}$. On the other hand, W_{hi}, W_{hf}, W_{ho}, and W_{hg} are the weight matrices of each of the four layers for their connection to the previous short-term state, $h_{(t-1)}$.

Finally, b_i, b_f, b_o, and b_g are the bias terms for each of the four layers. TensorFlow initializes b_f to a vector full of 1's instead of 0's. This prevents it from forgetting everything at the beginning of training.

GRU cell

There are many other variants of the LSTM cell. One particularly popular variant is the **Gated Recurrent Unit (GRU)** cell. Kyunghyun Cho and others proposed the GRU cell in a 2014 paper that also introduced the autoencoder network we mentioned earlier.

Technically, a GRU cell is a simplified version of an LSTM cell where both the state vectors are merged into a single vector called $h_{(t)}$. A single gate controller controls both the forget gate and the input gate. If the gate controller's output is 1, the input gate is open and the forget gate is closed.

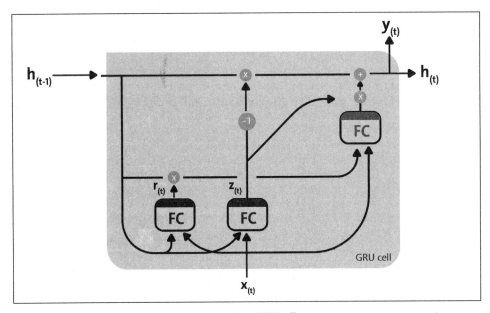

Figure 12: A GRU cell

On the other hand, if it's output is 0, the opposite happens. Whenever a memory must be stored, the location where it will be stored is erased first, which is actually a frequent variant of the LSTM cell in and of itself. The second simplification is that since the full state vector is output at every time step, there is no output gate. However, a new gate controller controls which part of the previous state will be shown to the main layer.

The following equations are used to do the GRU computations of a cell's long-term state, its short-term state, and its output at each time step for a single instance:

$$z_{(t)} = \sigma\left(W_{xz}^T . x_{(t)} + W_{hz}^T . h_{(t-1)}\right)$$

$$r_{(t)} = \sigma\left(W_{xr}^T . x_{(t)} + W_{hr}^T . h_{(t-1)}\right)$$

$$g_{(t)} = \tanh\left(W_{xg}^T . x_{(t)} + W_{hg}^T . \left(r_{(t)} \otimes h_{(t-1)}\right)\right)$$

$$h_{(t)} = \left(1 - z_{(t)}\right) \otimes h_{(t-1)} + z_{(t)} \otimes g_{(t)}$$

Creating a GRU cell in TensorFlow is straightforward. Here is an example:

```
gru_cell = tf.nn.rnn_cell.GRUCell(num_units=n_neurons)
```

These simplifications are not a weakness of this type of architecture; it seems to perform successfully. The LSTM or GRU cells are one of the main reasons behind the success of RNNs in recent years, in particular for applications in **NLP**.

We will see examples of using LSTM in this chapter, but the next section shows an example of using an RNN for spam/ham text classification.

Implementing an RNN for spam prediction

In this section, we will see how to implement an RNN in TensorFlow to predict spam/ham from texts.

Data description and preprocessing

The popular spam dataset from the UCI ML repository will be used, which can be downloaded from http://archive.ics.uci.edu/ml/machine-learning-databases/00228/smsspamcollection.zip.

The dataset contains texts from several emails, some of which were marked as spam. Here we will train a model that will learn to distinguish between spam and non-spam emails using only the text of the email. Let's get started by importing the required libraries and model:

```
import os
import re
import io
```

```
import requests
import numpy as np
import matplotlib.pyplot as plt
import tensorflow as tf
from zipfile import ZipFile
from tensorflow.python.framework import ops
import warnings
```

Additionally, we can stop printing the warning produced by TensorFlow if you want:

```
warnings.filterwarnings("ignore")
os.environ['TF_CPP_MIN_LOG_LEVEL'] = '3'
ops.reset_default_graph()
```

Now, let's create the TensorFlow session for the graph:

```
sess = tf.Session()
```

The next task is to set the RNN parameters:

```
epochs = 300
batch_size = 250
max_sequence_length = 25
rnn_size = 10
embedding_size = 50
min_word_frequency = 10
learning_rate = 0.0001
dropout_keep_prob = tf.placeholder(tf.float32)
```

Let's manually download the dataset and store it in a `text_data.txt` file in the `temp` directory. First, we set the path:

```
data_dir = 'temp'
data_file = 'text_data.txt'
if not os.path.exists(data_dir):
    os.makedirs(data_dir)
```

Now, we directly download the dataset in zipped format:

```
if not os.path.isfile(os.path.join(data_dir, data_file)):
    zip_url = 'http://archive.ics.uci.edu/ml/machine-learning-databases/00228/smsspamcollection.zip'
    r = requests.get(zip_url)
    z = ZipFile(io.BytesIO(r.content))
    file = z.read('SMSSpamCollection')
```

We still need to format the data:

```
text_data = file.decode()
text_data = text_data.encode('ascii',errors='ignore')
text_data = text_data.decode().split('\n')
```

Now, store in it the directory mentioned earlier in a text file:

```
    with open(os.path.join(data_dir, data_file), 'w') as
file_conn:
        for text in text_data:
            file_conn.write("{}\n".format(text))
else:
    text_data = []
    with open(os.path.join(data_dir, data_file), 'r') as
file_conn:
        for row in file_conn:
            text_data.append(row)
    text_data = text_data[:-1]
```

Let's split the words that have a word length of at least 2:

```
text_data = [x.split('\t') for x in text_data if len(x)>=1]
[text_data_target, text_data_train] = [list(x) for x in
zip(*text_data)]
```

Now we create a text cleaning function:

```
def clean_text(text_string):
    text_string = re.sub(r'([^\s\w]|_|[0-9])+', '', text_string)
    text_string = " ".join(text_string.split())
    text_string = text_string.lower()
    return(text_string)
```

We call the preceding method to clean the text:

```
text_data_train = [clean_text(x) for x in text_data_train]
```

Now we need to do one of the most important tasks, which is creating word embedding –changing text into numeric vectors:

```
vocab_processor =
tf.contrib.learn.preprocessing.VocabularyProcessor(max_sequence_
length, min_frequency=min_word_frequency)
text_processed =
np.array(list(vocab_processor.fit_transform(text_data_train)))
```

Recurrent Neural Networks

Now let's shuffle to make the dataset balance:

```
text_processed = np.array(text_processed)
text_data_target = np.array([1 if x=='ham' else 0 for x in
text_data_target])
shuffled_ix = np.random.permutation(np.arange(len(text_data_target)))
x_shuffled = text_processed[shuffled_ix]
y_shuffled = text_data_target[shuffled_ix]
```

Now that we have shuffled the data, we can split the data into a training and testing set:

```
ix_cutoff = int(len(y_shuffled)*0.75)
x_train, x_test = x_shuffled[:ix_cutoff], x_shuffled[ix_cutoff:]
y_train, y_test = y_shuffled[:ix_cutoff], y_shuffled[ix_cutoff:]
vocab_size = len(vocab_processor.vocabulary_)
print("Vocabulary size: {:d}".format(vocab_size))
print("Training set size: {:d}".format(len(y_train)))
print("Test set size: {:d}".format(len(y_test)))
```

Following is the output of the preceding code:

```
>>>
Vocabulary size: 933
Training set size: 4180
Test set size: 1394
```

Before we start training, let's create placeholders for our TensorFlow graph:

```
x_data = tf.placeholder(tf.int32, [None, max_sequence_length])
y_output = tf.placeholder(tf.int32, [None])
```

Let's create the embedding:

```
embedding_mat = tf.get_variable("embedding_mat",
shape=[vocab_size, embedding_size], dtype=tf.float32,
initializer=None, regularizer=None, trainable=True, collections=None)
embedding_output = tf.nn.embedding_lookup(embedding_mat, x_data)
```

Now it's time to construct our RNN. The following code defines the RNN cell:

```
cell = tf.nn.rnn_cell.BasicRNNCell(num_units = rnn_size)
output, state = tf.nn.dynamic_rnn(cell, embedding_output,
dtype=tf.float32)
output = tf.nn.dropout(output, dropout_keep_prob)
```

Now let's define the way to get the output from our RNN sequence:

```
output = tf.transpose(output, [1, 0, 2])
last = tf.gather(output, int(output.get_shape()[0]) - 1)
```

Next, we define the weights and the biases for the RNN:

```
weight = bias = tf.get_variable("weight", shape=[rnn_size, 2],
dtype=tf.float32, initializer=None, regularizer=None,
trainable=True, collections=None)
bias = tf.get_variable("bias", shape=[2], dtype=tf.float32,
initializer=None, regularizer=None, trainable=True,
collections=None)
```

The logits output is then defined. It uses both the weight and the bias from the preceding code:

```
logits_out = tf.nn.softmax(tf.matmul(last, weight) + bias)
```

Now we define the losses for each prediction so that later on, they can contribute to the loss function:

```
losses =
tf.nn.sparse_softmax_cross_entropy_with_logits_v2(logits=logits_ou
t, labels=y_output)
```

We then define the loss function:

```
loss = tf.reduce_mean(losses)
```

We now define the accuracy of each prediction:

```
accuracy = tf.reduce_mean(tf.cast(tf.equal(tf.argmax(logits_out, 1),
tf.cast(y_output, tf.int64)), tf.float32))
```

We then create the `training_op` with `RMSPropOptimizer`:

```
optimizer = tf.train.RMSPropOptimizer(learning_rate)
train_step = optimizer.minimize(loss)
```

Now let's initialize all the variables using the `global_variables_initializer()` method:

```
init_op = tf.global_variables_initializer()
sess.run(init_op)
```

Additionally, we can create some empty lists to keep track of the training loss, testing loss, training accuracy, and the testing accuracy in each epoch:

```
train_loss = []
test_loss = []
train_accuracy = []
test_accuracy = []
```

Now we are ready to perform the training, so let's get started. The workflow of the training goes as follows:

- Shuffle the training data
- Select the training set and calculate generations
- Run training step for each batch
- Run loss and accuracy of training
- Run the evaluation steps.

The following codes includes all of the aforementioned steps:

```
shuffled_ix = np.random.permutation(np.arange(len(x_train)))
x_train = x_train[shuffled_ix]
y_train = y_train[shuffled_ix]
num_batches = int(len(x_train)/batch_size) + 1

for i in range(num_batches):
    min_ix = i * batch_size
    max_ix = np.min([len(x_train), ((i+1) * batch_size)])
    x_train_batch = x_train[min_ix:max_ix]
    y_train_batch = y_train[min_ix:max_ix]
    train_dict = {x_data: x_train_batch, y_output: \
y_train_batch, dropout_keep_prob:0.5}
    sess.run(train_step, feed_dict=train_dict)
    temp_train_loss, temp_train_acc = sess.run([loss,\
                    accuracy], feed_dict=train_dict)
train_loss.append(temp_train_loss)
train_accuracy.append(temp_train_acc)
test_dict = {x_data: x_test, y_output: y_test, \
dropout_keep_prob:1.0}
temp_test_loss, temp_test_acc = sess.run([loss, accuracy], \
                feed_dict=test_dict)
test_loss.append(temp_test_loss)
test_accuracy.append(temp_test_acc)
```

```
    print('Epoch: {}, Test Loss: {:.2}, Test Acc: {:.2}'.
format(epoch+1, temp_test_loss, temp_test_acc))
    print('\nOverall accuracy on test set (%):
{}'.format(np.mean(temp_test_acc)*100.0))
```

Following is the output of the preceding code:

```
>>>
Epoch: 1, Test Loss: 0.68, Test Acc: 0.82
Epoch: 2, Test Loss: 0.68, Test Acc: 0.82
Epoch: 3, Test Loss: 0.67, Test Acc: 0.82
...
Epoch: 997, Test Loss: 0.36, Test Acc: 0.96
Epoch: 998, Test Loss: 0.36, Test Acc: 0.96
Epoch: 999, Test Loss: 0.35, Test Acc: 0.96
Epoch: 1000, Test Loss: 0.35, Test Acc: 0.96
Overall accuracy on test set (%): 96.19799256324768
```

Well done! The accuracy of the RNN is above 96%, which is outstanding. Now let's observe how the loss propagates across each iteration and over time:

```
epoch_seq = np.arange(1, epochs+1)
plt.plot(epoch_seq, train_loss, 'k--', label='Train Set')
plt.plot(epoch_seq, test_loss, 'r-', label='Test Set')
plt.title('RNN training/test loss')
plt.xlabel('Epochs')
plt.ylabel('Loss')
plt.legend(loc='upper left')
plt.show()
```

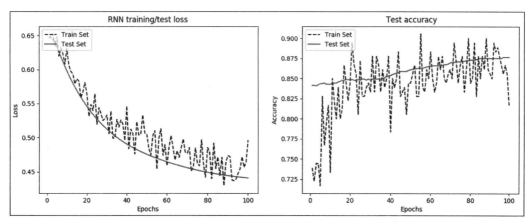

Figure 13: a) RNN training and test loss per epoch b) test accuracy per epoch

We also plot the accuracy over time:

```
plt.plot(epoch_seq, train_accuracy, 'k--', label='Train Set')
plt.plot(epoch_seq, test_accuracy, 'r-', label='Test Set')
plt.title('Test accuracy')
plt.xlabel('Epochs')
plt.ylabel('Accuracy')
plt.legend(loc='upper left')
plt.show()
```

The next application uses time series data for predictive modeling. We will also see how to develop more complex RNNs called LSTM networks.

Developing a predictive model for time series data

RNNs, specifically LSTM models, is often a difficult topic to understand. Time series prediction is a useful application of RNNs because of temporal dependencies in the data. Time series data is abundantly available online. In this section, we will see an example of using an LSTM for handling time series data. Our LSTM network will be able to predict the number of airline passengers in the future.

Description of the dataset

The dataset that I will be using is data about international airline passengers from 1949 to 1960. The dataset can be downloaded from

`https://datamarket.com/data/set/22u3/international-airline-passengers-monthly-totals-in#!ds=22u3&display=line`. The following screenshot shows the metadata of the international airline passengers:

Chapter 6

Dataset title	International airline passengers: monthly totals in thousands. Jan 49 – Dec 60
Last updated	1 Feb 2014, 19:52
Last updated by source	20 Jun 2012
Provider	Time Series Data Library
Provider source	Box & Jenkins (1976)
Source URL	http://datamarket.com/data/list/?q=provider:tsdl
Units	Thousands of passengers
Dataset metrics	144 fact values in 1 timeseries.
Time granularity	Month
Time range	Jan 1949 – Dec 1960
Language	English
License	Default open license
License summary	This data release is licensed as follows: You may copy and redistribute the data. You may make derivative works from the data. You may use the data for commercial purposes. You may not sublicense the data when redistributing it. You may not redistribute the data under a different license. Source attribution on any use of this data: Must refer source.
Description	Transport and tourism, Source: Box & Jenkins (1976), in file: data/airpass, Description: International airline passengers: monthly totals in thousands. Jan 49 – Dec 60

Figure 14: Metadata of the international airline passengers (source: `https://datamarket.com/`)

You can download the data by choosing the **Export** tab and then selecting **CSV (,)** in the **Export** group. You will have to edit the CSV file manually to remove the header line, as well as the additional footer line. I have downloaded and saved the data file named `international-airline-passengers.csv`. The following graph is a nice plot of the time series data:

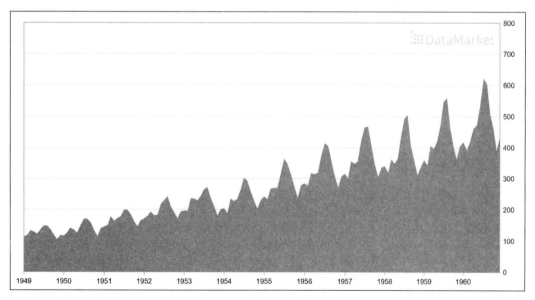

Figure 15: International airline passengers: monthly totals in thousands from Jan 49 – Dec 60

Pre-processing and exploratory analysis

Now let's load the original dataset and see some facts. At first, we load the time series as follows (see `time_series_preprocessor.py`):

```
import csv
import numpy as np
```

Here, we can see the signature of `load_series()`, which is a user-defined method that loads the time series and normalizes it:

```
def load_series(filename, series_idx=1):
    try:
        with open(filename) as csvfile:
            csvreader = csv.reader(csvfile)
            data = [float(row[series_idx]) for row in csvreader if len(row) > 0]
            normalized_data = (data - np.mean(data)) / np.std(data)
        return normalized_data
    except IOError:
        Print("Error occurred")

        return None
```

Now let's invoke the preceding method to load the time series and print (issue `$ python3 plot_time_series.py` on Terminal) the number of series in the dataset:

```
import csv
import numpy as np
import matplotlib.pyplot as plt
import time_series_preprocessor as tsp
timeseries = tsp.load_series('international-airline-passengers.csv')
print(timeseries)
```

The following is the output of the preceding code:

```
>>>
[-1.40777884 -1.35759023 -1.24048348 -1.26557778 -1.33249593 -1.21538918
 -1.10664719 -1.10664719 -1.20702441 -1.34922546 -1.47469699 -1.35759023
…..
  2.85825285  2.72441656  1.9046693   1.5115252   0.91762667  1.26894693]
```

```
print(np.shape(timeseries))
```

>>>

144

That means there are `144` entries in the time series. Let's plot the time series:

```
plt.figure()
plt.plot(timeseries)
plt.title('Normalized time series')
plt.xlabel('ID')
plt.ylabel('Normalized value')
plt.legend(loc='upper left')
plt.show()
```

The following is the output of the preceding code:

>>>

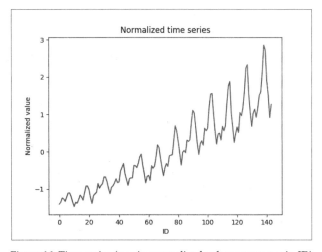

Figure 16: Time series (y-axis, normalized value versus x-axis, ID)

Once we have loaded the time series dataset, the next task is to prepare the training set. Since we will be evaluating the model multiple times to predict future values, we will split the data into training and testing. To be more specific, the `split_data()` function divides the dataset into two components for training and testing, 75% for training and 25% for testing:

```
def split_data(data, percent_train):
    num_rows = len(data)
    train_data, test_data = [], []
    for idx, row in enumerate(data):
```

```
            if idx < num_rows * percent_train:
                train_data.append(row)
            else:
                test_data.append(row)
    return train_data, test_data
```

LSTM predictive model

Once we have our dataset ready, we can train the predictor by loading the data in an acceptable format. For this step, I have written a Python script called `TimeSeriesPredictor.py`, which starts by importing the necessary library and modules (issue `$ python3 TimeSeriesPredictor.py` command on Terminal for this script):

```
import numpy as np
import tensorflow as tf
from tensorflow.python.ops import rnn, rnn_cell
import time_series_preprocessor as tsp
import matplotlib.pyplot as plt
```

Next, we define the hyperparameters for the LSTM network (tune it accordingly):

```
input_dim = 1
seq_size = 5
hidden_dim = 5
```

We now define the weight variables (no biases) and input placeholders:

```
W_out = tf.get_variable("W_out", shape=[hidden_dim, 1],
dtype=tf.float32, initializer=None, regularizer=None,
trainable=True, collections=None)
b_out = tf.get_variable("b_out", shape=[1], dtype=tf.float32,
initializer=None, regularizer=None, trainable=True,
collections=None)
x = tf.placeholder(tf.float32, [None, seq_size, input_dim])
y = tf.placeholder(tf.float32, [None, seq_size])
```

The next task is to construct the LSTM network. The following method, `LSTM_Model()`, takes three parameters, as follows:

- x: Inputs of size [T, batch_size, input_size]
- W: A matrix of fully-connected output layer weights
- b: A vector of fully-connected output layer biases

Now let's see the signature of the method:

```
def LSTM_Model():
    cell = rnn_cell.BasicLSTMCell(hidden_dim)
    outputs, states = rnn.dynamic_rnn(cell, x, dtype=tf.float32)
    num_examples = tf.shape(x)[0]
    W_repeated = tf.tile(tf.expand_dims(W_out, 0), [num_examples, 1, 1])
    out = tf.matmul(outputs, W_repeated) + b_out
    out = tf.squeeze(out)
    return out
```

Additionally, we create three empty lists to store the training loss, test loss, and the step:

```
train_loss = []
test_loss = []
step_list = []
```

The next method, called `train()`, is used to train the LSTM network:

```
def trainNetwork(train_x, train_y, test_x, test_y):
    with tf.Session() as sess:
        tf.get_variable_scope().reuse_variables()
        sess.run(tf.global_variables_initializer())
        max_patience = 3
        patience = max_patience
        min_test_err = float('inf')
        step = 0
        while patience > 0:
            _, train_err = sess.run([train_op, cost], feed_dict={x: train_x, y: train_y})
            if step % 100 == 0:
                test_err = sess.run(cost, feed_dict={x: test_x, y: test_y})
                print('step: {}\t\ttrain err: {}\t\ttest err: {}'.format(step, train_err, test_err))
                train_loss.append(train_err)
                test_loss.append(test_err)
                step_list.append(step)
                if test_err < min_test_err:
                    min_test_err = test_err
                    patience = max_patience
                else:
                    patience -= 1
```

Recurrent Neural Networks

```
        step += 1
        save_path = saver.save(sess, 'model.ckpt')
        print('Model saved to {}'.format(save_path))
```

The next task is to create the cost optimizer and instantiate `training_op`:

```
cost = tf.reduce_mean(tf.square(LSTM_Model()- y))
train_op = tf.train.AdamOptimizer(learning_rate=0.003).minimize(cost)
```

Additionally, here we have an auxiliary op called saving the model:

```
saver = tf.train.Saver()
```

Now that we have created the model, the next method, called `testLSTM()`, is used to test the prediction power of the model on the test set:

```
def testLSTM(sess, test_x):
        tf.get_variable_scope().reuse_variables()
        saver.restore(sess, 'model.ckpt')
        output = sess.run(LSTM_Model(), feed_dict={x: test_x})
        return output
```

To plot the predicted results, we have a function called `plot_results()`. The signature is as follows:

```
def plot_results(train_x, predictions, actual, filename):
    plt.figure()
    num_train = len(train_x)
    plt.plot(list(range(num_train)), train_x, color='b', label='training data')
    plt.plot(list(range(num_train, num_train + len(predictions))), predictions, color='r', label='predicted')
    plt.plot(list(range(num_train, num_train + len(actual))), actual, color='g', label='test data')
    plt.legend()
    if filename is not None:
        plt.savefig(filename)
    else:
        plt.show()
```

Model evaluation

To evaluate the model, we have a method called `main()` that actually invokes the preceding methods to create and train the LSTM network. The workflow of the code is as following:

1. Load the data
2. Slide a window through the time series data to construct the training dataset
3. Do the same window sliding strategy to construct the test dataset
4. Train a model on the training dataset
5. Visualize the model's performance

Let's see the signature of the method:

```
def main():
    data = tsp.load_series('international-airline-passengers.csv')
    train_data, actual_vals = tsp.split_data(data=data,
percent_train=0.75)
    train_x, train_y = [], []
    for i in range(len(train_data) - seq_size - 1):
        train_x.append(np.expand_dims(train_data[i:i+seq_size],
axis=1).tolist())
        train_y.append(train_data[i+1:i+seq_size+1])
    test_x, test_y = [], []
    for i in range(len(actual_vals) - seq_size - 1):
        test_x.append(np.expand_dims(actual_vals[i:i+seq_size],
axis=1).tolist())
        test_y.append(actual_vals[i+1:i+seq_size+1])
    trainNetwork(train_x, train_y, test_x, test_y)
    with tf.Session() as sess:
        predicted_vals = testLSTM(sess, test_x)[:,0]
        # Following prediction results of the model given ground
truth values
        plot_results(train_data, predicted_vals, actual_vals,
'ground_truth_predition.png')
        prev_seq = train_x[-1]
        predicted_vals = []
        for i in range(1000):
            next_seq = testLSTM(sess, [prev_seq])
            predicted_vals.append(next_seq[-1])
            prev_seq = np.vstack((prev_seq[1:], next_seq[-1]))
        # Following predictions results where only the training
data was given
        plot_results(train_data, predicted_vals, actual_vals,
'prediction_on_train_set.png')
>>>
```

Finally, we call the `main()` method to perform the training. Once the training is completed, it further plots the prediction results of the model consisting of ground truth values versus predictions results, where only the training data was given:

>>>

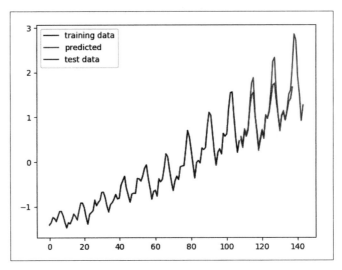

Figure 17: The results of the model on the ground truth values

The next graph shows the prediction results on the training data. This procedure has less information available, but it still did a good job of matching the trends in the data:

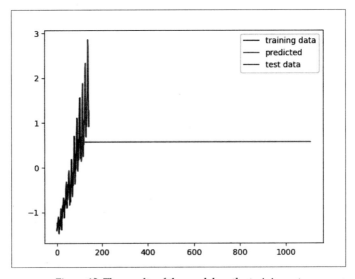

Figure 18: The results of the model on the training set

The following method helps us plot the training and the test error:

```
def plot_error():
    # Plot training loss over time
    plt.plot(step_list, train_loss, 'r--', label='LSTM training loss per iteration', linewidth=4)
    plt.title('LSTM training loss per iteration')
    plt.xlabel('Iteration')
    plt.ylabel('Training loss')
    plt.legend(loc='upper right')
    plt.show()

    # Plot test loss over time
    plt.plot(step_list, test_loss, 'r--', label='LSTM test loss per iteration', linewidth=4)
    plt.title('LSTM test loss per iteration')
    plt.xlabel('Iteration')
    plt.ylabel('Test loss')
    plt.legend(loc='upper left')
    plt.show()
```

Now we call the preceding method as follows:

```
plot_error()
>>>
```

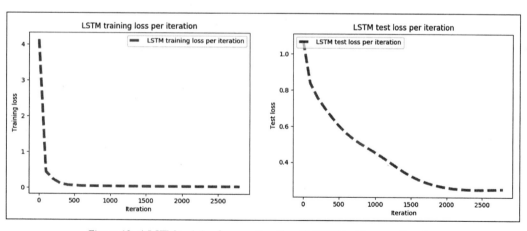

Figure 19: a) LSTM training loss per iteration, b) LSTM test loss per iteration

We can use a time series predictor to reproduce realistic fluctuations in data. Now you can prepare your own dataset and do some other predictive analytics. The next example is about sentiment analysis from the product and movie review dataset. We will also see how to develop a more complex RNN using an LSTM network.

An LSTM predictive model for sentiment analysis

Sentiment analysis is one of the most widely used tasks in NLP. An LSTM network can be used to classify short texts into desired categories, a classification problem. For example, a set of tweets can be categorized as either positive or negative. In this section, we will see such an example.

Network design

The implemented LSTM network will have three layers: an embedding layer, an RNN layer, and a softmax layer. A high-level view of this can be seen in the following diagram. Here, I summarize the functionalities of all of the layers:

- **Embedding layer**: We will see an example in *Chapter 8, Advanced TensorFlow Programming* that shows that text datasets cannot be fed to **Deep Neural Networks** (**DNNs**) directly, so an additional layer called an embedding layer is required. For this layer, we transform each input, which is a tensor of k words, into a tensor of k N-dimensional vectors. This is called word embedding, where N is the embedding size. Every word will be associated with a vector of weights that needs to be learned during the training process. You can gain more insight into word embedding at vector representations of words.
- **RNN layer**: Once we have constructed the embedding layer, there will be a new layer called the RNN layer, which is made out of LSTM cells with a dropout wrapper. LSTM weights need to be learned during the training process, as described in the previous sections. The RNN layer is unrolled dynamically (as shown in figure 4), taking k word embeddings as input and outputting k M-dimensional vectors, where **M** is the hidden size of the LSTM cells.
- **Softmax or sigmoid layer**: The RNN layer's output is averaged across k time steps, obtaining a single tensor of size M. Finally, a softmax layer, for example, is used to compute classification probabilities.

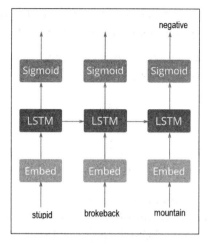

Figure 20: The high-level view of the LSTM network for sentiment analysis

We will see later how cross-entropy can be used as the loss function, and RMSProp is the optimizer that minimizes it.

LSTM model training

The UMICH SI650 – sentiment classification dataset (with duplication removed) contains data about product and movie reviews donated by the University of Michigan can be downloaded from https://inclass.kaggle.com/c/si650winter11/data. Unwanted or special characters have been cleaned, before getting, the tokens (see the data.csv file).

The following script also removes stop words (see data_preparation.py). Some samples are given that are labeled as either negative or positive (1 is positive and 0 is negative):

Sentiment	SentimentText
1	The Da Vinci Code book is just awesome.
1	I liked the Da Vinci Code a lot.
0	OMG, I HATE BROKEBACK MOUNTAIN.
0	I hate Harry Potter.

Table 1: A sample of the sentiment dataset

Now, let's see a step-by-step example of training the LSTM network for this task. At first, we import the necessary modules and packages (execute the `train.py` file):

```python
from data_preparation import Preprocessing
from lstm_network import LSTM_RNN_Network
import tensorflow as tf
import pickle
import datetime
import time
import os
import matplotlib.pyplot as plt
```

In the preceding import declarations, `data_preparation` and `lstm_network` are two helper Python scripts that are used for dataset preparation and network design. We will see more details later shortly. Now let's define parameters for the LSTM:

```python
data_dir = 'data/' # Data directory containing 'data.csv'
stopwords_file = 'data/stopwords.txt' # Path to stopwords file
n_samples= None # Set n_samples=None to use the whole dataset

# Directory where TensorFlow summaries will be stored'
summaries_dir= 'logs/'
batch_size = 100 #Batch size
train_steps = 1000 #Number of training steps
hidden_size= 75 # Hidden size of LSTM layer
embedding_size = 75 # Size of embeddings layer
learning_rate = 0.01
test_size = 0.2
dropout_keep_prob = 0.5 # Dropout keep-probability
sequence_len = None # Maximum sequence length
validate_every = 100 # Step frequency to validate
```

I believe the preceding parameters are self-explanatory. The next task is to prepare summaries to be used by the TensorBoard:

```python
summaries_dir = '{0}/{1}'.format(summaries_dir, datetime.datetime.now().strftime('%d_%b_%Y-%H_%M_%S'))
train_writer = tf.summary.FileWriter(summaries_dir + '/train')
validation_writer = tf.summary.FileWriter(summaries_dir + '/validation')
```

Now let's prepare the model directory:

```
model_name = str(int(time.time()))
model_dir = '{0}/{1}'.format(checkpoints_root, model_name)
if not os.path.exists(model_dir):
    os.makedirs(model_dir)
```

Next, let's prepare the data and build a TensorFlow graph (see the data_preparation.py file):

```
data_lstm = Preprocessing(data_dir=data_dir,
            stopwords_file=stopwords_file,
            sequence_len=sequence_len,
            test_size=test_size,
            val_samples=batch_size,
            n_samples=n_samples,
            random_state=100)
```

In the preceding code segment, `Preprocessing` is a class continuing (see data_preparation.py for detail) several function and constructor that help us pre-process the training and testing set in order to train the LSTM network. Here, I have provided the code for each function and its functionality.

The constructor of this class initializes the data pre-processor. This class provides an interface to load, pre-process, and split the data into training, validation, and testing sets. It takes the following parameters:

- `data_dir`: A data directory containing the dataset file, `data.csv`, with columns called `SentimentText` and `Sentiment`.
- `stopwords_file`: Optional. If provided, it discards each stop word from the original data.
- `sequence_len`: Optional. If `m` is the maximum sequence length in the dataset, it's required that `sequence_len >= m`. If `sequence_len` is `None`, then it'll be automatically assigned to `m`.
- `n_samples`: Optional. It's the number of samples to load from the dataset (which is useful for large datasets). If `n_samples` is `None`, then the whole dataset will be loaded (be careful; if the dataset is large it may take a while to pre-process every sample).
- `test_size`: Optional. `0<test_size<1`. It represents the proportion of the dataset to include in the testing set (the default is `0.2`).
- `val_samples`: Optional but can be used to represent the absolute number of validations samples (the default is `100`).

Recurrent Neural Networks

- `random_state`: This is an optional parameter for the random seed used for splitting data into training, testing, and validation sets (the default is 0).
- `ensure_preprocessed`: Optional. If `ensure_preprocessed=True`, it ensures that the dataset is already pre-processed (the default is `False`).

The code for the constructor is as follows:

```
def __init__(self, data_dir, stopwords_file=None,
sequence_len=None, n_samples=None, test_size=0.2, val_samples=100,
random_state=0, ensure_preprocessed=False):
        self._stopwords_file = stopwords_file
        self._n_samples = n_samples
        self.sequence_len = sequence_len
        self._input_file = os.path.join(data_dir, 'data.csv')
        self._preprocessed_file=os.path.join(data_dir,
"preprocessed_"+str(n_samples)+ ".npz")
        self._vocab_file = os.path.join(data_dir,
 "vocab_" + str(n_samples) + ".pkl")
        self._tensors = None
        self._sentiments = None
        self._lengths = None
        self._vocab = None
        self.vocab_size = None

        # Prepare data
        if os.path.exists(self._preprocessed_file)
 and os.path.exists(self._vocab_file):
            print('Loading preprocessed files ...')
            self.__load_preprocessed()
        else:
            if ensure_preprocessed:
                raise ValueError('Unable to find
 preprocessed files.')
            print('Reading data ...')
            self.__preprocess()
        # Split data in train, validation and test sets
        indices = np.arange(len(self._sentiments))
        x_tv, self._x_test, y_tv, self._y_test,
 tv_indices, test_indices = train_test_split(
            self._tensors,
            self._sentiments,
            indices,
            test_size=test_size,
            random_state=random_state,
```

```
                stratify=self._sentiments[:, 0])
                self._x_train,self._x_val,self._y_train,
self._y_val,train_indices,val_indices= train_test_split(x_tv,
y_tv, tv_indices, test_size=val_samples,random_state = random_state,
                stratify=y_tv[:, 0])
        self._val_indices = val_indices
        self._test_indices = test_indices
        self._train_lengths = self._lengths[train_indices]
        self._val_lengths = self._lengths[val_indices]
        self._test_lengths = self._lengths[test_indices]
        self._current_index = 0
        self._epoch_completed = 0
```

Now let's see the signature of the preceding method. We start with the `_preprocess()` method, which loads data from `data_dir/data.csv`, pre-processes each sample loaded, and stores intermediate files to avoid pre-processing later. The workflow is as follows:

1. Load the data
2. Clean the sample text
3. Prepare the vocabulary dictionary
4. Remove the most uncommon words (they are probably grammar mistakes), encode the samples into tensors, and pad each tensor with zeros according to `sequence_len`
5. Save intermediate files
6. Store sample lengths for future use

Now let's take a look at the following code block, which represents the preceding workflow:

```
def __preprocess(self):
    data = pd.read_csv(self._input_file, nrows=self._n_samples)
    self._sentiments = np.squeeze(data.as_matrix(columns=['Sentiment']))
    self._sentiments = np.eye(2)[self._sentiments]
    samples = data.as_matrix(columns=['SentimentText'])[:, 0]
    samples = self.__clean_samples(samples)
    vocab = dict()
    vocab[''] = (0, len(samples))    # add empty word
    for sample in samples:
        sample_words = sample.split()
```

```python
            for word in list(set(sample_words)):  # distinct words
                value = vocab.get(word)
                if value is None:
                    vocab[word] = (-1, 1)
                else:
                    encoding, count = value
                    vocab[word] = (-1, count + 1)
        sample_lengths = []
        tensors = []
        word_count = 1
        for sample in samples:
            sample_words = sample.split()
            encoded_sample = []
            for word in list(set(sample_words)):  # distinct words
                value = vocab.get(word)
                if value is not None:
                    encoding, count = value
                    if count / len(samples) > 0.0001:
                        if encoding == -1:
                            encoding = word_count
                            vocab[word] = (encoding, count)
                            word_count += 1
                        encoded_sample += [encoding]
                    else:
                        del vocab[word]
            tensors += [encoded_sample]
            sample_lengths += [len(encoded_sample)]
        self.vocab_size = len(vocab)
        self._vocab = vocab
        self._lengths = np.array(sample_lengths)
        self.sequence_len, self._tensors =
    self.__apply_to_zeros(tensors, self.sequence_len)
        with open(self._vocab_file, 'wb') as f:
            pickle.dump(self._vocab, f)
        np.savez(self._preprocessed_file, tensors=self._tensors,
    lengths=self._lengths, sentiments=self._sentiments)
```

Next, we invoke the preceding method and load the intermediate files, avoiding data pre-processing:

```python
def __load_preprocessed(self):
    with open(self._vocab_file, 'rb') as f:
        self._vocab = pickle.load(f)
    self.vocab_size = len(self._vocab)
```

```
load_dict = np.load(self._preprocessed_file)
self._lengths = load_dict['lengths']
self._tensors = load_dict['tensors']
self._sentiments = load_dict['sentiments']
self.sequence_len = len(self._tensors[0])
```

Once we have the pre-processed dataset, the next task is to clean the samples. The workflow is as follows:

1. Prepare regex patterns.
2. Clean each sample.
3. Restore HTML characters.
4. Remove @users and URLs.
5. Transform to lowercase.
6. Remove punctuation symbols.
7. Replace CC(C+) (a character occurring more than twice in a row) with C.
8. Remove stop words.

Now let's write the above steps programatically. For this, we have the following function:

```
def __clean_samples(self, samples):
    print('Cleaning samples ...')
    ret = []
    reg_punct = '[' + re.escape(''.join(string.punctuation)) + ']'
    if self._stopwords_file is not None:
        stopwords = self.__read_stopwords()
        sw_pattern = re.compile(r'\b(' + '|'.join(stopwords) + r')\b')
    for sample in samples:
        text = html.unescape(sample)
        words = text.split()
        words = [word for word in words if not word.startswith('@') and not word.startswith('http://')]
        text = ' '.join(words)
        text = text.lower()
        text = re.sub(reg_punct, ' ', text)
        text = re.sub(r'([a-z])\1{2,}', r'\1', text)
        if stopwords is not None:
            text = sw_pattern.sub('', text)
        ret += [text]
    return ret
```

The `__apply_to_zeros()` method returns the `padding_length` used and a NumPy array of padded tensors. First, it finds the maximum length, m, and ensures that m>=`sequence_len`. Then it pads the list with zeros according to `sequence_len`:

```
def __apply_to_zeros(self, lst, sequence_len=None):
    inner_max_len = max(map(len, lst))
    if sequence_len is not None:
        if inner_max_len > sequence_len:
            raise Exception('Error: Provided sequence length is not sufficient')
        else:
            inner_max_len = sequence_len
    result = np.zeros([len(lst), inner_max_len], np.int32)
    for i, row in enumerate(lst):
        for j, val in enumerate(row):
            result[i][j] = val
    return inner_max_len, result
```

The next task is to remove all the stop words (which are provided in the `data/StopWords.txt file`). This method returns the stop words list:

```
def __read_stopwords(self):
    if self._stopwords_file is None:
        return None
    with open(self._stopwords_file, mode='r') as f:
        stopwords = f.read().splitlines()
    return stopwords
```

The `next_batch()` method takes `batch_size>0` as the number of samples that'll be included, returns batch size samples (`text_tensor`, `text_target`, `text_length`) after completing the epoch, and randomly shuffles the training samples:

```
def next_batch(self, batch_size):
    start = self._current_index
    self._current_index += batch_size
    if self._current_index > len(self._y_train):
        self._epoch_completed += 1
        ind = np.arange(len(self._y_train))
        np.random.shuffle(ind)
        self._x_train = self._x_train[ind]
        self._y_train = self._y_train[ind]
        self._train_lengths = self._train_lengths[ind]
        start = 0
        self._current_index = batch_size
```

```
        end = self._current_index
    return self._x_train[start:end], self._y_train[start:end],
self._train_lengths[start:end]
```

The next method, called get_val_data(), is then used to get the validation set to be used during the training period. It takes the original text and returns the validation data. By default, it returns the original_text (original_samples, text_tensor, text_target, text_length), or otherwise returns text_tensor, text_target, text_length:

```
    def get_val_data(self, original_text=False):
        if original_text:
            data = pd.read_csv(self._input_file,
nrows=self._n_samples)
            samples = data.as_matrix(columns=['SentimentText'])[:, 0]
            return samples[self._val_indices], self._x_val, self._y_val,
self._val_lengths
        return self._x_val, self._y_val, self._val_lengths
```

Finally, we have an additional method called get_test_data(), which is used to prepare the testing set that will be used during the model evaluation period:

```
    def get_test_data(self, original_text=False):
        if original_text:
            data = pd.read_csv(self._input_file,
nrows=self._n_samples)
            samples = data.as_matrix(columns=['SentimentText'])[:,
0]
            return samples[self._test_indices], self._x_test, self._y_
test, self._test_lengths
        return self._x_test, self._y_test, self._test_lengths
```

Now we prepare the data so that the LSTM network can feed it:

```
lstm_model = LSTM_RNN_Network(hidden_size=[hidden_size],
                              vocab_size=data_lstm.vocab_size,
                              embedding_size=embedding_size,
                              max_length=data_lstm.sequence_len,
                              learning_rate=learning_rate)
```

In the preceding code segment, `LSTM_RNN_Network` is a class containing several functions and constructors that help us create the LSTM network. The upcoming constructor builds a TensorFlow LSTM model. It takes the following parameters:

- `hidden_size`: An array holding the number of units in an LSTM cell of rnn layers
- `vocab_size`: The vocabulary size in the sample
- `embedding_size`: Words will be encoded using a vector of this size
- `max_length`: The maximum length of an input tensor
- `n_classes`: The number of classification classes
- `learning_rate`: The learning rate of the RMSProp algorithm
- `random_state`: The random state for dropout

The code for the constructor is as follows:

```
def __init__(self, hidden_size, vocab_size, embedding_size,
max_length, n_classes=2, learning_rate=0.01, random_state=None):
    # Build TensorFlow graph
    self.input = self.__input(max_length)
    self.seq_len = self.__seq_len()
    self.target = self.__target(n_classes)
    self.dropout_keep_prob = self.__dropout_keep_prob()
    self.word_embeddings = self.__word_embeddings(self.input,
vocab_size, embedding_size, random_state)
    self.scores = self.__scores(self.word_embeddings,
self.seq_len, hidden_size, n_classes, self.dropout_keep_prob,
                    random_state)
    self.predict = self.__predict(self.scores)
    self.losses = self.__losses(self.scores, self.target)
    self.loss = self.__loss(self.losses)
    self.train_step = self.__train_step(learning_rate, self.loss)
    self.accuracy = self.__accuracy(self.predict, self.target)
    self.merged = tf.summary.merge_all()
```

The next function is called `_input()`, and it takes a parameter called param `max_length`, which is the maximum length of an input tensor. It then returns an input placeholder with the shape `[batch_size, max_length]` for the TensorFlow computation:

```
def __input(self, max_length):
    return tf.placeholder(tf.int32, [None, max_length],
name='input')
```

Next, the _seq_len() function returns a sequence length placeholder with the shape [batch_size]. It holds each tensor's real length in a given batch, allowing a dynamic sequence length:

```
def __seq_len(self):
    return tf.placeholder(tf.int32, [None], name='lengths')
```

The next function is called _target(). It takes a parameter called param n_classes, which contains the number of classification classes. Finally, it returns the target placeholder with the shape [batch_size, n_classes]:

```
def __target(self, n_classes):
    return tf.placeholder(tf.float32, [None, n_classes], name='target')
```

_dropout_keep_prob() returns a placeholder holding the dropout keep probability to reduce the overfitting:

```
def __dropout_keep_prob(self):
    return tf.placeholder(tf.float32, name='dropout_keep_prob')
```

The _cell() method is used to build a LSTM cell with a dropout wrapper. It takes the following parameters:

- hidden_size: It is the number of units in the LSTM cell
- dropout_keep_prob: This indicates the tensor holding the dropout keep probability
- seed: It is an optional value that ensures the reproducibility of the computation for the random state for the dropout wrapper.

Finally, it returns an LSTM cell with a dropout wrapper:

```
def __cell(self, hidden_size, dropout_keep_prob, seed=None):
    lstm_cell = tf.nn.rnn_cell.LSTMCell(hidden_size, state_is_tuple=True)
    dropout_cell = tf.nn.rnn_cell.DropoutWrapper(lstm_cell, input_keep_prob=dropout_keep_prob, output_keep_prob = dropout_keep_prob, seed=seed)
    return dropout_cell
```

Once we have created the LSTM cells, we can create the embedding of the input tokens. For this, `__word_embeddings()` does the trick. It builds an embedding layer with the shape `[vocab_size, embedding_size]`, with input parameters such as x, which is the input with the shape `[batch_size, max_length]`. The `vocab_size` is the vocabulary size, that is, the number of possible words that may appear in a sample. The `embedding_size` is the words that will be represented using a vector of this size and seed is optional, but it ensures the random state for the embedding initialization.

Finally, it returns the embedding lookup tensor with the shape `[batch_size, max_length, embedding_size]`:

```
def __word_embeddings(self, x, vocab_size, embedding_size, seed=None):
    with tf.name_scope('word_embeddings'):
        embeddings = tf.get_variable("embeddings", shape=[vocab_size, embedding_size], dtype=tf.float32, initializer=None, regularizer=None, trainable=True, collections=None)
        embedded_words = tf.nn.embedding_lookup(embeddings, x)
    return embedded_words
```

The `__rnn_layer()` method creates the LSTM layer. It takes several input parameters, which are described here:

- `hidden_size`: This is the number of units in the LSTM cell
- `x`: This is the input with shape
- `seq_len`: This is the sequence length tensor with shape
- `dropout_keep_prob`: This is the tensor holding the dropout keep probability
- `variable_scope`: This is the name of the variable scope (the default layer is rnn_layer)
- `random_state`: This is the random state for the dropout wrapper

Finally, it returns the outputs with the shape `[batch_size, max_seq_len, hidden_size]`:

```
def __rnn_layer(self, hidden_size, x, seq_len, dropout_keep_prob, variable_scope=None, random_state=None):
    with tf.variable_scope(variable_scope, default_name='rnn_layer'):
        lstm_cell = self.__cell(hidden_size, dropout_keep_prob, random_state)
        outputs, _ = tf.nn.dynamic_rnn(lstm_cell, x, dtype=tf.float32, sequence_length=seq_len)
    return outputs
```

The _score() method is used to compute the network output. It takes several input parameters, as follows:

- embedded_words: This is the embedding lookup tensor with the shape [batch_size, max_length, embedding_size]
- seq_len: This is the sequence length tensor with the shape [batch_size]
- hidden_size: This is an array holding the number of units in the LSTM cell in each RNN layer
- n_classes: This is the number of classification classes
- dropout_keep_prob: This is the tensor holding the dropout keep probability
- random_state: This is an optional parameter, but it can be used to ensure the random state for the dropout wrapper

Finally, the _score() method returns the linear activation of each class with the shape [batch_size, n_classes]:

```
def __scores(self, embedded_words, seq_len, hidden_size,
n_classes, dropout_keep_prob, random_state=None):
    outputs = embedded_words
    for h in hidden_size:
        outputs = self.__rnn_layer(h, outputs, seq_len,
dropout_keep_prob)
    outputs = tf.reduce_mean(outputs, axis=[1])
    with tf.name_scope('final_layer/weights'):
        w = tf.get_variable("w", shape=[hidden_size[-1],
            n_classes], dtype=tf.float32, initializer=None,
                regularizer=None, trainable=True,
                collections=None)
        self.variable_summaries(w, 'final_layer/weights')
    with tf.name_scope('final_layer/biases'):
        b = tf.get_variable("b", shape=[n_classes],
            dtype=tf.float32, initializer=None,
            regularizer=None,trainable=True, collections=None)
        self.variable_summaries(b, 'final_layer/biases')
    with tf.name_scope('final_layer/wx_plus_b'):
        scores = tf.nn.xw_plus_b(outputs, w, b, name='scores')
        tf.summary.histogram('final_layer/wx_plus_b', scores)
    return scores
```

The `_predict()` method takes scores as the linear activation of each class with the shape `[batch_size, n_classes]` and returns softmax (to normalize the score in a scale of `[0, 1]`) activations with the shape `[batch_size, n_classes]`:

```
def __predict(self, scores):
    with tf.name_scope('final_layer/softmax'):
        softmax = tf.nn.softmax(scores, name='predictions')
        tf.summary.histogram('final_layer/softmax', softmax)
    return softmax
```

The `_losses()` method returns the cross-entropy losses (since softmax is used as the activation function) with the shape `[batch_size]`. It also takes two parameters, such as scores, as the linear activation of each class with the shape `[batch_size, n_classes]` and the target tensor with the shape `[batch_size, n_classes]`:

```
def __losses(self, scores, target):
    with tf.name_scope('cross_entropy'):
        cross_entropy = tf.nn.softmax_cross_entropy_with_logits_v2(logits=scores, labels=target, name='cross_entropy')
    return cross_entropy
```

The `_loss()` function computes and returns the mean cross-entropy loss. It takes only one parameter, called losses, which indicates the cross-entropy losses with the shape `[batch_size]` and is computed by the previous function:

```
def __loss(self, losses):
    with tf.name_scope('loss'):
        loss = tf.reduce_mean(losses, name='loss')
        tf.summary.scalar('loss', loss)
    return loss
```

Now, `_train_step()` computes and returns the `RMSProp` training step operation. It takes two parameters, `learning_rate`, which is the learning rate for the `RMSProp` optimizer; and the mean cross-entropy loss computed by the previous function:

```
def __train_step(self, learning_rate, loss):
    return tf.train.RMSPropOptimizer(learning_rate).minimize(loss)
```

When it is time for performance evaluation, the _accuracy() function computes the accuracy of the classification. It takes three parameters, predict, which the softmax activation is having the shape [batch_size, n_classes]; and the target tensor with the shape [batch_size, n_classes] and the mean accuracy obtained in the current batch:

```
def _accuracy(self, predict, target):
    with tf.name_scope('accuracy'):
        correct_pred = tf.equal(tf.argmax(predict, 1), tf.argmax(target, 1))
        accuracy = tf.reduce_mean(tf.cast(correct_pred, tf.float32), name='accuracy')
        tf.summary.scalar('accuracy', accuracy)
    return accuracy
```

The next function is called initialize_all_variable() and, as you may be able to guess, it initializes all variables:

```
def initialize_all_variables(self):
    return tf.global_variables_initializer()
```

Finally, we have a static method called variable_summaries(), which attaches a lot of summaries to a tensor for the TensorBoard visualization. It takes the following parameters:

```
var: is the variable to summarize
mean: mean of the summary name.
```

The signature is given below:

```
@staticmethod
def variable_summaries(var, name):
    with tf.name_scope('summaries'):
        mean = tf.reduce_mean(var)
        tf.summary.scalar('mean/' + name, mean)
        with tf.name_scope('stddev'):
            stddev = tf.sqrt(tf.reduce_mean(tf.square(var - mean)))
        tf.summary.scalar('stddev/' + name, stddev)
        tf.summary.scalar('max/' + name, tf.reduce_max(var))
        tf.summary.scalar('min/' + name, tf.reduce_min(var))
        tf.summary.histogram(name, var)
```

Recurrent Neural Networks

Now we need to create a TensorFlow session before we can train the model:

```
sess = tf.Session()
```

Let's initialize all the variables:

```
init_op = tf.global_variables_initializer()
sess.run(init_op)
```

Then we save the TensorFlow model for future use:

```
saver = tf.train.Saver()
```

Now let's prepare the training set:

```
x_val, y_val, val_seq_len = data_lstm.get_val_data()
```

Now we should write the logs of the TensorFlow graph computation:

```
train_writer.add_graph(lstm_model.input.graph)
```

Additionally, we can create some empty lists to hold the training loss, validation loss, and the steps so that we can see them graphically:

```
train_loss_list = []
val_loss_list = []
step_list = []
sub_step_list = []
step = 0
```

Now we start the training. In each step, we record the training error. The validation errors are recorded in each sub-step:

```
for i in range(train_steps):
    x_train, y_train, train_seq_len = data_lstm.next_batch(batch_size)
    train_loss, _, summary = sess.run([lstm_model.loss, lstm_model.train_step, lstm_model.merged],
                                        feed_dict={lstm_model.input: x_train,
                                                    lstm_model.target: y_train,
                                                    lstm_model.seq_len: train_seq_len,
                                                    lstm_model.dropout_keep_prob:dropout_keep_prob})
    train_writer.add_summary(summary, i)   # Write train summary for step i (TensorBoard visualization)
```

```
            train_loss_list.append(train_loss)
            step_list.append(i)
                print('{0}/{1} train loss: {2:.4f}'.format(i + 1,
    FLAGS.train_steps, train_loss))
            if (i + 1) %validate_every == 0:
                val_loss, accuracy, summary = sess.run([lstm_model.loss,
    lstm_model.accuracy, lstm_model.merged],
                                                        feed_dict={lstm_model.
    input: x_val,
                                                                    lstm_model.
    target: y_val,
                                                                    lstm_model.
    seq_len: val_seq_len,
                                                                    lstm_model.
    dropout_keep_prob: 1})
                validation_writer.add_summary(summary, i)
                print('   validation loss: {0:.4f} (accuracy
    {1:.4f})'.format(val_loss, accuracy))
                step = step + 1
                val_loss_list.append(val_loss)
                sub_step_list.append(step)
```

The following is the output of the preceding code:

```
>>>

1/1000 train loss: 0.6883
2/1000 train loss: 0.6879
3/1000 train loss: 0.6943

99/1000 train loss: 0.4870
100/1000 train loss: 0.5307
validation loss: 0.4018 (accuracy 0.9200)
...
199/1000 train loss: 0.1103
200/1000 train loss: 0.1032
validation loss: 0.0607 (accuracy 0.9800)
...
299/1000 train loss: 0.0292
300/1000 train loss: 0.0266
```

Recurrent Neural Networks

```
validation loss: 0.0417 (accuracy 0.9800)
...
998/1000 train loss: 0.0021
999/1000 train loss: 0.0007
1000/1000 train loss: 0.0004
validation loss: 0.0939 (accuracy 0.9700)
```

The preceding code prints the training and validation error. When the training is over, the model will be saved to the checkpoint directory that has a unique id:

```
checkpoint_file = '{}/model.ckpt'.format(model_dir)
save_path = saver.save(sess, checkpoint_file)
print('Model saved in: {0}'.format(model_dir))
```

Following is the output of the preceding code:

```
>>>
Model saved in checkpoints/1517781236
```

The checkpoint directory will produce at least three files:

- `config.pkl` contains parameters used to train the model.
- `model.ckpt` contains the weights of the model.
- `model.ckpt.meta` contains the TensorFlow graph definition.

Let's see how the training went, that is, what were the training and the validation losses like:

```
# Plot loss over time
plt.plot(step_list, train_loss_list, 'r--', label='LSTM training loss per iteration', linewidth=4)
plt.title('LSTM training loss per iteration')
plt.xlabel('Iteration')
plt.ylabel('Training loss')
plt.legend(loc='upper right')
plt.show()

# Plot accuracy over time
plt.plot(sub_step_list, val_loss_list, 'r--', label='LSTM validation loss per validating interval', linewidth=4)
plt.title('LSTM validation loss per validation interval')
plt.xlabel('Validation interval')
plt.ylabel('Validation loss')
plt.legend(loc='upper left')
plt.show()
```

The following is the output of the preceding code:

>>>

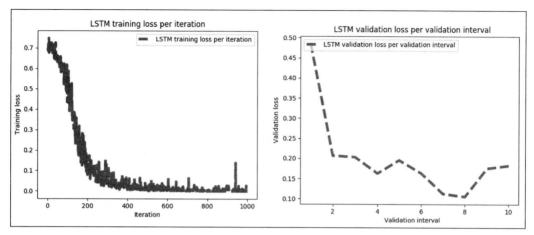

Figure 21: a) LSTM training loss per iteration on test set, b) LSTM validation loss per validation interval

If we examine the preceding graphs, it is clear that the training went pretty well in both the training phase and the validation phase with only 1,000 steps. However, readers should increase the training step, tune the hyperparameters, and see how it goes.

Visualizing through TensorBoard

Now let's observe the TensorFlow computational graph on TensorBoard. Simply execute the following command and access TensorBoard at `localhost:6006/`:

```
tensorboard --logdir /home/logs/
```

Recurrent Neural Networks

The graph tab shows the execution graph, including the gradients used, `loss_op`, the accuracy, the final layer, the optimizer used (in our case it's `RMSPro`), the LSTM layer (that is, RNN layer), the embedding layer, and `save_op`:

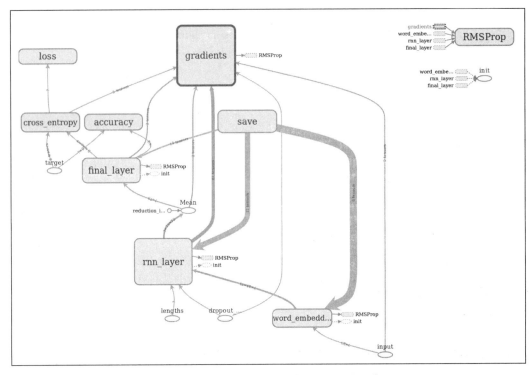

Figure 22: The execution graph on TensorBoard

The execution graph shows that the computations we have done for this LSTM-based classifier for sentiment analysis are quite transparent. We can also observe the validation, training losses, accuracies, and the operations in the layers:

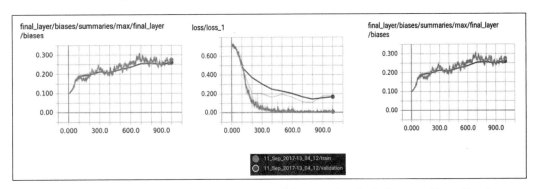

Figure 23: Validation, training losses, accuracies and the operations in the layers on TensorBoard

LSTM model evaluation

We have trained and saved our LSTM model. We can easily restore the trained model and do some evaluation. We need to prepare the testing set and use the previously trained TensorFlow model to make predictions on it. Let's do this straight away. First, we load the required models:

```
import tensorflow as tf
from data_preparation import Preprocessing
   import pickle
```

Then we load to show the checkpoint directory where the model was saved. For our case, it was checkpoints/1505148083.

> For this step, execute the predict.py script with the following command:
>
> **$ python3 predict.py --checkpoints_dir checkpoints/1517781236**

```
# Change this path based on output by 'python3 train.py'
checkpoints_dir = 'checkpoints/1517781236'

ifcheckpoints_dir is None:
    raise ValueError('Please, a valid checkpoints directory
is required (--checkpoints_dir <file name>)')
```

Now load the testing dataset and prepare it to evaluate the model:

```
data_lstm = Preprocessing(data_dir=data_dir,
              stopwords_file=stopwords_file,
              sequence_len=sequence_len,
              n_samples=n_samples,
              test_size=test_size,
              val_samples=batch_size,
              random_state=random_state,
              ensure_preprocessed=True)
```

In the preceding code, use the following parameters exactly as we did in the training step:

```
data_dir = 'data/' # Data directory containing 'data.csv'
stopwords_file = 'data/stopwords.txt' # Path to stopwords file.
sequence_len = None # Maximum sequence length
n_samples= None # Set n_samples=None to use the whole dataset
test_size = 0.2
batch_size = 100 #Batch size
random_state = 0 # Random state used for data splitting. Default
is 0
```

The workflow for this evaluation method is as follows:

1. First, import the meta graph and evaluate the model using the testing data
2. Create the TensorFlow session for the computation
3. Import the graph and restore its weights
4. Recover the input/output tensors
5. Perform the prediction
6. Finally, we print the accuracy and the result on the simple testing set

Step 1 has already been completed previously. This code does steps 2 to 5:

```
original_text, x_test, y_test, test_seq_len = data_lstm.get_test_data(original_text=True)
graph = tf.Graph()
with graph.as_default():
    sess = tf.Session()
    print('Restoring graph ...')
    saver = tf.train.import_meta_graph("{}/model.ckpt.meta".format(FLAGS.checkpoints_dir))
    saver.restore(sess, ("{}/model.ckpt".format(checkpoints_dir)))
    input = graph.get_operation_by_name('input').outputs[0]
    target = graph.get_operation_by_name('target').outputs[0]
    seq_len = graph.get_operation_by_name('lengths').outputs[0]
    dropout_keep_prob = graph.get_operation_by_name('dropout_keep_prob').outputs[0]
    predict = graph.get_operation_by_name('final_layer/softmax/predictions').outputs[0]
    accuracy = graph.get_operation_by_name('accuracy/accuracy').outputs[0]
    pred, acc = sess.run([predict, accuracy],
                        feed_dict={input: x_test,
                                   target: y_test,
                                   seq_len: test_seq_len,
                                   dropout_keep_prob: 1})
    print("Evaluation done.")
```

The following is the output of the preceding code:

```
>>>
Restoring graph ...
The evaluation was done.
```

Well done! The training is finished, so let's print the results:

```
print('\nAccuracy: {0:.4f}\n'.format(acc))
for i in range(100):
    print('Sample: {0}'.format(original_text[i]))
    print('Predicted sentiment: [{0:.4f}, {1:.4f}]'.format(pred[i,
0], pred[i, 1]))
    print('Real sentiment: {0}\n'.format(y_test[i]))
```

Following is the output of the preceding code:

```
>>>
Accuracy: 0.9858

Sample: I loved the Da Vinci code, but it raises many theological
questions most of which are very absurd...
Predicted sentiment: [0.0000, 1.0000]
Real sentiment: [0. 1.]
…
Sample: I'm sorry I hate to read Harry Potter, but I love the movies!
Predicted sentiment: [1.0000, 0.0000]
Real sentiment: [1. 0.]
…
Sample: I LOVE Brokeback Mountain...
Predicted sentiment: [0.0002, 0.9998]
Real sentiment: [0. 1.]
…
Sample: We also went to see Brokeback Mountain which totally SUCKED!!!
Predicted sentiment: [1.0000, 0.0000]
Real sentiment: [1. 0.]
```

The accuracy is above 98%. This is brilliant! However, you could try to iterate the training for even higher iterations with tuned hyperparameters, and you might get even higher accuracy. I leave this up to the readers.

In the next section, we will see how to develop a more advanced ML project using LSTM, which is called human activity recognition using smartphones dataset. In short, our ML model will be able to classify human movement into six categories: walking, walking upstairs, walking downstairs, sitting, standing, and laying.

Human activity recognition using LSTM model

The **Human Activity Recognition** (**HAR**) database was built by taking measurements from 30 participants who performed **activities of daily living** (**ADL**) while carrying a waist-mounted smartphone with embedded inertial sensors. The objective is to classify their activities into one of the six categories mentioned previously.

Dataset description

The experiments were carried out on a group of 30 volunteers within an age range of 19-48 years. Each person accomplished six activities (walking, walking upstairs, walking downstairs, sitting, standing, and laying) while wearing a Samsung Galaxy S II smartphone on their waist. Using an accelerometer and a gyroscope, the author captured 3-axial linear acceleration and 3-axial angular velocity at a constant rate of 50 Hz.

Only two sensors, an accelerometer, and gyroscope, were used. The sensor signals were pre-processed by applying noise filters and then sampled in fixed-width sliding windows of 2.56 sec with a 50% overlap. This gives 128 readings per window. The gravitational and body motion components from the sensor acceleration signal were separated via a Butterworth low-pass filter into body acceleration and gravity.

> For more information, please refer to this paper: *Davide Anguita, Alessandro Ghio, Luca Oneto, Xavier Parra, and Jorge L. Reyes-Ortiz, A Public Domain Dataset for Human Activity Recognition Using Smartphones* and *21st European Symposium on Artificial Neural Networks, Computational Intelligence and Machine Learning, ESANN 2013. Bruges, Belgium 24-26, April 2013.*

For simplicity, the gravitational force was assumed to have only a few low-frequency components. Therefore, a filter of 0.3 Hz cut-off frequency was used. From each window, a feature vector was found by calculating variables from the time and frequency domain.

The experiments have been video-recorded to facilitate manually labeling of the data. The dataset has been randomly partitioned into two sets, where 70% of the volunteers were selected for the training data and 30% for the testing data. When I explore the dataset, both the training and testing set have the following file structure:

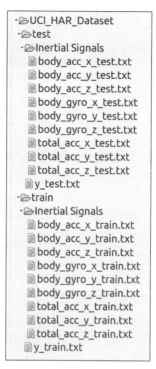

Figure 24: HAR dataset file structure

For each record in the dataset, the following is provided:

- Triaxial acceleration from the accelerometer and the estimated body acceleration
- Triaxial angular velocity from the gyroscope sensor
- A 561-feature vector with time and frequency domain variables
- Its activity label
- An identifier of the subject who carried out the experiment

Therefore, we know the problem that needs to be addressed. It is time to explore the technology and the related challenges.

Workflow of the LSTM model for HAR

The overall algorithm has the following workflow:

1. Load the data.
2. Define the hyperparameters.
3. Set up the LSTM model using imperative programming and the hyperparameters.
4. Apply batch-wise training. That is, pick a batch of data, feed it to the model, then, after some iterations, evaluate the model and print the batch loss and the accuracy.
5. Output the chart for the training and test errors.

The above steps can be followed and constructed a pipeline:

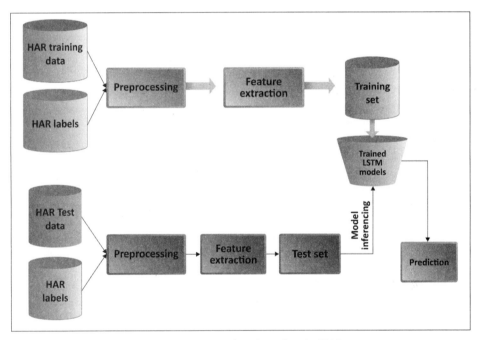

Figure 25: An LSTM-based pipeline for HAR

Implementing an LSTM model for HAR

First, we import the required packages and modules:

```
import numpy as np
import matplotlib
import matplotlib.pyplot as plt
import tensorflow as tf
from sklearn import metrics
from tensorflow.python.framework import ops
import warnings
import random
warnings.filterwarnings("ignore")
os.environ['TF_CPP_MIN_LOG_LEVEL'] = '3'
```

As stated earlier, `INPUT_SIGNAL_TYPES` contain some useful constants. They are separate normalized input features for the neural network:

```
INPUT_SIGNAL_TYPES = [
    "body_acc_x_",
    "body_acc_y_",
    "body_acc_z_",
    "body_gyro_x_",
    "body_gyro_y_",
    "body_gyro_z_",
    "total_acc_x_",
    "total_acc_y_",
    "total_acc_z_"
]
```

The labels are defined in another array – that is output classes used to learn how to classify:

```
LABELS = [
    "WALKING",
    "WALKING_UPSTAIRS",
    "WALKING_DOWNSTAIRS",
    "SITTING",
    "STANDING",
    "LAYING"
]
```

Recurrent Neural Networks

We are now assuming that you have already downloaded the HAR dataset from `https://archive.ics.uci.edu/ml/machine-learning-databases/00240/UCI HAR Dataset.zip` and put in a folder named `UCIHARDataset` (or you can choose a suitable name that sounds better). Additionally, we need to provide the paths to the training and the testing set:

```
DATASET_PATH = "UCIHARDataset/"
print("\n" + "Dataset is now located at: " + DATASET_PATH)

TRAIN = "train/"
TEST = "test/"
```

Then we load and map the data from each `.txt` file based on the input signal type defined by the `INPUT_SIGNAL_TYPES` array in the Array [Array [Array [Float]]] format. X denotes the neural network's training and testing inputs:

```
def load_X(X_signals_paths):
    X_signals = []

    for signal_type_path in X_signals_paths:
        file = open(signal_type_path, 'r')
        # Read dataset from disk, dealing with text files' syntax
        X_signals.append(
            [np.array(serie, dtype=np.float32) for serie in [
                row.replace('  ', ' ').strip().split(' ') for row in file
            ]]
        )
        file.close()

    return np.transpose(np.array(X_signals), (1, 2, 0))

X_train_signals_paths = [
    DATASET_PATH + TRAIN + "Inertial Signals/" + signal + "train.txt" for signal in INPUT_SIGNAL_TYPES
]
X_test_signals_paths = [
    DATASET_PATH + TEST + "Inertial Signals/" + signal + "test.txt" for signal in INPUT_SIGNAL_TYPES
]

X_train = load_X(X_train_signals_paths)
X_test = load_X(X_test_signals_paths)
```

Then we load y, the labels for the neural network's training and testing outputs:

```
def load_y(y_path):
    file = open(y_path, 'r')
    # Read dataset from disk, dealing with text file's syntax
    y_ = np.array(
        [elem for elem in [
            row.replace('  ', ' ').strip().split(' ') for row in file
        ]],
        dtype=np.int32
    )
    file.close()

    # We subtract 1 to each output class for 0-based indexing
    return y_ - 1

y_train_path = DATASET_PATH + TRAIN + "y_train.txt"
y_test_path = DATASET_PATH + TEST + "y_test.txt"

y_train = load_y(y_train_path)
y_test = load_y(y_test_path)
```

Let's look at some of the dataset's statistics, such as the number of training series (as described earlier, this is with a 50% overlap between each series), the number of test series, the number of timesteps per series, and the number of input parameters per timestep:

```
training_data_count = len(X_train)
test_data_count = len(X_test)
n_steps = len(X_train[0])
n_input = len(X_train[0][0])
print("Number of training series: "+ trainingDataCount)
print("Number of test series: "+ testDataCount)
print("Number of timestep per series: "+ nSteps)
print("Number of input parameters per timestep: "+ nInput)
```

The following is the output of the preceding code:

```
>>>
Number of training series: 7352
Number of test series: 2947
Number of timestep per series: 128
Number of input parameters per timestep: 9
```

Recurrent Neural Networks

Now let's define some core parameter definitions for the training. The whole neural network's structure could be summarized by enumerating those parameters and the fact an LSTM is used:

```
n_hidden = 32 # Hidden layer num of features
n_classes = 6 # Total classes (should go up, or should go down)

learning_rate = 0.0025
lambda_loss_amount = 0.0015
training_iters = training_data_count * 300 #Iterate 300 times
batch_size = 1500
display_iter = 30000 # to show test set accuracy during training
```

We have defined all the core parameters and network parameters. These are random choices. I did not do hyperparameter tuning but still it worked well. Therefore, I would suggest tuning these hyperparameters using grid-search techniques. There are many online materials available.

Nevertheless, before we construct the LSTM network and start the training, let's print some debugging information to make sure that the execution does not stop halfway through:

```
print("Some useful info to get an insight on dataset's shape and normalization:")
print("(X shape, y shape, every X's mean, every X's standard deviation)")
print(X_test.shape, y_test.shape, np.mean(X_test), np.std(X_test))
print("The dataset is therefore properly normalized, as expected, but not yet one-hot encoded.")
```

The following is the output of the preceding code:

```
>>>
Some useful info to get an insight on dataset's shape and normalization:
(X shape, y shape, every X's mean, every X's standard deviation)
(2947, 128, 9) (2947, 1) 0.0991399 0.395671
```

The dataset is therefore properly normalized, as expected, but not yet one-hot encoded.

Now that the training dataset is in corrected and normalized order, it is time to construct the LSTM network. The following function returns a TensorFlow LSTM network from the given parameters. Moreover, two LSTM cells are stacked together, which adds depth to the neural network:

```
def LSTM_RNN(_X, _weights, _biases):
    _X = tf.transpose(_X, [1,0,2])# permute n_steps & batch_size
    _X = tf.reshape(_X, [-1, n_input])
    _X = tf.nn.relu(tf.matmul(_X, _weights['hidden']) + _biases['hidden'])
    _X = tf.split(_X, n_steps, 0)
    lstm_cell_1 = tf.nn.rnn_cell.BasicLSTMCell(n_hidden, forget_bias=1.0, state_is_tuple=True)
    lstm_cell_2 = tf.nn.rnn_cell.BasicLSTMCell(n_hidden, forget_bias=1.0, state_is_tuple=True)
    lstm_cells = tf.nn.rnn_cell.MultiRNNCell([lstm_cell_1, lstm_cell_2], state_is_tuple=True)
    outputs, states = tf.contrib.rnn.static_rnn(lstm_cells, _X, dtype=tf.float32)
    lstm_last_output = outputs[-1]
    return tf.matmul(lstm_last_output, _weights['out']) + _biases['out']
```

If we look at the preceding code snippet carefully, we can see that we get the last time step's output feature for a "many to one" style classifier. Now, the question is what is a many-to-one RNN classifier? Well, similar to figure 5, we accept a time series of feature vectors (one vector per time step) and convert them to a probability vector in the output for classification.

Now that we have been able to construct our LSTM network, we need to prepare the training dataset into a batch. The following function fetches a batch_size amount of data from (X|y)_train data:

```
def extract_batch_size(_train, step, batch_size):
    shape = list(_train.shape)
    shape[0] = batch_size
    batch_s = np.empty(shape)
    for i in range(batch_size):
        index = ((step-1)*batch_size + i) % len(_train)
        batch_s[i] = _train[index]
    return batch_s
```

After that, we need to encode output labels from number indexes to binary categories. Then we perform training steps with `batch_size`. For example, `[[5], [0], [3]]` needs to be converted into a shape similar to `[[0, 0, 0, 0, 0, 1], [1, 0, 0, 0, 0, 0], [0, 0, 0, 1, 0, 0]]`. Well, we can do this with one-hot-encoding. The following method does exactly the same transformation:

```
def one_hot(y_):
    y_ = y_.reshape(len(y_))
    n_values = int(np.max(y_)) + 1
    return np.eye(n_values)[np.array(y_, dtype=np.int32)]
```

Excellent! Our dataset is ready, so we can start building the network. First, we create two separate placeholders for the input and the labels:

```
x = tf.placeholder(tf.float32, [None, n_steps, n_input])
y = tf.placeholder(tf.float32, [None, n_classes])
```

We then create the required weight vectors:

```
weights = {
    'hidden': tf.Variable(tf.random_normal([n_input, n_hidden])),
    'out': tf.Variable(tf.random_normal([n_hidden, n_classes], mean=1.0))
}
```

Then we create the required bias vectors:

```
biases = {
    'hidden': tf.Variable(tf.random_normal([n_hidden])),
    'out': tf.Variable(tf.random_normal([n_classes]))
}
```

Then we build the model by passing the input tensor, the weight vector, and the bias vector as follows:

```
pred = LSTM_RNN(x, weights, biases)
```

Additionally, we also need to compute the `cost` op, the regularization, the optimizer, and the evaluation. We use L2 loss for regularization, that prevents this overkill neural network to over fit issue in the training:

```
l2 = lambda_loss_amount * sum(tf.nn.l2_loss(tf_var) for tf_var in tf.trainable_variables())

cost = tf.reduce_mean(tf.nn.softmax_cross_entropy_with_logits_v2(labels=y, logits=pred)) + l2
```

```
optimizer =
tf.train.AdamOptimizer(learning_rate=learning_rate).minimize(cost)
correct_pred = tf.equal(tf.argmax(pred,1), tf.argmax(y,1))
accuracy = tf.reduce_mean(tf.cast(correct_pred, tf.float32))
```

Great! So far, everything has been fine. Now we are ready to train the neural network. First, we create some lists to hold some training's performance:

```
test_losses = []
test_accuracies = []
train_losses = []
train_accuracies = []
```

Then we create a TensorFlow session, launch the graph, and initialize the global variables:

```
sess =
tf.InteractiveSession(config=tf.ConfigProto(log_device_placement=
False))
init = tf.global_variables_initializer()
sess.run(init)
```

Then we perform training steps with `batch_size` amount of example data in each loop. We first fit the training using the batch of data, and then we evaluate the network only at a few steps for faster training. Additionally, we evaluate on the testing set (no learning happens here, just evaluation for diagnosis). Finally, we print the result:

```
step = 1
while step * batch_size <= training_iters:
    batch_xs = extract_batch_size(X_train, step, batch_size)
    batch_ys = one_hot(extract_batch_size(y_train, step,
batch_size))
    _, loss, acc = sess.run(
        [optimizer, cost, accuracy],
        feed_dict={
            x: batch_xs,
            y: batch_ys
        }
    )
    train_losses.append(loss)
    train_accuracies.append(acc)
    if (step*batch_size % display_iter == 0) or (step == 1) or
(step * batch_size > training_iters):
        print("Training iter #" + str(step*batch_size) + \
":   Batch Loss = " + "{:.6f}".format(loss) + \
```

```python
            ", Accuracy = {}".format(acc))
        loss, acc = sess.run(
            [cost, accuracy],
            feed_dict={
                x: X_test,
                y: one_hot(y_test)
            }
        )
        test_losses.append(loss)
        test_accuracies.append(acc)
        print("PERFORMANCE ON TEST SET: " + \
              "Batch Loss = {}".format(loss) + \
              ", Accuracy = {}".format(acc))
    step += 1
print("Optimization Finished!")
one_hot_predictions, accuracy, final_loss = sess.run(
    [pred, accuracy, cost],
    feed_dict={
        x: X_test,
        y: one_hot(y_test)
    })
test_losses.append(final_loss)
test_accuracies.append(accuracy)

print("FINAL RESULT: " + \
      "Batch Loss = {}".format(final_loss) + \
      ", Accuracy = {}".format(accuracy))
```

The following is the output of the preceding code:

```
>>>
Training iter #1500:   Batch Loss = 3.266330, Accuracy = 0.15733332931995392

PERFORMANCE ON TEST SET: Batch Loss = 2.6498606204986572, Accuracy = 0.15473362803459167

Training iter #30000:   Batch Loss = 1.538126, Accuracy = 0.6380000114440918
...
PERFORMANCE ON TEST SET: Batch Loss = 0.5507552623748779, Accuracy = 0.8924329876899719

Optimization Finished!

FINAL RESULT: Batch Loss = 0.6077192425727844, Accuracy = 0.8686800003051758
```

Well done! The training went well. However, a visual overview would be more useful:

```
indep_train_axis = np.array(range(batch_size,
(len(train_losses)+1)*batch_size, batch_size))
plt.plot(indep_train_axis, np.array(train_losses),     "b--",
label="Train losses")
plt.plot(indep_train_axis, np.array(train_accuracies), "g--",
label="Train accuracies")
indep_test_axis = np.append(
    np.array(range(batch_size, len(test_losses)*display_iter,
    display_iter)[:-1]),
    [training_iters])
plt.plot(indep_test_axis, np.array(test_losses),     "b-", label="Test
losses")
plt.plot(indep_test_axis, np.array(test_accuracies), "g-", label="Test
accuracies")
plt.title("Training session's progress over iterations")
plt.legend(loc='upper right', shadow=True)
plt.ylabel('Training Progress (Loss or Accuracy values)')
plt.xlabel('Training iteration')
plt.show()
```

The following is the output of the preceding code:

>>>

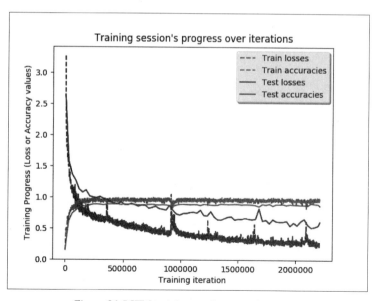

Figure 26: LSTM training sessions over iterations

We will need to compute other performance metrics, such as `accuracy`, `precision`, `recall`, and `f1 measure`:

```
predictions = one_hot_predictions.argmax(1)
print("Testing Accuracy: {}%".format(100*accuracy))
print("")
print("Precision: {}%".format(100*metrics.precision_score(y_test,
predictions, average="weighted")))
print("Recall: {}%".format(100*metrics.recall_score(y_test,
predictions, average="weighted")))
print("f1_score: {}%".format(100*metrics.f1_score(y_test,
predictions, average="weighted")))
```

The following is the output of the preceding code:

```
>>>
Testing Accuracy: 89.51476216316223%
Precision: 89.65053428376297%
Recall: 89.51476077366813%
f1_score: 89.48593061935716%
```

Since the problem that we are approaching is a multiclass classification, drawing the confusion matrix make sense:

```
print("")
print ("Showing Confusion Matrix")

cm = metrics.confusion_matrix(y_test, predictions)
df_cm = pd.DataFrame(cm, LABELS, LABELS)
plt.figure(figsize = (16,8))
plt.ylabel('True label')
plt.xlabel('Predicted label')
sn.heatmap(df_cm, annot=True, annot_kws={"size": 14}, fmt='g',
linewidths=.5)
plt.show()
```

The following is the output of the preceding code:

>>>

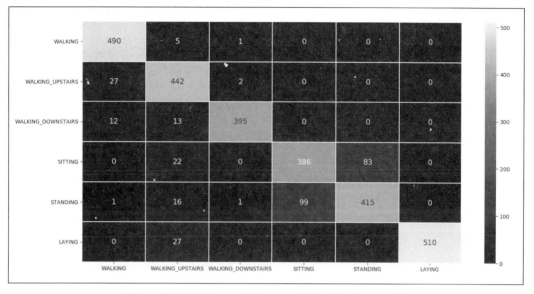

Figure 27: Multiclass confusion matrix (predicted vs actual)

In the confusion matrix, the training and testing data are not equally distributed amongst classes, so it is normal that more than a sixth of the data is correctly classified in the last category. Having said that, we have managed to achieve a prediction accuracy of about 87%. We will see more analysis soon. It could have been higher, but the training was done on the CPU, so it has low precision and, of course, takes a long time. Therefore, I would recommend that you train on a GPU instead to get a better result. In addition, tuning hyperparameters could be a good option.

Summary

LSTM networks are equipped with special hidden units, called memory cells, whose purpose is to remember the previous input for a long time. These cells take, at each instant of time, the previous state and the current input of the network as input. By combining them with the current contents of memory, and deciding what to keep and what to delete from memory with a gating mechanism by other units, LSTM has proved to be very useful and an effective way of learning long-term dependency.

In this chapter, we discussed RNNs. We saw how to make predictions with data that has a high temporal dependency. We saw how to develop several real-life predictive models that make the predictive analytics easier using RNNs and the different architectural variants. We started the chapter with a theoretical background of RNNs.

Then we looked at a few examples that showed a systematic way of implementing predictive models for image classification, sentiment analysis of movies and products, and spam prediction for NLP. Then we saw how to develop predictive models for time series data. Finally, we saw a more advanced application of RNNs for human activity recognition, and we observed a classification accuracy of about 87%.

DNNs are structured in a uniform manner so that, at each layer of the network, thousands of identical artificial neurons perform the same computation. Therefore, the architecture of a DNN fits quite well with the kinds of computation that a GPU can efficiently perform. GPUs have additional advantages over CPUs; these include having more computational units and having a higher bandwidth for memory retrieval.

Furthermore, in many deep learning applications that require a lot of computational effort, the graphics-specific capabilities of GPUs can be exploited to further speed up calculations. In the next chapter, we will see how to make the training faster, more accurate, and even distributed among nodes.

7
Heterogeneous and Distributed Computing

A computation expressed using TensorFlow can be executed with little or no changes on a wide variety of heterogeneous systems, ranging from mobile devices such as phones and tablets up to large-scale distributed systems of hundreds of machines and thousands of computational devices, such as GPU cards.

In this chapter, we explore this fundamental topic on TensorFlow. In particular, we shall consider the possibility of executing TensorFlow models on GPU cards and distributed systems.

GPUs have additional advantages over CPUs, including having more computational units and having a higher bandwidth for memory retrieval. Furthermore, in many deep learning applications that require a lot of computational effort, GPU graphics specific capabilities can be exploited to further speed up calculations.

At the same time, a distributed computing strategy can be useful if you have to handle a very large dataset to train your model.

The chapter introduces the following topics:

- GPGPU computing
- The GPU programming model
- The TensorFlow GPU setup
- Distributed computing
- The distributed TensorFlow setup

GPGPU computing

There are several reasons that led to deep learning (DL) being developed and placed at the center of attention in the field of machine learning (ML) in the recent decades.

One reason, perhaps the main one, is surely represented by the progress in hardware, with the availability of new processors, such as Graphics Processing Units (GPUs), which have greatly reduced the time needed to train networks, reducing the time by 10 or even 20 times.

In fact, since the connections between the individual neurons have a numerically estimated weight, and since networks learn by calibrating the weights appropriately, increasing network complexity would cause high computing power, and high computing power can be handled by GPU.

The GPGPU history

GPGPU is an acronym that stands for **General Purpose Computing on Graphics Processing Units**. It recognizes the trend of employing GPU technology for applications other than graphics. Until 2006, the graphics API OpenGL and DirectX standards were the only ways to program with a GPU. Any attempt to execute arbitrary calculations on the GPU was subject to the programming restrictions of those APIs.

GPUs are designed to produce a color for each pixel on the screen using programmable arithmetic units called **pixel shaders**. Programmers realized that if the input was numerical data with a different meaning from pixel colors, then they could program the pixel shader to perform arbitrary computations.

There were memory limitations because the programs could only receive a handful of input color and texture units as input data. It was almost impossible to predict how a GPU would handle floating-point data (if it had been able to process them), so many scientific calculations could not use the GPU.

Anyone who wanted to resolve a numerical problem would have to learn OpenGL or DirectX, the only ways to communicate with the GPU.

Chapter 7

The CUDA architecture

In 2006, NVIDIA presented the first GPU to support DirectX 10. The *GeForce 8800GTX* was also the first GPU to use the CUDA architecture. This architecture included several new components designed specifically for GPU computing and that aimed to remove the limitations that prevented previous GPUs from being used for non-graphical calculations. In fact, the execution units on the GPU could read and write arbitrary memory as well as access a cache maintained in software called **shared memory**.

These architectural features that were added made a CUDA GPU excel at general-purpose calculations as well as in traditional graphics tasks. The following diagram summarizes the division of space between the various components of a GPU and a CPU. As you can see, a GPU devotes more transistors to data processing; it is a *highly parallel*, *multithreaded*, and *manycore* processor:

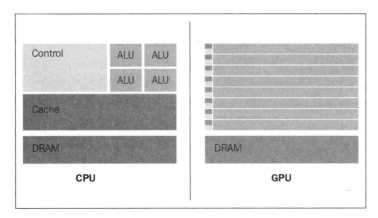

Figure 1: CPU versus GPU architecture

Note that almost all the space on the GPU chip is dedicated to the ALU, rather than cache and control, making it suitable for repetitive calculations on large amounts of data. The GPU accesses local memory and is connected to the system, the CPU, via a bus (currently, the PCI Express).

The graphics chip consists of a series of multiprocessors, the **streaming multiprocessor** (**SM**). The number of multiprocessors depends on the specific characteristics and the performance class of each GPU.

Each multiprocessor is in turn formed of stream processors (or cores). Each of these processors can perform basic arithmetic operations on integer or floating point numbers with single and double precision.

The GPU programming model

At this point, it is necessary to introduce some basic concepts to clarify the CUDA programming model.

The first distinction is between host and device.

The code executed on the host side is a part of code executed on the CPU, and this will also include the RAM and the hard disk. The code executed on the device is the code that is automatically loaded on the graphics card and runs on the latter.

Another important concept is the kernel; it stands for a function performed on the device and launched from the host. The code defined in the kernel will be performed in parallel by an array of threads.

The following schema summarizes how the GPU programming model works:

- The running program will have source code to run on the CPU and code to run on the GPU
- The CPU and GPU have separated memories
- The data is transferred from the CPU to the GPU to be computed
- The data output from GPU computation is copied back to CPU memory

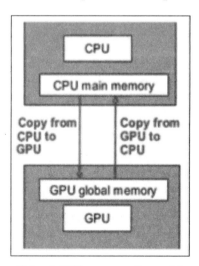

Figure 2: GPU programming model

The TensorFlow GPU setup

To use TensorFlow with NVIDIA GPUs, the first step is to install the CUDA Toolkit.

 To know more, visit https://developer.nvidia.com/cuda-downloads.

Once the CUDA Toolkit is installed, you must download the cuDNN v5.1 library for Linux from https://developer.nvidia.com/cudnn.

cuDNN is a library that helps accelerate deep learning frameworks, such as TensorFlow and Theano. Here's a brief explanation from the NVIDIA website:

> "The NVIDIA CUDA® Deep Neural Network library (cuDNN) is a GPU-accelerated library of primitives for deep neural networks. cuDNN provides highly tuned implementations for standard routines such as forward and backward convolution, pooling, normalization, and activation layers. cuDNN is part of the NVIDIA Deep Learning SDK."

Before installing it, you'll need to register on NVIDIA's Accelerated Computing Developer Program. Once you're registered, log in and download cuDNN 5.1 to your local computer.

Once it is downloaded, decompress the files and copy them into the CUDA Toolkit directory (we've assumed here that the directory is /usr/local/cuda/):

```
$ sudo tar -xvf cudnn-8.0-linux-x64-v5.1-rc.tgz -C /usr/local
```

Update TensorFlow

We're assuming you'll be using TensorFlow to build your deep neural network models. Simply update TensorFlow via pip with the upgrade flag.

We suppose you're currently using TensorFlow 0.11:

```
pip install – upgrade https://storage.googleapis.com/tensorflow/linux/gpu/tensorflow-0.10.0rc0-cp27-none-linux_x86_64.whl
```

Now you should have everything you need to run a model using your GPU.

GPU representation

In TensorFlow, the supported devices are represented as strings:

- `"/cpu:0"`: The CPU of your machine
- `"/gpu:0"`: The GPU of your machine, if you have one
- `"/gpu:1"`: The second GPU of your machine, and so on

The execution flow gives priority when an operation is assigned to a GPU device.

Using a GPU

To use a GPU in your TensorFlow program, just type the following:

```
with tf.device("/gpu:0"):
```

Then you need to do the setup operations. This line of code will create a new context manager, telling TensorFlow to perform those actions on the GPU.

Let's consider the following example, in which we want to execute the following sum of two large matrices: $A^n + B^n$.

Define the basic imports:

```
import numpy as np
import tensorflow as tf
import datetime
```

We can configure a TensorFlow program to find out which devices your operations and tensors are assigned to. To do this, we'll create a session with the following `log_device_placement` parameter set to `True`:

```
log_device_placement = True
```

Then we set the `n` parameter, which is the number of multiplications to perform:

```
n=10
```

Then we build two random large matrixes. We use NumPy's `rand` function to perform this operation:

```
A = np.random.rand(10000, 10000).astype('float32')
B = np.random.rand(10000, 10000).astype('float32')
```

`A` and `B` will each be of size 10000x10000.

The following arrays will be used to store the results:

```
c1 = []
c2 = []
```

Next, we define the kernel matrix multiplication function, that will be performed by the GPU:

```
def matpow(M, n):
    if n == 1:
        return M
    else:
        return tf.matmul(M, matpow(M, n-1))
```

As we previously explained, we must configure the GPU and the GPU with the operations to perform:

The GPU will compute the A^n and B^n operations and store results in c1:

```
with tf.device('/gpu:0'):
    a = tf.placeholder(tf.float32, [10000, 10000])
    b = tf.placeholder(tf.float32, [10000, 10000])
    c1.append(matpow(a, n))
    c1.append(matpow(b, n))
```

The addition of all elements in c1 ($A^n + B^n$) is performed by the CPU, so we define the following:

```
with tf.device('/cpu:0'):
  sum = tf.add_n(c1)
```

The datetime class allows us to evaluate the computational time:

```
t1_1 = datetime.datetime.now()
with tf.Session(config=tf.ConfigProto\
        (log_device_placement=log_device_placement)) as sess:
    sess.run(sum, {a:A, b:B})
t2_1 = datetime.datetime.now()
```

The computational time is then displayed:

```
print("GPU computation time: " + str(t2_1-t1_1))
```

On my laptop, using a GeForce 840M graphic card, the result is as follows:

GPU computation time: 0:00:13.816644

GPU memory management

In some cases, it is desirable for the process to only allocate a subset of the available memory, or to only grow the memory usage as it is needed by the process. TensorFlow provides two config options on the session to control this.

The first is the `allow_growth` option, which attempts to allocate only as much GPU memory based on runtime allocations: it starts out allocating very little memory, and as sessions get run and more GPU memory is needed, we extend the amount of GPU memory needed by the TensorFlow process.

Note that we do not release memory, since that can lead to even worse memory fragmentation. To turn this option on, set the option in `ConfigProto` as follows:

```
config = tf.ConfigProto()
config.gpu_options.allow_growth = True
session = tf.Session(config=config, ...)
```

The second method is the `per_process_gpu_memory_fraction` option, which determines the fraction of the overall amount of memory that each visible GPU should be allocated. For example, you can tell TensorFlow to only allocate 40% of the total memory of each GPU as follows:

```
config = tf.ConfigProto()
config.gpu_options.per_process_gpu_memory_fraction = 0.4
session = tf.Session(config=config, ...)
```

This is useful if you want to truly limit the amount of GPU memory available to the TensorFlow process.

Assigning a single GPU on a multi-GPU system

If you have more than one GPU in your system, the GPU with the lowest ID will be selected by default. If you would like to run your session on a different GPU, you will need to specify the preference explicitly.

For example, we can try to change the GPU assignation in the previous code:

```
with tf.device('/gpu:1'):
    a = tf.placeholder(tf.float32, [10000, 10000])
    b = tf.placeholder(tf.float32, [10000, 10000])
    c1.append(matpow(a, n))
    c1.append(matpow(b, n))
```

In this way, we are telling gpu1 to execute the kernel function. If the device we have specified does not exist (as in my case), you will get `InvalidArgumentError`:

```
InvalidArgumentError (see above for traceback): Cannot assign a device
to node 'Placeholder_1': Could not satisfy explicit device specification
'/device:GPU:1' because no devices matching that specification are
registered in this process; available devices: /job:localhost/replica:0/
task:0/cpu:0
     [[Node: Placeholder_1 = Placeholder[dtype=DT_FLOAT, shape=[100,100],
_device="/device:GPU:1"] ()]]
```

If you would like TensorFlow to automatically choose an existing and supported device to run the operations if the specified one doesn't exist, you can set `allow_soft_placement` to True in the configuration option when creating the session.

Again, we set '/gpu:1' for the following node:

```
with tf.device('/gpu:1'):
    a = tf.placeholder(tf.float32, [10000, 10000])
    b = tf.placeholder(tf.float32, [10000, 10000])
    c1.append(matpow(a, n))
    c1.append(matpow(b, n))
```

Then we build a `Session` with the following `allow_soft_placement` parameter set to `True`:

```
with tf.Session(config=tf.ConfigProto\
                (allow_soft_placement=True,\
                log_device_placement=log_device_placement))\
                as sess:
```

In this way, when we run the session, no `InvalidArgumentError` will be displayed. We'll get a correct result, in this case, with a little delay:

```
GPU computation time: 0:00:15.006644
```

The source code for GPU with soft placement

Here's the complete source code, just for clarity:

```
import numpy as np
import tensorflow as tf
import datetime

log_device_placement = True
```

```
n = 10

A = np.random.rand(10000, 10000).astype('float32')
B = np.random.rand(10000, 10000).astype('float32')

c1 = []
c2 = []


def matpow(M, n):
    if n == 1:
        return M
    else:
        return tf.matmul(M, matpow(M, n-1))

with tf.device('/gpu:0'):
    a = tf.placeholder(tf.float32, [10000, 10000])
    b = tf.placeholder(tf.float32, [10000, 10000])
    c1.append(matpow(a, n))
    c1.append(matpow(b, n))

with tf.device('/cpu:0'):
    sum = tf.add_n(c1)

t1_1 = datetime.datetime.now()
with tf.Session(config=tf.ConfigProto\
                (allow_soft_placement=True,\
                log_device_placement=log_device_placement))\
                as sess:
    sess.run(sum, {a:A, b:B})
t2_1 = datetime.datetime.now()
```

Using multiple GPUs

If you would like to run TensorFlow on multiple GPUs, you can construct your model by assigning a specific chink of code to a GPU. For example, if we have two GPUs, we can split the previous code as follows, assigning the first matrix computation to the first GPU:

```
with tf.device('/gpu:0'):
    a = tf.placeholder(tf.float32, [10000, 10000])
    c1.append(matpow(a, n))
```

Chapter 7

The second matrix computation is assigned to the second GPU:

```
with tf.device('/gpu:1'):
    b = tf.placeholder(tf.float32, [10000, 10000])
    c1.append(matpow(b, n))
```

The CPU will manage the results. Also, note that we used the shared c1 array to collect them:

```
with tf.device('/cpu:0'):
    sum = tf.add_n(c1)
```

In the following code snippet, we provide a concrete example of management of two GPUs:

```
import numpy as np
import tensorflow as tf
import datetime

log_device_placement = True
n = 10

A = np.random.rand(10000, 10000).astype('float32')
B = np.random.rand(10000, 10000).astype('float32')

c1 = []

def matpow(M, n):
    if n == 1:
        return M
    else:
        return tf.matmul(M, matpow(M, n-1))

#FIRST GPU
with tf.device('/gpu:0'):
    a = tf.placeholder(tf.float32, [10000, 10000])
    c1.append(matpow(a, n))

#SECOND GPU
with tf.device('/gpu:1'):
    b = tf.placeholder(tf.float32, [10000, 10000])
    c1.append(matpow(b, n))


with tf.device('/cpu:0'):
```

[319]

```
        sum = tf.add_n(c1)

    t1_1 = datetime.datetime.now()
    with tf.Session(config=tf.ConfigProto\
                 (allow_soft_placement=True,\
                log_device_placement=log_device_placement))\
                 as sess:
         sess.run(sum, {a:A, b:B})
    t2_1 = datetime.datetime.now()
```

Distributed computing

DL models have to be trained on a large amount of data to improve their performance. However, training a deep network with millions of parameters may take days, or even weeks. In *Large Scale Distributed Deep Networks, Dean et al.* proposed two paradigms, namely *model parallelism* and *data parallelism*, which allow us to train and serve a network model on multiple physical machines. In the following section, we introduce these paradigms with a focus on distributed TensorFlow capabilities.

Model parallelism

Model parallelism gives every processor the same data but applies a different model to it. If the network model is too big to fit into one machine's memory, different parts of the model can be assigned to different machines. A possible model parallelism approach is to have the first layer on a machine (node 1), the second layer on the second machine (node 2), and so on. Sometimes this is not the optimal approach, because the last layer has to wait for the first layer's computation to complete during the forward step, and the first layer has to wait for the deepest layers during the backpropagation step. Only if the model is parallelizable (such as GoogleNet) can this happen on different machines without coming across such a bottleneck:

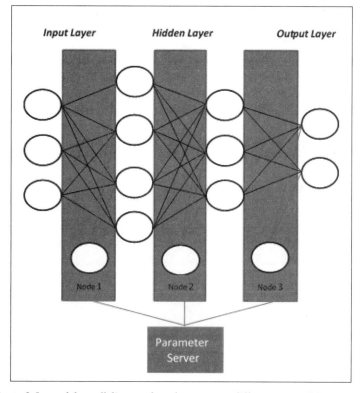

Figure 3: In model parallelism, each node computes different parts of the network

People who train neural networks may have been the originators of the term *model parallelism* almost 20 years ago, because they have different models of neural network to train and test, and multiple layers within a network that can be trained with the same data.

Data parallelism

Data parallelism means the application of a single instruction to multiple data items. It is the ideal workload for a SIMD (single instruction, multiple data) computer architecture, the oldest and simplest form of parallel processing on electronic digital computers.

In this approach, the network model fits in one machine, called the *parameter server*, while most of the computational work is done by multiple machines called *workers*:

- **Parameter server**: This is a CPU where you store the variables you need in the workers. In my case, this is where I defined the weights variables needed for my networks.
- **Workers**: This is where we do most of our computationally intensive work.

Each worker is in charge of reading, computing, and updating the model parameters, and sending them to the parameter server:

- In the **forward pass**, the worker *takes variables* from the parameter server, do something with them on our workers.
- In the **backward pass**, the worker *sends the current state* back to the parameter server, which does an update operation and gives us the new weights to try out:

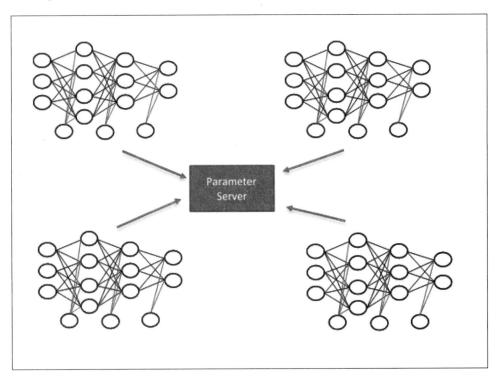

Figure 4: In the data parallelism model, each node computes all the parameters

Two main options are possible for data parallelism:

- **Synchronous training**: All the workers read the parameters at the same time, compute a training operation, and wait for all the others to be done. Then the gradients will be averaged and a single update will be sent to the parameter server. Hence, at any point in time, the workers will all be aware of the same values for the graph parameters.
- **Asynchronous training**: The workers will read from the parameter server(s) asynchronously, compute their training operation, and send asynchronous updates. At any point in time, two different workers might be aware of different values for the graph parameters.

The distributed TensorFlow setup

In this section, we will explore the mechanisms through which computation in TensorFlow can be distributed. The first step in running distributed TensorFlow is to specify the architecture of the cluster using tf.train.ClusterSpec:

```
import tensorflow as tf

cluster = tf.train.ClusterSpec({"ps": ["localhost:2222"],\
                                "worker": ["localhost:2223",\
                                           "localhost:2224"]})
```

Nodes are typically divided into two jobs: parameter servers (ps), which host variables, and workers, which perform heavy computation. In the preceding code, we have one parameter server and two workers, as well as the IP address and port of each node.

Then we have to build a tf.train.Server for each parameter server and worker, previously defined:

```
ps = tf.train.Server(cluster, job_name="ps", task_index=0)

worker0 = tf.train.Server(cluster,\
                          job_name="worker", task_index=0)
worker1 = tf.train.Server(cluster,\
                          job_name="worker", task_index=1)
```

The tf.train.Server object contains a set of local devices, a set of connections to other tasks in its tf.train.ClusterSpec, and a tf.Session that can use these to perform a distributed computation. It is created to allow for connectivity between devices.

Heterogeneous and Distributed Computing

Next, we assign model variables to workers using the following command:

```
tf.device :

with tf.device("/job:ps/task:0"):
    a = tf.constant(3.0, dtype=tf.float32)
    b = tf.constant(4.0)
```

Copy these instructions into a file named `main.py`.

In two separate files, `worker0.py` and `worker1.py`, we must define the workers. In `worker0.py`, multiply the two variables `a` and `b` and print out the result:

```
import tensorflow as tf
from main import *

with tf.Session(worker0.target) as sess:
    init = tf.global_variables_initializer()
    add_node = tf.multiply(a,b)
    sess.run(init)
    print(sess.run(add_node))
```

In `worker1.py`, first change the value of `a` and then multiply the two variables `a` and `b`:

```
import tensorflow as tf
from main import *

with tf.Session(worker1.target) as sess:
    init = tf.global_variables_initializer()
    a = tf.constant(10.0, dtype=tf.float32)
    add_node = tf.multiply(a,b)
    sess.run(init)
    a = add_node
    print(sess.run(add_node))
```

To execute this example, first run the `main.py` file from the command prompt.

You should get a result like this:

```
>python main.py

Found device 0 with properties:
name: GeForce 840M
major: 5 minor: 0 memoryClockRate (GHz) 1.124
pciBusID 0000:08:00.0
```

```
       Total memory: 2.00GiB
       Free memory: 1.66GiB

           Started server with target: grpc://localhost:2222
```

Then we can run the workers:

```
> python worker0.py

Found device 0 with properties:
name: GeForce 840M
major: 5 minor: 0 memoryClockRate (GHz) 1.124
pciBusID 0000:08:00.0
Total memory: 2.00GiB
Free memory: 1.66GiB


    Start master session 83740f48d039c97d with config:
    12.0
> python worker1.py

Found device 0 with properties:
name: GeForce 840M
major: 5 minor: 0 memoryClockRate (GHz) 1.124
pciBusID 0000:08:00.0
Total memory: 2.00GiB
Free memory: 1.66GiB


    Start master session 3465f63a4d9feb85 with config:
    40.0
```

Summary

In this chapter, we had a quick look at two fundamentals topics related to optimizing the computation of DNNs.

The first topic explained how to use GPUs and TensorFlow to implement DNNs. They are structured in a very uniform manner so that, at each layer of the network, thousands of identical artificial neurons perform the same computation. Hence, the architecture of a DNN fits quite well with the kinds of computation that a GPU can efficiently perform.

The second topic introduced distributed computing. This was initially used to perform very complex calculations that could not be completed by a single machine. Likewise, analyzing large amounts of data quickly by splitting this task among different nodes appears to be the best strategy when faced with such a big challenge.

At the same time, DL problems can be exploited using distributed computing. DL computations can be divided into multiple activities (tasks); each of them will be given a fraction of data and will return a result that will have to be recomposed with the results provided by the other activities. Alternatively, in most complex situations, a different calculation algorithm can be assigned to each machine.

Finally, in the last example, we showed how computation in TensorFlow can be distributed.

8
Advanced TensorFlow Programming

The development of **deep learning** (**DL**) networks requires rapid prototyping when testing new models. For this reason, several TensorFlow-based libraries have been built, which abstract many programming concepts and provides high-level building blocks.

We'll describe the main characteristics of each library with an application example.

This chapter covers the following high-level TensorFlow APIs and their overviews:

- tf.estimator
- TFLearn
- Pretty Tensor
- Keras

tf.estimator

tf.estimator is a high-level TensorFlow API for creating and training models by encapsulating the functionalities for training, evaluating, predicting and exporting. TensorFlow recently re-branded and released the TF Learn package within TensorFlow under a new name, TF Estimator, probably to avoid confusion with the TFLearn package from tflearn.org.

tf.estimator allows developers to easily extend the package and implement new ML algorithms by using the existing modular components and TensorFlow's low-level APIs, which serve as the building blocks of ML algorithms. Some examples of these building blocks are evaluation metrics, layers, losses, and optimizers.

The main features provided by tf.estimator are described in the next sections.

Estimators

An estimator is a rule that calculates an estimate of a given quantity. Estimators are used to train and evaluate TensorFlow models. Each estimator is an implementation of a particular type of ML algorithm. They currently support regression and classification problems. A list of the available estimators includes *LinearRegressor/Classifier, DNNRegressor/Classifier, DNNLinearCombinedRegressor/Classifier, TensorForestEstimator, SVM, LogisticRegressor,* and a *generic* estimator that can be used to construct a custom model for either classification or regression problems. This provides a wide range of state-of-art ML algorithms, as well as the building blocks users need to construct their own algorithms.

Graph actions

Graph actions contain all the complicated logic for *distributed training, inference,* and the *evaluation* of a model. They are built on top of TensorFlow's low-level APIs; these complexities are hidden away from users so that they can focus on using the simplified interface to conduct their research. Estimators can then be distributed using multiple machines and devices, and all extended estimators get this functionality for free.

For example, *tf.estimator.RunConfig* specifies the runtime configurations for an Estimator run, providing the required parameters such as the number of cores to be used and the amount of GPU memory to be used. It also contains a *ClusterConfig* that specifies the configuration for a distributed run. It configures the tasks, clusters, master nodes, parameter servers, and everything else.

Parsing resources

Similar to libraries such as pandas, a high-level DataFrame module was included in tf.estimator to facilitate many common data reading/parsing tasks from resources such as TensorFlow.

Flower predictions

To illustrate the basic functionalities of the tf.estimator module, we will start by building a basic deep neural network (DNN) model and training it on the Iris dataset with the aim of predicting flower species based on sepal/petal geometry. The Iris dataset contains 150 rows of data, comprising 50 samples from each of three related **Iris** species: *Iris Setosa*, *Iris Virginica*, and *Iris Versicolor*. Each row contains the following data for each flower sample: sepal length, sepal width, petal length, petal width, and flower species. Flower species are represented as integers, with **0** denoting *Iris Setosa*, **1** denoting *Iris Versicolor*, and **2** denoting *Iris Virginica*:

Figure 1: Iris dataset

The example described in this section, `premade_estimator.py`, is downloadable from `https://github.com/tensorflow/models/blob/master/samples/core/get_started/premade_estimator.py`.

To fetch the training data, `iris_data.py` is used, downloadable from https://github.com/tensorflow/models/blob/master/samples/core/get_started/iris_data.py.

The Iris dataset is randomized and split into two separate CSVs; the first is the training set of 120 samples (`iris_training.csv`):

```
TRAIN_URL = "http://download.tensorflow.org/data/iris_training.csv"
```

The second is the testing set of 30 samples (`iris_test.csv`):

```
TEST_URL = "http://download.tensorflow.org/data/iris_test.csv"
```

Here you find the feature fields:

```
CSV_COLUMN_NAMES = ['SepalLength', 'SepalWidth',
                    'PetalLength', 'PetalWidth', 'Species']
```

Here you find the species to classify:

```
SPECIES = ['Setosa', 'Versicolor', 'Virginica']
```

Training and testing data are loaded using the `iris_data.load_data()` function:

```
(train_x, train_y), (test_x, test_y) = iris_data.load_data()
```

tf.estimator offers a variety of predefined estimators that can be used to run training and evaluation operations on input data.

Here we'll configure a **DNN Classifier** model to fit the Iris data. Using tf.estimator, we instantiate a tf.estimator.DNNClassifier with just a couple of lines of code. The code above defines the model's features, which specifies the datatypes for the features in the dataset:

```
my_feature_columns = []
for key in train_x.keys():
  my_feature_columns.append(tf.feature_column.numeric_column(key=key))
```

All the feature data is continuous, so tf.feature_column.numeric_column is the appropriate function to use to construct the feature columns. There are four features in the dataset (sepal width, sepal height, petal width, and petal height), so the shape must be set to [4] to hold all the data.

Now let's build a classifier, using the DNNClassifier model:

```
classifier = tf.estimator.DNNClassifier(
        feature_columns=my_feature_columns,
        hidden_units=[10, 10],
        n_classes=3)
```

DNNClassifier model uses the following arguments:

- `feature_columns= my_feature_columns`: The set of feature columns defined previously
- `hidden_units=[10, 10]`: Two hidden layers, containing 10 and 10 neurons
- `n_classes=3`: Three target classes, representing the three Iris species

Define the input pipeline (`input_fn`) and train the data using the `train` method. The number of training steps is 1000:

```
classifier.train(
        input_fn=lambda:iris_data.train_input_fn(train_x, train_y,
                                                args.batch_size),
        steps=args.train_steps)
```

The model's accuracy is evaluated using the `evaluate` method:

```
eval_result = classifier.evaluate(
        input_fn=lambda:iris_data.eval_input_fn(test_x, test_y,
                                                args.batch_size))

print('\nTest set accuracy: {accuracy:0.3f}\n'.format(**eval_result))
```

Like the method `train`, `evaluate` takes an input function that builds its input pipeline. `evaluate` returns a `dict` with the evaluation results.

The code example (`premade_estimator.py`) outputs training logs followed by some predictions against the test set:

```
INFO:tensorflow:loss = 120.53493, step = 1
INFO:tensorflow:global_step/sec: 437.609
INFO:tensorflow:loss = 14.973656, step = 101 (0.291 sec)
INFO:tensorflow:global_step/sec: 369.482
INFO:tensorflow:loss = 8.025629, step = 201 (0.248 sec)
INFO:tensorflow:global_step/sec: 267.963
INFO:tensorflow:loss = 7.3872843, step = 301 (0.364 sec)
INFO:tensorflow:global_step/sec: 337.761
```

```
INFO:tensorflow:loss = 7.1775312, step = 401 (0.260 sec)
INFO:tensorflow:global_step/sec: 684.081
INFO:tensorflow:loss = 6.1282234, step = 501 (0.146 sec)
INFO:tensorflow:global_step/sec: 686.175
INFO:tensorflow:loss = 7.441858, step = 601 (0.146 sec)
INFO:tensorflow:global_step/sec: 731.402
INFO:tensorflow:loss = 4.633889, step = 701 (0.137 sec)
INFO:tensorflow:global_step/sec: 687.698
INFO:tensorflow:loss = 8.395943, step = 801 (0.145 sec)
INFO:tensorflow:global_step/sec: 687.174
INFO:tensorflow:loss = 6.0668287, step = 901 (0.146 sec)
INFO:tensorflow:Saving checkpoints for 1000 into C:\Users\GIANCA~1\
AppData\Local\Temp\tmp9yaobdrg\model.ckpt.
INFO:tensorflow:Loss for final step: 7.467471.
INFO:tensorflow:Starting evaluation at 2018-03-03-14:11:13
INFO:tensorflow:Restoring parameters from C:\Users\GIANCA~1\AppData\
Local\Temp\tmp9yaobdrg\model.ckpt-1000
INFO:tensorflow:Finished evaluation at 2018-03-03-14:11:14
INFO:tensorflow:Saving dict for global step 1000: accuracy =
0.96666664, average_loss = 0.060853884, global_step = 1000, loss =
1.8256165

Test set accuracy: 0.967

INFO:tensorflow:Restoring parameters from C:\Users\GIANCA~1\AppData\
Local\Temp\tmp9yaobdrg\model.ckpt-1000
```

We can use the trained model to predict the species of an iris flower based on some *unlabeled measurements*.

Let's consider the following flower samples:

```
expected = ['Setosa', 'Versicolor', 'Virginica']
    predict_x = {
        'SepalLength': [5.1, 5.9, 6.9],
        'SepalWidth': [3.3, 3.0, 3.1],
        'PetalLength': [1.7, 4.2, 5.4],
        'PetalWidth': [0.5, 1.5, 2.1],
    }
```

As with training and evaluation, we make predictions using a single function call, via the predict method:

```
predictions = classifier.predict(
    input_fn=lambda:iris_data.eval_input_fn(predict_x,
                                        labels=None,
        batch_size=args.batch_size))

for pred_dict, expec in zip(predictions, expected):
    template = ('\nPrediction is "{}" ({:.1f}%), expected "{}"')
    class_id = pred_dict['class_ids'][0]
    probability = pred_dict['probabilities'][class_id]
    print(template.format(iris_data.SPECIES[class_id],
                            100 * probability, expec))
```

The preceding code yields the following output:

```
Prediction is "Setosa" (99.8%), expected "Setosa"

Prediction is "Versicolor" (99.8%), expected "Versicolor"

Prediction is "Virginica" (97.4%), expected "Virginica"
```

TFLearn

TFLearn is a library that wraps a lot of new TensorFlow APIs with the nice and familiar scikit-learn API.

TensorFlow is all about a building and executing graphs. This is a very powerful concept, but it is also cumbersome to start with.

Looking under the hood of TF.Learn, we just used three parts:

- **layers**: A set of advanced TensorFlow functions that allow us to easily build complex graphs, from fully connected layers, convolution, and batch norm to losses and optimization.
- **graph_actions**: A set of tools to perform training, evaluating, and running inference on TensorFlow graphs.
- **Estimator**: This packages everything into a class that follows scikit-learn interface and provides a way to easily build and train custom TensorFlow models.

Advanced TensorFlow Programming

Installation

To install TFLearn, the easiest way is to run the following command:

```
pip install git+https://github.com/tflearn/tflearn.git
```

For the latest stable version, use this command:

```
pip install tflearn
```

Otherwise, you can also install it from source by running the following (from the source folder):

```
python setup.py install
```

Titanic survival predictor

In this tutorial, we will learn to use TFLearn and TensorFlow to model the chance of survival of passengers on the Titanic using their personal information (such as gender and age). To tackle this classic ML task, we are going to build a DNN classifier.

Let's take a look at the dataset (TFLearn will automatically download it for you).

For each passenger, the following information is provided:

survived	Survived (0 = No; 1 = Yes)
pclass	Passenger Class (1 = st; 2 = nd; 3 = rd)
name	Name
sex	Sex
age	Age
sibsp	Number of Siblings/Spouses Aboard
parch	Number of Parents/Children Aboard
ticket	Ticket Number
fare	Passenger Fare

Here are some examples from the dataset:

survived	pclass	name	sex	age	sibsp	parch	ticket	fare
1	1	Aubart, Mme. Leontine Pauline	female	24	0	0	PC 17477	69.3000
0	2	Bowenur, Mr. Solomon	male	42	0	0	211535	13.0000
1	3	Baclini, Miss. Marie Catherine	female	5	2	1	2666	19.2583
0	3	Youseff, Mr. Gerious	male	45.5	0	0	2628	7.2250

There are two classes in our task: not survived (class = 0) and survived (class = 1). The passenger data has 8 features. The Titanic dataset is stored in a CSV file, so we can use the TFLearn load_csv() function to load the data from the file into a Python list. We specify the target_column argument to indicate that our labels (survived or not) are located in the first column (id: 0). The functions will return a tuple: (data, labels).

Let's start with importing the NumPy and TFLearn libraries:

```
import numpy as np
import tflearn as tfl
```

Download the Titanic dataset:

```
from tflearn.datasets import titanic
titanic.download_dataset('titanic_dataset.csv')
```

Load the CSV file, and indicate that the first column represents labels:

```
from tflearn.data_utils import load_csv
data, labels = load_csv('titanic_dataset.csv', target_column=0,
                        categorical_labels=True, n_classes=2)
```

Data needs some preprocessing before it is ready to be used in our DNN classifier. We must delete the column fields that won't help us with our analysis. We discard the *name* and *ticket* fields, because we estimate that a passenger's name and ticket are not related with their chance of surviving:

```
def preprocess(data, columns_to_ignore):
```

The preprocessing phase starts by descending the id and delete columns:

```
    for id in sorted(columns_to_ignore, reverse=True):
        [r.pop(id) for r in data]
    for i in range(len(data)):
```

The *sex* field is converted to float (to be manipulated):

```
        data[i][1] = 1. if data[i][1] == 'female' else 0.
    return np.array(data, dtype=np.float32)
```

As already described, the *name* and *ticket* fields will be ignored by the analysis:

```
to_ignore=[1, 6]
```

Then we call the preprocess procedure:

```
data = preprocess(data, to_ignore)
```

Next, we specify the shape of our input data. The input sample has a total of 6 features, and we will process samples in batches to save memory, so our data input shape is [None, 6]. The None parameter means an unknown dimension, so we can change the total number of samples that are processed in a batch:

```
net = tfl.input_data(shape=[None, 6])
```

Finally, we build a 3-layer neural network with this simple sequence of statements:

```
net = tfl.fully_connected(net, 32)
net = tfl.fully_connected(net, 32)
net = tfl.fully_connected(net, 2, activation='softmax')
net = tfl.regression(net)
```

TFLearn provides a model wrapper, DNN, that automatically performs neural network classifier tasks:

```
model = tfl.DNN(net)
```

We will run it for 10 epochs with a batch size of 16:

```
model.fit(data, labels, n_epoch=10, batch_size=16,
show_metric=True)
```

When we run the model, we should get the following output:

```
Training samples: 1309
Validation samples: 0
--
Training Step: 82  | total loss: 0.64003
| Adam | epoch: 001 | loss: 0.64003 - acc: 0.6620 -- iter: 1309/1309
--
Training Step: 164  | total loss: 0.61915
| Adam | epoch: 002 | loss: 0.61915 - acc: 0.6614 -- iter: 1309/1309
--
Training Step: 246  | total loss: 0.56067
| Adam | epoch: 003 | loss: 0.56067 - acc: 0.7171 -- iter: 1309/1309
--
Training Step: 328  | total loss: 0.51807
| Adam | epoch: 004 | loss: 0.51807 - acc: 0.7799 -- iter: 1309/1309
--
Training Step: 410  | total loss: 0.47475
| Adam | epoch: 005 | loss: 0.47475 - acc: 0.7962 -- iter: 1309/1309
--
Training Step: 574  | total loss: 0.48988
| Adam | epoch: 007 | loss: 0.48988 - acc: 0.7891 -- iter: 1309/1309
--
```

```
Training Step: 656   | total loss: 0.55073
| Adam | epoch: 008 | loss: 0.55073 - acc: 0.7427 -- iter: 1309/1309
--
Training Step: 738   | total loss: 0.50242
| Adam | epoch: 009 | loss: 0.50242 - acc: 0.7854 -- iter: 1309/1309
--
Training Step: 820   | total loss: 0.41557
| Adam | epoch: 010 | loss: 0.41557 - acc: 0.8110 -- iter: 1309/1309
--
```

The model accuracy is around 81%, which means that it can predict the correct outcome (that is, whether the passenger survived or not) for 81% of the passengers.

PrettyTensor

PrettyTensor allows the developer to wrap TensorFlow operations to quickly chain any number of layers to define neural networks. Coming up is simple example of Pretty Tensor's capabilities: we wrap a standard TensorFlow object, pretty, into a library-compatible object; then we feed it through three fully connected layers, and we finally output a softmax distribution:

```
pretty = tf.placeholder([None, 784], tf.float32)
softmax = (prettytensor.wrap(examples)
  .fully_connected(256, tf.nn.relu)
  .fully_connected(128, tf.sigmoid)
  .fully_connected(64, tf.tanh)
  .softmax(10))
```

The PrettyTensor installation is very simple. You can just use the `pip` installer:

`sudo pip install prettytensor`

Chaining layers

PrettyTensor has three modes of operation that share the ability to chain methods.

Normal mode

In normal mode, every time a method is called, a new PrettyTensor is created. This allows easy chaining, and you can still use any particular object multiple times. This makes it easy to branch your network.

Sequential mode

In sequential mode, an internal variable – the head – keeps track of the most recent output tensor, thus allowing call chains to be broken into multiple statements.

Here is a quick example:

```
seq = pretty_tensor.wrap(input_data).sequential()
seq.flatten()
seq.fully_connected(200, activation_fn=tf.nn.relu)
seq.fully_connected(10, activation_fn=None)
result = seq.softmax(labels, name=softmax_name))
```

Branch and join

Complex networks can be built using the first-class `branch` and `join` methods:

- `branch` creates a separate PrettyTensor object that points to the current head when it is called, and this allows the user to define a separate tower that either ends in a regression target, ends in output, or rejoins the network. Rejoining allows the user to define composite layers like inception.
- `join` is used to join multiple inputs or to rejoin a composite layer.

Digit classifier

In this example, we'll define and train a two-layer model and a convolutional model in the style of LeNet 5:

```
import tensorflow as tf
import prettytensor as pt
from prettytensor.tutorial import data_utils

tf.app.flags.DEFINE_string('save_path',\
                            None, \
                            'Where to save the model checkpoints.')

FLAGS = tf.app.flags.FLAGS

BATCH_SIZE = 50
EPOCH_SIZE = 60000
TEST_SIZE = 10000
```

Since we are feeding our data as NumPy arrays, we need to create placeholders in the graph. These must then be fed using the feed `dict` statement:

```
image_placeholder = tf.placeholder\
                (tf.float32, [BATCH_SIZE, 28, 28, 1])

labels_placeholder = tf.placeholder\
                (tf.float32, [BATCH_SIZE, 10])
```

Next, we create the `multilayer_fully_connected` function. The first two layers are fully connected (`100` neurons) and the last layer is a `softmax` result layer. As you can see, chaining layers is a very simple operation:

```
def multilayer_fully_connected(images, labels):
    images = pt.wrap(images)
    with pt.defaults_scope\
            (activation_fn=tf.nn.relu,l2loss=0.00001):

        return (images.flatten().\
                fully_connected(100).\
                fully_connected(100).\
                softmax_classifier(10, labels))
```

Now we'll build a multilayer convolutional network: the architecture is similar to LeNet 5's. Please change this so you can experiment with other architectures:

```
def lenet5(images, labels):
    images = pt.wrap(images)
    with pt.defaults_scope\
            (activation_fn=tf.nn.relu, l2loss=0.00001):

        return (images.conv2d(5, 20).\
                max_pool(2, 2).\
                conv2d(5, 50).\
                max_pool(2, 2).\
                flatten().\
                fully_connected(500).\
                softmax_classifier(10, labels))
```

Depending on the chosen model, we may have a 2-layer classifier (`multilayer_fully_connected`) or a convolutional classifier (`lenet5`):

```
def make_choice():
    var = int(input('(1) = multy layer model    (2) = lenet 5 '))
    print(var)
    if var == 1:
```

```
        result = multilayer_fully_connected\
                (image_placeholder,labels_placeholder)
        run_model(result)
    elif var == 2:
        result = lenet5\
                (image_placeholder,labels_placeholder)
        run_model(result)
    else:
        print ('incorrect input value')
```

Finally, we define the accuracy of the chosen model:

```
    def run_model(result):
        accuracy = result.softmax.evaluate_classifier\
                    (labels_placeholder,phase=pt.Phase.test)
```

Next, we build the training and testing sets:

```
        train_images, train_labels = data_utils.mnist(training=True)
        test_images, test_labels = data_utils.mnist(training=False)
```

We will use a gradient descent optimizer procedure and apply it to the graph. The pt.apply_optimizer function adds regularization losses and sets up a step counter:

```
        optimizer = tf.train.GradientDescentOptimizer(0.01)
        train_op = pt.apply_optimizer\
                    (optimizer,losses=[result.loss])
```

We can set a save_path in the running session to automatically save the progress every so often. Otherwise, the model will be lost at the end of the session:

```
    runner = pt.train.Runner(save_path=FLAGS.save_path)
    with tf.Session():
        for epoch in range(0,10):
```

Shuffle the training data:

```
            train_images, train_labels = \
                data_utils.permute_data\
                ((train_images, train_labels))

            runner.train_model(train_op,result.\
                    loss,EPOCH_SIZE,\
                    feed_vars=(image_placeholder,\
                            labels_placeholder),\
                    feed_data=pt.train.\
                    feed_numpy(BATCH_SIZE,\
```

Chapter 8

```
                                train_images,\
                                train_labels),\
                   print_every=100)

        classification_accuracy = runner.evaluate_model\
                                (accuracy,\
                                TEST_SIZE,\
                                feed_vars=(image_placeholder,\
                                        labels_placeholder),\
                                feed_data=pt.train.\
                                feed_numpy(BATCH_SIZE,\
                                        test_images,\
                                        test_labels))
        print("epoch" , epoch + 1)
        print("accuracy", classification_accuracy )


if __name__ == '__main__':
    make_choice()
```

Running the example, we have to choose the model to train:

```
(1) = multylayer model    (2) = lenet 5
```

By selecting the `multylayer model`, we should have an accuracy of 95.5 %:

```
Extracting /tmp/data\train-images-idx3-ubyte.gz
Extracting /tmp/data\train-labels-idx1-ubyte.gz
Extracting /tmp/data\t10k-images-idx3-ubyte.gz
Extracting /tmp/data\t10k-labels-idx1-ubyte.gz
epoch 1
accuracy [0.8969]
epoch 2
accuracy [0.914]
epoch 3
accuracy [0.9188]
epoch 4
accuracy [0.9306]
epoch 5
accuracy [0.9353]
epoch 6
accuracy [0.9384]
epoch 7
accuracy [0.9445]
epoch 8
```

```
accuracy [0.9472]
epoch 9
accuracy [0.9531]
epoch 10
accuracy [0.9552]
```

While for the Lenet5 we should have an accuracy of 98.8 %:

```
Extracting /tmp/data\train-images-idx3-ubyte.gz
Extracting /tmp/data\train-labels-idx1-ubyte.gz
Extracting /tmp/data\t10k-images-idx3-ubyte.gz
Extracting /tmp/data\t10k-labels-idx1-ubyte.gz

epoch 1
accuracy [0.9686]
epoch 2
accuracy [0.9755]
epoch 3
accuracy [0.983]
epoch 4
accuracy [0.9841]
epoch 5
accuracy [0.9844]
epoch 6
accuracy [0.9863]
epoch 7
accuracy [0.9862]
epoch 8
accuracy [0.9877]
epoch 9
accuracy [0.9855]
epoch 10
accuracy [0.9886]
```

Keras

Keras is an open source neural network library that is written in Python. It focuses on being minimal, modular, and extensible, and was designed in order to enable fast experimentation with DNNs.

This library, whose primary author and maintainer is a Google engineer named François Chollet, was developed as part of the research effort of project ONEIROS (Open-ended Neuro-Electronic Intelligent Robot Operating System).

Keras was developed following these design principles:

- **Modularity**: A model is understood as a sequence or a graph of standalone, fully-configurable modules that can be plugged together with as few restrictions as possible. *Neural layers*, *cost functions*, *optimizers*, *initialization schemes*, and *activation functions* are all standalone modules that can be combined to create new models.
- **Minimalism**: Each module must be short (a few lines of code) and simple. The source code should be transparent upon the dirt reading.
- **Extensibility**: New modules are simple to add (like new classes and functions), and existing modules provide examples on which to base new modules. Being able to easily create new modules allows total expressiveness, making Keras suitable for advanced research.

Keras, is available both in the embedded version as a TensorFlow API, and as a library:

- tf.keras from https://www.tensorflow.org/api_docs/python/tf/keras
- Keras v 2.1.4 (please see at https://keras.io for updates and installation guide)

In the following sections we will see how to use both the first and the second implementation.

Keras programming models

The core data structure of Keras is a model, which is a way to organize layers. There are two types of model:

- **Sequential**: This is just a linear stack of layers used to implement simple models
- **Functional APIs**: These are used for more complex architectures, such as models with multiple output and directed acyclic graphs

Sequential model

In this section, we'll quickly explain how sequential models work by showing you the code. Let's start by importing and building the Keras `Sequential` model using the TensorFlow APIs:

```
import tensorflow as tf
from tensorflow.python.keras.models import Sequential
model = Sequential()
```

Once we have defined a model we can add one or more layers. The stacking operation is provided by the `add()` statement:

```
from keras.layers import Dense, Activation
```

For example, let's add a first fully connected neural network layer and the activation function:

```
model.add(Dense(output_dim=64, input_dim=100))
model.add(Activation("relu"))
```

Then we add a second `softmax` layer:

```
model.add(Dense(output_dim=10))
model.add(Activation("softmax"))
```

If the model looks fine, we must `compile()` the model, specifying the loss function and the optimizer to be used:

```
model.compile(loss='categorical_crossentropy',\
              optimizer='sgd',\
              metrics=['accuracy'])
```

We can now configure our optimizer. Keras tries to make programming reasonably simple, allowing the user to be fully in control when they need to.

Once compiled, the model must fit the data:

```
model.fit(X_train, Y_train, nb_epoch=5, batch_size=32)
```

Alternatively, we can feed batches to our model manually:

```
model.train_on_batch(X_batch, Y_batch)
```

Once it is trained, we can use our model to make predictions on new data:

```
classes = model.predict_classes(X_test, batch_size=32)
proba = model.predict_proba(X_test, batch_size=32)
```

Sentiment classification of movie reviews

In this example, we apply the Keras sequential model to a sentiment analysis problem. Sentiment analysis is the act of deciphering the opinions contained in a written or spoken text. The main purpose of this technique is to identify the sentiment (or polarity) of a lexical expression, which may have a neutral, positive, or negative connotation. The problem that we want to solve is the IMDB movie review sentiment classification problem: each movie review is a variable sequence of words, and the sentiment (positive or negative) of each movie review must be classified.

The problem is very complex because the sequences can vary in length and contain a large vocabulary of input symbols. The solution requires the model to learn long-term dependencies between symbols in the input sequence.

The IMDB dataset contains 25,000 highly polar movie reviews (good or bad) for training, and the same amount again for testing. The data was collected by Stanford researchers and was used in a 2011 paper where a 50-50 split of the data was used for training and testing. In this paper, an accuracy of 88.89% was achieved.

Once we have defined our problem, we are ready to develop a sequential LSTM model to classify the sentiment of movie reviews. We can quickly develop a LSTM for the IMDB problem and achieve good accuracy. Let's start off by importing the classes and functions required for this model and initializing the random number generator to a constant value to ensure we can easily reproduce the results.

In this example we are using the embedded Keras in TensorFlow APIs:

```
import numpy
from tensorflow.python.keras.models import Sequential
from tensorflow.python.keras.datasets import imdb
from tensorflow.python.keras.layers import Dense
from tensorflow.python.keras.layers import LSTM
from tensorflow.python.keras.layers import Embedding
from tensorflow.python.keras.preprocessing import sequence
numpy.random.seed(7)
```

We load the IMDB dataset. We are restricting the dataset to the top 5,000 words. We also split the dataset into training (50%) and testing (50%) sets.

Keras provides built-in access to the IMDB dataset. The `imdb.load_data()` function allows you to load the dataset in a format that is ready for use in neural networks and DL models. The words have been replaced by integers that indicate the ordered frequency of each word in the dataset. The sentences in each review therefore comprise a sequence of integers.

Here's the code:

```
top_words = 5000
(X_train, y_train), (X_test, y_test) = \
                   imdb.load_data(num_words=top_words)
```

Next, we need to truncate and pad the input sequences so that they are all the same length for modeling. The model will learn the zero values that carry no information, so the sequences are not the same length in terms of content, but the vectors need to be the same length to be computed in Keras. The sequence length in each review varies, so we restricted each review to 500 words, truncating long reviews and padding the shorter reviews with zero values:

Let's see:

```
max_review_length = 500
X_train = sequence.pad_sequences\
            (X_train, maxlen=max_review_length)
X_test = sequence.pad_sequences\
            (X_test, maxlen=max_review_length)
```

We can now *define*, *compile*, and *fit* our LSTM model.

To resolve the sentiment classification problem, we'll use the *word embedding* technique. It consists of representing words in a continuous vector space, which is an area in which the words that are semantically similar are mapped to neighboring points. Word embedding is based on the *distributional hypothesis*, which states that the words that appear in a given context must share the same semantic meaning. Each movie review will then be mapped into a real vector domain, where the similarity between words in terms of meaning translates to closeness in the vector space. Keras provides a convenient way to convert positive integer representations of words into a word embedding by using an embedding layer.

Here, we define the length of the embedding vector and the model:

```
embedding_vector_length = 32
model = Sequential()
```

The first layer is the *embedded layer*. It uses 32 length vectors to represent each word:

```
model.add(Embedding(top_words, \
            embedding_vector_length,\
            input_length=max_review_length))
```

The next layer is the *LSTM* layer with 100 memory units. Finally, because this is a classification problem, we use a dense output layer with a single neuron and a `sigmoid` *activation* function to make predictions about the classes (good and bad) in the problem:

```
model.add(LSTM(100))
model.add(Dense(1, activation='sigmoid'))
```

Chapter 8

Because it is a binary classification problem, we use `binary_crossentropy` as the loss function, while the optimizer used here is the `adam` optimization algorithm (we encountered it in a previous TensorFlow implementation):

```
model.compile(loss='binary_crossentropy',\
              optimizer='adam',\
              metrics=['accuracy'])
print(model.summary())
```

We only fit three epochs, because the model quickly overfits. A batch size of 64 reviews is used to space out weight updates:

```
model.fit(X_train, y_train, \
          validation_data=(X_test, y_test),\
          num_epochs=3, \
          batch_size=64)
```

Then, we estimate the model's performance on unseen reviews:

```
scores = model.evaluate(X_test, y_test, verbose=0)
print("Accuracy: %.2f%%" % (scores[1]*100))
```

Running this example produces the following output:

```
Epoch 1/3
16750/16750 [==============================] - 107s - loss: 0.5570 - acc: 0.7149
Epoch 2/3
16750/16750 [==============================] - 107s - loss: 0.3530 - acc: 0.8577
Epoch 3/3
16750/16750 [==============================] - 107s - loss: 0.2559 - acc: 0.9019

Accuracy: 86.79%
```

You can see that this simple LSTM with little tuning achieves near state-of-the-art results on the IMDB problem. Importantly, this is a template that you can use to apply LSTM networks to your own sequence classification problems.

Functional API

To build complex networks, the *functional approach*, which we will describe here, turns out to be very useful. As shown in *Chapter 4, Convolutional Neural Networks*, the most popular neural networks (AlexNET, VGG, and so on) consist of one or more neural mini-networks repeated several times. Functional API consists of considering a neural network as a function that we can call several times. This approach turns out to be computationally advantageous because in order to build a neural network, even a complex one, just a few lines of code are needed.

In the following examples, we are using the Keras v2.1.4 from `https://keras.io`.

Let's see how it works. First, you need to import the `Model` module:

```
from keras.models import Model
```

The first thing to do is to specify the input for the model. Let's declare a tensor of shape 28×28×1 using the `Input()` function:

```
from keras.layers import Input
digit_input = Input(shape=(28, 28,1))
```

This is one of the notable differences between sequential models and Functional APIs. So, using the `Conv2D` and `MaxPooling2D` APIs, we build a convolutional layer:

```
x = Conv2D(64, (3, 3))(digit_input)

x = Conv2D(64, (3, 3))(x)

x = MaxPooling2D((2, 2))(x)

out = Flatten()(x)
```

Note that the variable x specifies the variable to which the layer is applied. Finally, we define the model by specifying the input and output:

```
vision_model = Model(digit_input, out)
```

Of course, we will also need to specify the *loss, optimizer*, and so on using the fit and compile methods, in the same way as we did for the sequential models.

SqueezeNet

In this example, we introduce a small CNN architecture called **SqueezeNet** that achieves AlexNet-level accuracy on ImageNet with 50 times fewer parameters. This architecture is inspired by the inception module of GoogleNet and was published in the paper : *SqueezeNet: AlexNet-level accuracy with 50x fewer parameters and < 1MB model size*, downloadable from the following link: `http://arxiv.org/pdf/1602.07360v2.pdf`.

Chapter 8

The idea behind SqueezeNet is to reduce the number of parameters we have to deal with using a compression scheme. This strategy reduces the number of parameters using fewer filters. This is done by feeding squeeze layers into what they refer to as expand layers. These two layers compose the so-called **Fire Module** as shown in the following diagram:

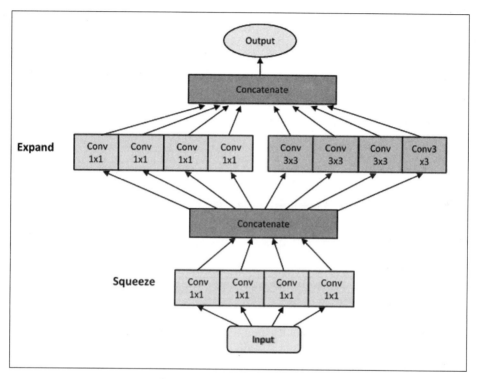

Figure 2: SqueezeNet Fire Module

`fire_module` is composed of 1×1 convolution filters followed by a *ReLU* operation:

```
x = Convolution2D(squeeze,(1,1),padding='valid', name='fire2/
squeeze1x1')(x)
x = Activation('relu', name='fire2/relu_squeeze1x1')(x)
```

The expand part has two portions: `left` and `right`.

The `left` part uses 1×1 convolutions and is called expand 1×1:

```
left = Conv2D(expand, (1, 1), padding='valid', name=s_id + exp1x1)(x)
left = Activation('relu', name=s_id + relu + exp1x1)(left)
```

The `right` part uses 3×3 convolutions and is called `expand3x3`. Both of these parts are followed by a ReLU layer:

```
right = Conv2D(expand, (3, 3), padding='same', name=s_id + exp3x3)(x)
right = Activation('relu', name=s_id + relu + exp3x3)(right)
```

The final output of the Fire Module is a *concatenation* of left and right:

```
x = concatenate([left, right], axis=channel_axis, name=s_id + 'concat')
```

Then, `fire_module` is used repeatedly to build the complete network, which looks like this:

```
x = Convolution2D(64,(3,3),strides=(2,2), padding='valid',\
                                    name='conv1')(img_input)
x = Activation('relu', name='relu_conv1')(x)
x = MaxPooling2D(pool_size=(3, 3), strides=(2, 2), name='pool1')(x)

x = fire_module(x, fire_id=2, squeeze=16, expand=64)
x = fire_module(x, fire_id=3, squeeze=16, expand=64)
x = MaxPooling2D(pool_size=(3, 3), strides=(2, 2), name='pool3')(x)

x = fire_module(x, fire_id=4, squeeze=32, expand=128)
x = fire_module(x, fire_id=5, squeeze=32, expand=128)
x = MaxPooling2D(pool_size=(3, 3), strides=(2, 2), name='pool5')(x)

x = fire_module(x, fire_id=6, squeeze=48, expand=192)
x = fire_module(x, fire_id=7, squeeze=48, expand=192)
x = fire_module(x, fire_id=8, squeeze=64, expand=256)
x = fire_module(x, fire_id=9, squeeze=64, expand=256)
x = Dropout(0.5, name='drop9')(x)

x = Convolution2D(classes, (1, 1), padding='valid', name='conv10')(x)
x = Activation('relu', name='relu_conv10')(x)
x = GlobalAveragePooling2D()(x)
x = Activation('softmax', name='loss')(x)
model = Model(inputs, x, name='squeezenet')
```

The following diagram shows the SqueezeNet architecture:

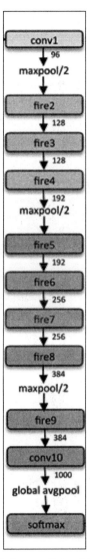

Figure 3: SqueezeNet architecture

You can download the Keras implementation of SqueezeNet (the `squeezenet.py` file) from the following link: https://github.com/rcmalli/keras-squeezenet.

Then we test the model on the following `squeeze_test.jpg` (227×227) image:

Figure 4: SqueezeNet test image

We do this by just using the following few lines of code:

```
import os
import numpy as np
import squeezenet as sq
from keras.applications.imagenet_utils import preprocess_input
from keras.applications.imagenet_utils import preprocess_input, decode_predictions
from keras.preprocessing import image


model = sq.SqueezeNet()
img = image.load_img('squeeze_test.jpg', target_size=(227, 227))
x = image.img_to_array(img)
x = np.expand_dims(x, axis=0)
x = preprocess_input(x)

preds = model.predict(x)
print('Predicted:', decode_predictions(preds))
```

As you can see, the results are very interesting:

```
Predicted: [[('n02504013', 'Indian_elephant', 0.64139527),
('n02504458', 'African_elephant', 0.22846894), ('n01871265', 'tusker',
0.12922771), ('n02397096', 'warthog', 0.00037213496), ('n02408429',
'water_buffalo', 0.00032306617)]]
```

Summary

In this chapter, we looked at some TensorFlow-based libraries for DL research and development. We introduced tf.estimator, which is a simplified interface for DL/ML, and is now part of TensorFlow and a high-level ML API that makes it easy to train, configure, and evaluate a variety of ML models. We used the estimator feature to implement a classifier for the Iris dataset.

We also had a look at the TFLearn library, which wraps a lot of TensorFlow APIs. In the example, we used TFLearn to estimate the chance of survival of passengers on the Titanic. To tackle this task, we built a DNN classifier.

Then, we introduced PrettyTensor, which allows TensorFlow operations to be wrapped to chain any number of layers. We implemented a convolutional model in the style of LeNet to quickly resolve the handwritten classification model.

Then we had a quick look at Keras, which is designed for minimalism and modularity, allowing the user to quickly define DL models. Using Keras, we have learned how to develop a simple single-layer LSTM model for the IMDB movie review sentiment classification problem. In the last example, we used Keras' functionality to build a SqueezeNet neural network starting from a pretrained inception model.

The next chapter introduces *reinforcement learning*. We'll explore the basic principles and algorithms of reinforcement learning. We'll also see some examples using TensorFlow and the OpenAI Gym framework, which is a powerful toolkit for developing and comparing reinforcement learning algorithms.

9
Recommendation Systems Using Factorization Machines

Factorization models are very popular in recommendation systems because they can be used to discover latent features underlying the interactions between two different kinds of entities. In this chapter, we will provide several examples of how to develop recommendation system for predictive analytics.

We will see the theoretical background of recommendation systems, such as matrix factorization. Later in the chapter, we will see how to use a collaborative approach to develop a movie recommendation system. Finally, will see how to use **Factorization Machines** (**FMs**) and improved versions of them to develop more robust recommendation systems.

In summary, the following topics will be covered in this chapter:

- Recommendation systems
- A movie recommendation system using the collaborative filtering approach
- K-means for clustering similar movies
- FM-based recommendation systems
- Using improved FMs for movie recommendation

Recommendation systems

Recommender techniques are nothing but information agents that try to predict items that users may be interested in and recommend the best ones to the target user. These techniques can be classified based on the information sources they use. For example, user features (age, gender, income, and location), item features (keywords and genres), user-item ratings (explicit ratings and transaction data), and other information about the user and item that are useful for the process of recommendation.

Thus, a **recommendation system** (otherwise known as a recommendation engine or RE) is a subclass of information filtering systems that help to predict the rating or preference, based on the rating provided by users for an item. In recent years, recommendation systems have become increasingly popular.

For example, at Amazon, the importance of suggesting the right item to the right user can be gauged by the fact that 35% of all sales are estimated to be generated by the recommendation engine. Therefore, Amazon is investing a large amount of talent and resources on getting better at AI – specifically "deep learning" technology – to make recommendation engines which learn and scale even more efficiently.

Consequently, they are being used in many areas, such as movies, music, news, books, research articles, search queries, social tags, products, jokes, restaurants, garments, financial services, life insurance, and online dating sites.

There are a couple of ways to develop REs to produce a list of recommendations. For example, collaborative and content-based filtering, knowledge-based, or the personality-based approach.

Collaborative filtering approaches

Using collaborative filtering approaches, an RE can be built based on a user's past behavior. Numerical ratings are given on consumed items. Sometimes, it can be based on the decisions made by other users who also have purchased the same items, using some widely-used data mining algorithms such as Apriori or FP-growth. In the following diagram, you can get some idea of different recommendation systems:

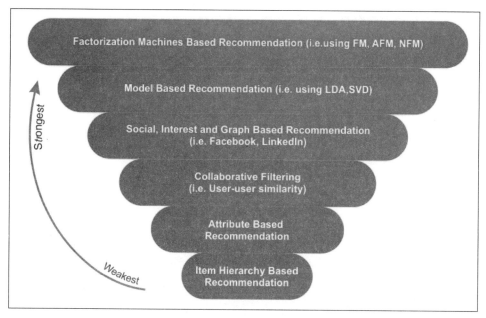

Figure 1: A comparative view of different recommendation systems

Collaborative filtering-based approaches often suffer from the following three problems:

- **Cold start**: Sometimes they can be stuck when a large amount of data about users is required to make a more accurate recommendation system.
- **Scalability**: A large amount of computation power is often necessary to calculate recommendations using a dataset with millions of users and products.
- **Sparsity**: This often happens with crowdsourced datasets when a huge number of items are sold on major e-commerce sites. All recommendation datasets are crowd-sourced in some sense. This is a general problem for almost all recommendation systems that have a sufficiently large number of items to offer to a sufficiently large number of users and need not be confined to e-commerce sites only.

  In this case, active users may rate only a small subset of the whole items sold, so even the most popular items have very few ratings. Accordingly, the user versus items matrix becomes very sparse. In other words, handling a large-scale sparse matrix is computationally very challenging.

To overcome these issues, a particular type of collaborative filtering algorithm uses matrix factorization, which is a low-rank matrix approximation technique. We will see an example of this later in the chapter.

Content-based filtering approaches

With content-based filtering approaches, a series of discrete characteristics of an item is utilized to recommend additional items with similar properties. Sometimes it is based on a description of the item and a profile of the user's preferences. These approaches try to recommend items that are similar to those that a user liked in the past, or that are being used currently.

A key issue with content-based filtering is whether the system is able to learn user preferences from their actions regarding one content source and use them with other content types. When this type of RE is deployed, it can then be used to predict items, or ratings for items, that the user may have an interest in.

Hybrid recommender systems

As you have seen, there are several pros and cons of using collaborative filtering and content-based filtering approaches. Therefore, to overcome the limitations of these two approaches, the recent trend has shown that a hybrid approach can be more effective and accurate. Sometimes, factorization approaches such as **Factorization Matrix** (**FM**) and **Singular Value Decomposition** (SVD) are used to make them robust. Hybrid approaches can be implemented in several ways:

- Content-based and collaborative predictions are computed separately and later are combined into one model. In this approach, FM and SVD are often used extensively.
- Content-based capabilities are added to a collaborative approach or vice versa. Again, FM and SVD are used for better prediction.

Netflix is a perfect example of using this hybrid approach to make recommendations to their subscribers. This site makes recommendations in two ways:

- **Collaborative filtering**: By comparing the watching and searching habits of similar users
- **Content-based filtering**: By recommending movies that share characteristics with films that a user has rated highly

Model-based collaborative filtering

Collaborative filtering methods are classified as memory-based i.e. user-based algorithm and model-based collaborative filtering (kernel mapping recommended). In the model-based collaborative filtering technique, users and products are described by a small set of factors, also called **latent factors** (**LFs**).

The LFs are then used to predict the missing entries. The **Alternating Least Squares** (**ALS**) algorithm is used to learn these latent factors. There are several advantages:

- Compared to a memory-based approach, a model-based approach can handle the sparsity of the original matrix better
- Using this approach, the resulting models become much smaller than the actual dataset, which imparts scalability to the overall system.
- Model-based systems are faster than memory-based systems because the resulting model is much smaller than what you need to query the whole dataset.
- Using this approach, it is relatively easy to avoid overfitting.

As a downside, the model-based approach is not flexible and adaptable because it is difficult to add data to the model. The quality of predictions depends on the way the model is built, but since this approach is inflexible, we cannot utilize all the data. This implies that we may not get high predictive accuracy.

Movie recommendation using collaborative filtering

In this section, we will see how to utilize collaborative filtering to develop a recommendation engine. However, before that let's discuss the utility matrix of preferences.

The utility matrix

In a collaborative filtering-based recommendation system, there are dimensions of entities: users and items (items refer to products, such as movies, games, and songs). As a user, you might have preferences for certain items. Therefore, these preferences must be extracted out of the data about items, users, or ratings. This data is often represented as a utility matrix, such as a user-item pair. This type of value can represent what is known about the degree of preference that the user has for a particular item.

The entry in the matrix can come from an ordered set. For example, integers 1-5 can be used to represent the number of stars that the user gave when rating items. We have already mentioned that users might not rate items very often, so most entries are unknown. Therefore, assigning 0 to unknown items would fail, which also means that the matrix is might be sparse. An unknown rating implies that we have no explicit information about the user's preference for the item.

Table 1 shows an example utility matrix. The matrix represents the rating users have given to movies on a 1-5 scale, with 5 being the highest rating. A blank entry represents the fact that the particular user has not provided any rating for that particular movie. HP1, HP2, and HP3 are acronyms for Harry Potter I, II, and III, TW stands for Twilight, and SW1, SW2, and SW3 for Star Wars episodes 1, 2, and 3. The letters A, B, C, and D represent the users:

	HP1	HP2	HP3	TW	SW1	SW2	SW3
A	4			5	1		
B	5	5	4				
C				2	4	5	
D		3					3

Table 1: Utility matrix (user versus movies matrix)

There are many blank entries for the user-movie pairs. This means that users have not rated those movies. In a real-life scenario, the matrix might be even sparser, with the typical user rating only a tiny fraction of all available movies. Using this matrix, the goal is to predict the blanks in the utility matrix. Now, let's see an example. Suppose we are curious to know whether user A would like SW2. It is difficult to work this out because there is not much data to work within the matrix in Table 1.

Thus, in practice, we might develop a movie recommendation engine to consider other properties of movies, such as the producer, director, leading actors, or even the similarity of their names. This way, we can compute the similarity of the movies SW1 and SW2. This similarity would drive us to conclude that since A did not like SW1, so they are unlikely to enjoy SW2 either.

However, this might not work for a larger dataset. Therefore, with much more data, we might observe that the people who rated both SW1 and SW2 were inclined to give them similar ratings. Finally, we can conclude that A would also give SW2 a low rating, similar to A's rating of SW1.

In the next section, we will see how to develop a movie recommendation engine using the collaborative filtering approach. We will see how to utilize this type of matrix.

> How to use the code repo: there are eight Python scripts in this code
> repo (that is, Deep Learning with TensorFlow_09_Codes/Collaborative
> Filtering/). First, execute the eda.py that performs an exploratory
> analysis of the dataset. Then, invoke the train.py script to perform
> the training. Finally, Test.py can be used for model inferencing and
> evaluation.
>
> Here is the brief functionality of each script:
>
>
>
> - eda.py: This is used for the exploratory analysis of the
>   MovieLens 1M dataset.
> - train.py: It performs the training as well as validation. Then
>   it prints the validation error. Finally, it creates the user-item
>   dense table.
> - Test.py: It restores the user vs item table generated in the
>   training. Then evaluates all the models.
> - run.py: It is used for model inferencing and does predictions.
> - kmean.py: It clusters similar movies.
> - main.py: It computes the top k movies, creates the user
>   rating, finds top k similar items, computes the user similarity,
>   computes the item correlation and computes the user Pearson
>   correlation.
> - readers.py: It reads the rating and movies data and performs
>   some preprocessing. Finally, it prepares the dataset for the
>   batch training.
> - model.py: It creates the model and computes the train/
>   validation loss.

The workflow can be described as follows:

1. First, we will train a model by using the available ratings.
2. Then we use the trained model to predict the missing ratings in the users versus movies matrix.
3. Then, with all the predicted ratings, a new user versus movie matrix will be constructed and saved in the form of a .pkl file.
4. Then, we use this matrix to make predictions of ratings for particular users.
5. Finally, we will train the K-means model to cluster related movies.

Description of the dataset

Before we start implementing the movie RE, let's look at the dataset that will be used. The MovieLens 1M dataset was downloaded from the MovieLens website at `http://files.grouplens.org/datasets/movielens/ml-1m.zip`.

I sincerely acknowledge and thank F. Maxwell Harper and Joseph A. Konstan for making the datasets available for use. The dataset was published in *MovieLens Dataset: History and Context. ACM Transactions on Interactive Intelligent Systems (TiiS) 5, 4, Article 19 (December 2015), 19 pages.*

There are three files in the dataset, and they relate to movies, ratings, and users. These files contain 1,000,209 anonymous ratings of approximately 3,900 movies made by 6,040 MovieLens users who joined MovieLens in 2000.

Ratings data

All the ratings are contained in the `ratings.dat` file and are in the following format - UserID::MovieID::Rating::Timestamp:

- UserIDs range between 1 and 6,040
- MovieIDs range between 1 and 3,952
- Ratings are made on a 5-star scale
- Timestamp is represented in seconds

Note that each user has rated at least 20 movies.

Movies data

Movie information is in the movies.dat file and is in the following format - MovieID::Title::Genres:

- Titles are identical to titles provided by IMDb (with the release year)
- Genres are pipe-separated (::), and each movie is categorized as action, adventure, animation, children's, comedy, crime, drama, war, documentary, fantasy, film-noir, horror, musical, mystery, romance, sci-fi, thriller, and western

Users data

User information is in the users.dat file and is in the following format - UserID::Gender::Age::Occupation::Zip-code.

All demographic information is provided voluntarily by the users and is not checked for accuracy. Only users who have provided some demographic information are included in this dataset. An M for male and F for female denotes gender.

Age is chosen from the following ranges:

- 1: Under 18
- 18: 18-24
- 25: 25-34
- 35: 35-44
- 45: 45-49
- 50: 50-55
- 56: 56+

Occupation is chosen from the following choices:

0: other, or not specified

1: academic/educator

2: artist

3: clerical/admin

4: college/grad student

5: customer service

6: doctor/health care

7: executive/managerial

8: farmer

9: homemaker

10: K-12 student

11: lawyer

12: programmer

13: retired

14: sales/marketing

15: scientist

16: self-employed

17: technician/engineer

18: tradesman/craftsman

19: unemployed

20: writer

Exploratory analysis of the MovieLens dataset

Here, we will see an exploratory description of the dataset before we start developing the RE. I am assuming that the reader has already downloaded the MovieLens 1m dataset from http://files.grouplens.org/datasets/movielens/ml-1m.zip and unzipped it in the input directory in this code repo. Now, for this, execute the $ python3 eda.py command on the terminal:

1. First, let's import the required libraries and packages:

   ```
   import matplotlib.pyplot as plt
   import seaborn as sns
   import pandas as pd
   import numpy as np
   ```

2. Now let's load the users, ratings, and movies dataset and create a pandas DataFrame:

   ```
   ratings_list = [i.strip().split("::") for i in open('Input/ratings.dat', 'r').readlines()]
   users_list = [i.strip().split("::") for i in open('Input/users.dat', 'r').readlines()]
   movies_list = [i.strip().split("::") for i in open('Input/movies.dat', 'r',encoding='latin-1').readlines()]
   ratings_df = pd.DataFrame(ratings_list, columns = ['UserID', 'MovieID', 'Rating', 'Timestamp'], dtype = int)
   movies_df = pd.DataFrame(movies_list, columns = ['MovieID', 'Title', 'Genres'])
   user_df=pd.DataFrame(users_list, columns=['UserID','Gender','Age', 'Occupation','ZipCode'])
   ```

Chapter 9

3. The next task is to convert the categorical columns, such as `MovieID`, `UserID`, and `Age`, into numerical values using the built-in `to_numeric()` pandas function:

   ```
   movies_df['MovieID'] = movies_df['MovieID'].apply(pd.to_numeric)
   user_df['UserID'] = user_df['UserID'].apply(pd.to_numeric)
   user_df['Age'] = user_df['Age'].apply(pd.to_numeric)
   ```

4. Let's see some examples from the user table:

   ```
   print("User table description:")
   print(user_df.head())
   print(user_df.describe())
   >>>
   User table description:
   UserID Gender  Age     Occupation ZipCode
       1       F    1             10   48067
       2       M   56             16   70072
       3       M   25             15   55117
       4       M   45              7   02460
       5       M   25             20   55455
                UserID          Age
   count   6040.000000  6040.000000
   mean    3020.500000    30.639238
   std     1743.742145    12.895962
   min        1.000000     1.000000
   25%     1510.750000    25.000000
   50%     3020.500000    25.000000
   75%     4530.250000    35.000000
   max     6040.000000    56.000000
   ```

5. Let's see some info from the rating dataset:

   ```
   print("Rating table description:")
   print(ratings_df.head())
   print(ratings_df.describe())
   >>>
   Rating table description:
   UserID  MovieID  Rating   Timestamp
        1     1193       5   978300760
        1      661       3   978302109
        1      914       3   978301968
        1     3408       4   978300275
        1     2355       5   978824291

               UserID           MovieID           Rating          Timestamp
   ```

[365]

```
count    1.000209e+06  1.000209e+06  1.000209e+06  1.000209e+06
mean     3.024512e+03  1.865540e+03  3.581564e+00  9.722437e+08
std      1.728413e+03  1.096041e+03  1.117102e+00  1.215256e+07
min      1.000000e+00  1.000000e+00  1.000000e+00  9.567039e+08
25%      1.506000e+03  1.030000e+03  3.000000e+00  9.653026e+08
50%      3.070000e+03  1.835000e+03  4.000000e+00  9.730180e+08
75%      4.476000e+03  2.770000e+03  4.000000e+00  9.752209e+08
max      6.040000e+03  3.952000e+03  5.000000e+00  1.046455e+09
```

6. Let's look at some info from the movie dataset:

```
>>>
print("Movies table description:")
print(movies_df.head())
print(movies_df.describe())
>>>
Movies table description:
    MovieID                    Title
Genres
0         1                 Toy Story (1995)
Animation|Children's|Comedy
1         2                 Jumanji (1995)
Adventure|Children's|Fantasy
2         3                 Grumpier Old Men (1995)
Comedy|Romance
3         4                 Waiting to Exhale (1995)
Comedy|Drama
4         5                 Father of the Bride Part II (1995)
Comedy
              MovieID
count    3883.000000
mean     1986.049446
std      1146.778349
min         1.000000
25%       982.500000
50%      2010.000000
75%      2980.500000
max      3952.000000
```

7. Now let's see the top five most rated movies:

```
print("Top ten most rated movies:")
print(ratings_df['MovieID'].value_counts().head())
>>>
Top 10 most rated movies with title and rating count:

American Beauty (1999)
3428
Star Wars: Episode IV - A New Hope (1977)
2991
Star Wars: Episode V - The Empire Strikes Back (1980)    2990
Star Wars: Episode VI - Return of the Jedi (1983)
2883
Jurassic Park (1993)
2672
Saving Private Ryan (1998)
2653
Terminator 2: Judgment Day (1991)
2649
Matrix, The (1999)
2590
Back to the Future (1985)
2583
Silence of the Lambs, The (1991)
2578
```

8. Now let's look at the movie rating distribution. For this, let's use a histogram plot, that demonstrates an important pattern where votes are distributed normally:

```
plt.hist(ratings_df.groupby(['MovieID'])['Rating'].mean().sort_values(axis=0, ascending=False))
plt.title("Movie rating Distribution")
plt.ylabel('Count of movies')
plt.xlabel('Rating');
plt.show()
>>>
```

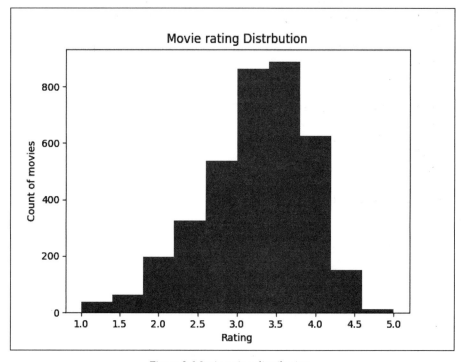

Figure 3: Movie rating distribution

9. Let's see how the ratings are distributed across different age groups:

```
user_df.Age.plot.hist()
plt.title("Distribution of users (by ages)")
plt.ylabel('Count of users')
plt.xlabel('Age');
plt.show()
>>>
```

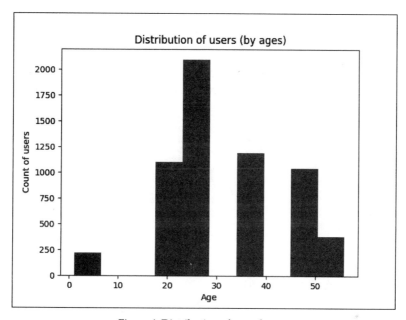

Figure 4: Distribution of users by age

10. Now let's see the highest-rated movie with a minimum of 150 ratings:

```
movie_stats = df.groupby('Title').agg({'Rating': [np.size, np.mean]})
print("Highest rated movie with minimum 150 ratings")
print(movie_stats.Rating[movie_stats.Rating['size'] > 150].sort_values(['mean'],ascending=[0]).head())
>>>
Top 5 and a highest rated movie with a minimum of 150 ratings
-------------------------------------------------------------
Title
size       mean
Seven Samurai (The Magnificent Seven)       628     4.560510
Shawshank Redemption, The (1994)                    2227    4.554558
Godfather, The (1972)                               2223
4.524966
```

```
Close Shave, A (1995)                                657
4.520548
Usual Suspects, The (1995)                          1783    4.517106
```

11. Let's look at gender bias in movie ratings, that is, how the movies' ratings compare by gender of the reviewer:

```
>>>
pivoted = df.pivot_table(index=['MovieID', 'Title'],
columns=['Gender'], values='Rating', fill_value=0)
print("Gender biasing towards movie rating")
print(pivoted.head())
```

12. We can now have a look at gender bias towards movie ratings and the difference between them, that is, how men and women rate the movies differently:

```
pivoted['diff'] = pivoted.M - pivoted.F
print(pivoted.head())
>>>
Gender                                                       F
M                       diff
MovieID Title
1  Toy Story (1995)                                     4.87817
4.130552            -0.057265
2  Jumanji (1995)                                       3.278409
3.175238            -0.103171
3  Grumpier Old Men (1995)         3.073529             2.994152
-0.079377
4  Waiting to Exhale (1995)        2.976471             2.482353
-0.494118
5  Father of the Bride Part II (1995) 3.212963          2.888298
-0.324665
```

13. From the preceding output, it is clear that in most cases, men provided higher ratings than women. Now that we have seen some info and statistics about the dataset, it is time to build our TensorFlow recommendation model.

Implementing a movie RE

In this example, we will see how to recommend the top k movies (where k is the number of movies), predict user ratings and recommend the top k similar items (where k is the number of items). Then we will see how to compute user similarity.

Then we will see the item-item correlation and user-user correlation using Pearson's correlation algorithm. Finally, we will see how to cluster similar movies using the K-means algorithm.

In other words, we will make a movie recommendation engine using the collaborative filtering approach and K-means to cluster similar movies.

Distance calculation: There are other ways to calculate the distance as well. For example:

1. Chebyshev distance can be used to measure the distance by considering only the most notable dimensions.
2. The Hamming distance algorithm can identify the difference between two strings.
3. Mahalanobis distance can be used to normalize the covariance matrix.
4. Manhattan distance is used to measure the distance by considering only axis-aligned directions.
5. The Haversine distance is used to measure the great-circle distances between two points on a sphere from the location.

Considering these distance-measuring algorithms, it is clear that the Euclidean distance algorithm would be the most appropriate to solve our purpose of distance calculation in the K-means algorithm

In summary, here is the workflow that will be used to develop this model:

1. First, train a model by using the available ratings.
2. Use that trained model to predict missing ratings in users versus movies matrix.
3. With all the predicted ratings, the users versus movies matrix become the trained users versus movies matrix, and we save both in the form of a `.pkl` file.
4. Then, we use the users versus movies matrix, or trained users versus movies matrix by the trained argument, for further processing.

Before training the model, the very first job is to prepare the training set by utilizing all of the available datasets.

Training the model with the available ratings

For this part, use the `train.py` script, which is dependent on other scripts. We will see the dependencies:

1. First, let's import necessary packages and modules:
    ```
    from collections import deque
    from six import next
    import readers
    ```

```
import os
import tensorflow as tf
import numpy as np
import model as md
import pandas as pd
import time
import matplotlib.pyplot as plt
```

2. Then we set the random seed for reproducibility:

   ```
   np.random.seed(12345)
   ```

3. The next task is to define the training parameters. Let's define the required data parameters, such as the location of the ratings dataset, the batch size, the dimension of SVD, the maximum epochs, and the checkpoint directory:

   ```
   data_file ="Input/ratings.dat"# Input user-movie-rating
   information file
   batch_size = 100 #Batch Size (default: 100)
   dims =15 #Dimensions of SVD (default: 15)
   max_epochs = 50 # Maximum epoch (default: 25)
   checkpoint_dir ="save/" #Checkpoint directory from training run
    val = True #True if Folders with files and False if single file
   is_gpu = True # Want to train model with GPU
   ```

4. We also need some other parameters, such as allowing soft placement and log device placement:

   ```
   allow_soft_placement = True #Allow device soft device placement
   log_device_placement=False #Log placement of ops on devices
   ```

5. We don't want to contaminate our fresh training with old metadata, or checkpoint and model files, so let's remove them if there are any:

   ```
   print("Start removing previous Files ...")
   if os.path.isfile("model/user_item_table.pkl"):
       os.remove("model/user_item_table.pkl")
   if os.path.isfile("model/user_item_table_train.pkl"):
       os.remove("model/user_item_table_train.pkl")
   if os.path.isfile("model/item_item_corr.pkl"):
       os.remove("model/item_item_corr.pkl")
   if os.path.isfile("model/item_item_corr_train.pkl"):
       os.remove("model/item_item_corr_train.pkl")
   if os.path.isfile("model/user_user_corr.pkl"):
       os.remove("model/user_user_corr.pkl")
   if os.path.isfile("model/user_user_corr_train.pkl"):
       os.remove("model/user_user_corr_train.pkl")
   if os.path.isfile("model/clusters.csv"):
   ```

```
        os.remove("model/clusters.csv")
    if os.path.isfile("model/val_error.pkl"):
        os.remove("model/val_error.pkl")
print("Done ...")
>>>
Start removing previous Files...
Done...
```

6. Then let's define the checkpoint directory. TensorFlow assumes this directory already exists, so we need to create it:

```
checkpoint_prefix = os.path.join(checkpoint_dir, "model")
if not os.path.exists(checkpoint_dir):
    os.makedirs(checkpoint_dir)
```

7. Before getting into the data, let's set the number of samples per batch, the dimension of the data, and the number of times the network sees all the training data:

```
batch_size =batch_size
dims =dims
max_epochs =max_epochs
```

8. Now let's specify the devices to be used for all TensorFlow computations, CPU or GPU:

```
if is_gpu:
    place_device = "/gpu:0"
else:
    place_device="/cpu:0"
```

9. Now we read the rating file with the delimiter, ::, through the get_data() function. A sample column consists of user ID, item ID, rating, and timestamp, for example, 3::1196::4::978297539. Then the above code does the purely integer-location based indexing for selection by position. After that, it splits the data into training and testing, 75% for training and 25% for testing. Finally, it uses the indices to separate the data and returns the data frame to use for the training:

```
def get_data():
    print("Inside get data ...")
    df = readers.read_file(data_file, sep="::")
    rows = len(df)
    df = df.iloc[np.random.permutation(rows)].reset_index(drop=True)
    split_index = int(rows * 0.75)
    df_train = df[0:split_index]
```

```
df_test = df[split_index:].reset_index(drop=True)
print("Done !!!")
print(df.shape)
return df_train, df_test,df['user'].max(),df['item'].max()
```

10. We then clip the limit of the values in an array: given an interval, values outside the interval are clipped to the edges of the interval. For example, if an interval of [0, 1] is specified, values smaller than 0 become 0, and values larger than 1 become 1:

```
def clip(x):
    return np.clip(x, 1.0, 5.0)
```

We then invoke the read_data() method to read data from the ratings file to build a TensorFlow model:

```
df_train, df_test,u_num,i_num = get_data()
>>>
Inside get data...
Done!!!
```

1. We then define the number of users in the dataset who rated the movies, and the number of movies in the dataset:

```
u_num = 6040 # Number of users in the dataset
i_num = 3952 # Number of movies in the dataset
```

2. Now let's generate the number of samples per batch:

```
samples_per_batch = len(df_train) // batch_size
print("Number of train samples %d, test samples %d, samples per batch %d" % (len(df_train), len(df_test), samples_per_batch))
>>>
Number of train samples 750156, test samples 250053, samples per batch 7501
```

3. Now, using a shuffle iterator, we generate random batches. In training, this helps to prevent a biased result as well as overfitting:

```
iter_train = readers.ShuffleIterator([df_train["user"], df_train["item"],df_train["rate"]], batch_size=batch_size)
```

4. For more on this class, refer to the readers.py script. For your convenience, here is the source of this class:

```
class ShuffleIterator(object):
    def __init__(self, inputs, batch_size=10):
        self.inputs = inputs
        self.batch_size = batch_size
        self.num_cols = len(self.inputs)
```

```
            self.len = len(self.inputs[0])
            self.inputs = np.transpose(np.vstack([np.array(self.
    inputs[i]) for i in range(self.num_cols)]))
        def __len__(self):
            return self.len
        def __iter__(self):
            return self
        def __next__(self):
            return self.next()
        def next(self):
            ids = np.random.randint(0, self.len, (self.batch_size,))
            out = self.inputs[ids, :]
            return [out[:, i] for i in range(self.num_cols)]
```

5. Then we sequentially generate one-epoch batches for testing (see `train.py`):

   ```
   iter_test = readers.OneEpochIterator([df_test["user"], df_
   test["item"], df_test["rate"]], batch_size=-1)
   ```

6. For more on this class, refer to the `readers.py` script. For your convenience, here is the source of this class:

   ```
   class OneEpochIterator(ShuffleIterator):
       def __init__(self, inputs, batch_size=10):
           super(OneEpochIterator, self).__init__(inputs, batch_
   size=batch_size)
           if batch_size > 0:
               self.idx_group = np.array_split(np.arange(self.len),
   np.ceil(self.len / batch_size))
           else:
               self.idx_group = [np.arange(self.len)]
           self.group_id = 0
       def next(self):
           if self.group_id >= len(self.idx_group):
               self.group_id = 0
               raise StopIteration
           out = self.inputs[self.idx_group[self.group_id], :]
           self.group_id += 1
           return [out[:, i] for i in range(self.num_cols)]
   ```

7. Now it's time to create the TensorFlow placeholders:

   ```
   user_batch = tf.placeholder(tf.int32, shape=[None], name="id_
   user")
   item_batch = tf.placeholder(tf.int32, shape=[None], name="id_
   item")
   rate_batch = tf.placeholder(tf.float32, shape=[None])
   ```

8. Now that our training set and the placeholders are ready to hold the batches of training values, it time to instantiate the model. For this, we use the `model()` method and use l2 regularization to avoid overfitting (see the `model.py` script):

```
infer, regularizer = md.model(user_batch, item_batch, user_num=u_num, item_num=i_num, dim=dims, device=place_device)
```

The `model()` method is as follows:

```
def model(user_batch, item_batch, user_num, item_num, dim=5, device="/cpu:0"):
    with tf.device("/cpu:0"):
        # Using a global bias term
        bias_global = tf.get_variable("bias_global", shape=[])

        # User and item bias variables: get_variable: Prefixes the name with the current variable
        # scope and performs reuse checks.
        w_bias_user = tf.get_variable("embd_bias_user", shape=[user_num])
        w_bias_item = tf.get_variable("embd_bias_item", shape=[item_num])

        # embedding_lookup: Looks up 'ids' in a list of embedding tensors
        # Bias embeddings for user and items, given a batch
        bias_user = tf.nn.embedding_lookup(w_bias_user, user_batch, name="bias_user")
        bias_item = tf.nn.embedding_lookup(w_bias_item, item_batch, name="bias_item")

        # User and item weight variables
        w_user = tf.get_variable("embd_user", shape=[user_num, dim],
                                initializer=tf.truncated_normal_initializer(stddev=0.02))
        w_item = tf.get_variable("embd_item", shape=[item_num, dim],
                                initializer=tf.truncated_normal_initializer(stddev=0.02))

        # Weight embeddings for user and items, given a batch
        embd_user = tf.nn.embedding_lookup(w_user, user_batch, name="embedding_user")
```

```
            embd_item = tf.nn.embedding_lookup(w_item, item_batch,
    name="embedding_item")

            # reduce_sum: Computes the sum of elements across
    dimensions of a tensor
            infer = tf.reduce_sum(tf.multiply(embd_user, embd_item),
    1)
            infer = tf.add(infer, bias_global)
            infer = tf.add(infer, bias_user)
            infer = tf.add(infer, bias_item, name="svd_inference")

            # l2_loss: Computes half the L2 norm of a tensor without
    the sqrt
            regularizer = tf.add(tf.nn.l2_loss(embd_user), tf.nn.l2_
    loss(embd_item), name="svd_regularizer")
        return infer, regularizer
```

9. Now let's define the training ops (see more in `models.py` script):

```
_, train_op = md.loss(infer, regularizer, rate_batch, learning_
rate=0.001, reg=0.05, device=place_device)
```

The `loss()` method is as follows:

```
def loss(infer, regularizer, rate_batch, learning_rate=0.1, reg=0.1,
device="/cpu:0"):
    with tf.device(device):
        cost_l2 = tf.nn.l2_loss(tf.subtract(infer, rate_batch))
        penalty = tf.constant(reg, dtype=tf.float32, shape=[],
name="l2")
        cost = tf.add(cost_l2, tf.multiply(regularizer, penalty))
        train_op = tf.train.FtrlOptimizer(learning_rate).
minimize(cost)
    return cost, train_op
```

1. Once we have instantiated the model and training ops, we can save the model for future use:

```
saver = tf.train.Saver()
init_op = tf.global_variables_initializer()
session_conf = tf.ConfigProto(
  allow_soft_placement=allow_soft_placement, log_device_
placement=log_device_placement)
```

2. Now we start training the model:

```
with tf.Session(config = session_conf) as sess:
    sess.run(init_op)
```

```python
        print("%s\t%s\t%s\t%s" % ("Epoch", "Train err", "Validation err", "Elapsed Time"))
        errors = deque(maxlen=samples_per_batch)
        train_error=[]
        val_error=[]
        start = time.time()

        for i in range(max_epochs * samples_per_batch):
            users, items, rates = next(iter_train)
            _, pred_batch = sess.run([train_op, infer], feed_dict={user_batch: users, item_batch: items, rate_batch: rates})
            pred_batch = clip(pred_batch)
            errors.append(np.power(pred_batch - rates, 2))

            if i % samples_per_batch == 0:
                train_err = np.sqrt(np.mean(errors))
                test_err2 = np.array([])
                for users, items, rates in iter_test:
                    pred_batch = sess.run(infer, feed_dict={user_batch: users, item_batch: items})
                    pred_batch = clip(pred_batch)
                    test_err2 = np.append(test_err2, np.power(pred_batch - rates, 2))
                end = time.time()

                print("%02d\t%.3f\t\t%.3f\t\t%.3f secs" % (i // samples_per_batch, train_err, np.sqrt(np.mean(test_err2)), end - start))
                train_error.append(train_err)
                val_error.append(np.sqrt(np.mean(test_err2)))
                start = end

    saver.save(sess, checkpoint_prefix)
    pd.DataFrame({'training error':train_error,'validation error':val_error}).to_pickle("val_error.pkl")
    print("Training Done !!!")

sess.close()
```

3. The preceding code carries out the training and saves the errors in a pickle file. Finally, it prints the training and validation error and the time taken:

```
>>>
Epoch      Train err      Validation err      Elapsed Time
00         2.816          2.812               0.118 secs
01         2.813          2.812               4.898 secs
...        ...            ...                 ...
48         2.770          2.767               1.618 secs
49         2.765          2.760               1.678 secs
```

Training Done!!!

The result is abridged, only a few steps have been shown. Now let's see these errors graphically:

```
error = pd.read_pickle("val_error.pkl")
error.plot(title="Training vs validation error (per epoch)")
plt.ylabel('Error/loss')
plt.xlabel('Epoch');
plt.show()
>>>
```

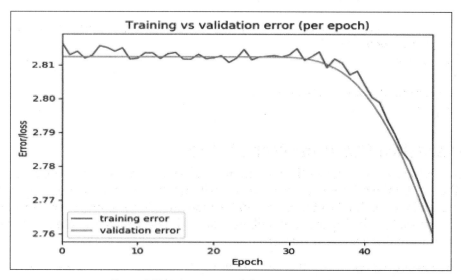

Figure 5: Training versus validation error per epoch

This graph shows that over time, both the training and the validation errors decrease, which means that we are walking in the correct direction. Nevertheless, you could still try to increase the steps and see if these two values can be further reduced, which means better accuracy.

Inferencing the saved model

The following code performs the model inferencing using the saved model and it prints the overall validation error:

```
if val:
    print("Validation ...")
    init_op = tf.global_variables_initializer()
    session_conf = tf.ConfigProto(
      allow_soft_placement=allow_soft_placement,
      log_device_placement=log_device_placement)
    with tf.Session(config = session_conf) as sess:
        new_saver = tf.train.import_meta_graph("{}.meta".format(checkpoint_prefix))
        new_saver.restore(sess, tf.train.latest_checkpoint(checkpoint_dir))
        test_err2 = np.array([])
        for users, items, rates in iter_test:
            pred_batch = sess.run(infer, feed_dict={user_batch: users, item_batch: items})
            pred_batch = clip(pred_batch)
            test_err2 = np.append(test_err2, np.power(pred_batch - rates, 2))
            print("Validation Error: ",np.sqrt(np.mean(test_err2)))
    print("Done !!!")
sess.close()
>>>
Validation Error:    2.14626890224
Done!!!
```

Generating the user-item table

The following method creates the user-item dataframe. It is used to create a trained DataFrame. All the missing values in the user-item table are filled in here using the SVD trained model. It takes the ratings dataframe and stores all the user ratings for all the movies. Finally, it generates a filled ratings dataframe, where the rows are the users and the columns are the items:

```
def create_df(ratings_df=readers.read_file(data_file, sep="::")):
    if os.path.isfile("model/user_item_table.pkl"):
        df=pd.read_pickle("user_item_table.pkl")
    else:
        df = ratings_df.pivot(index = 'user', columns ='item', values = 'rate').fillna(0)
```

```python
        df.to_pickle("user_item_table.pkl")

    df=df.T
    users=[]
    items=[]
    start = time.time()
    print("Start creating user-item dense table")
    total_movies=list(ratings_df.item.unique())

    for index in df.columns.tolist():
        #rated_movies=ratings_df[ratings_df['user']==index].drop(['st', 'user'], axis=1)
        rated_movie=[]
        rated_movie=list(ratings_df[ratings_df['user']==index].drop(['st', 'user'], axis=1)['item'].values)
        unseen_movies=[]
        unseen_movies=list(set(total_movies) - set(rated_movie))

        for movie in unseen_movies:
            users.append(index)
            items.append(movie)
    end = time.time()

    print(("Found in %.2f seconds" % (end-start)))
    del df
    rated_list = []

    init_op = tf.global_variables_initializer()
    session_conf = tf.ConfigProto(
      allow_soft_placement=allow_soft_placement,
      log_device_placement=log_device_placement)

    with tf.Session(config = session_conf) as sess:
        #sess.run(init_op)
        print("prediction started ...")
        new_saver = tf.train.import_meta_graph("{}.meta".format(checkpoint_prefix))
        new_saver.restore(sess, tf.train.latest_checkpoint(checkpoint_dir))
        test_err2 = np.array([])
        rated_list = sess.run(infer, feed_dict={user_batch: users, item_batch: items})
        rated_list = clip(rated_list)
        print("Done !!!")
```

Recommendation Systems Using Factorization Machines

```
        sess.close()

        df_dict={'user':users,'item':items,'rate':rated_list}
        df = ratings_df.drop(['st'],axis=1).append(pd.DataFrame(df_dict)).
pivot(index = 'user', columns ='item', values = 'rate').fillna(0)
        df.to_pickle("user_item_table_train.pkl")
        return df
```

Now let's invoke the preceding method to generate the user-item table as a pandas dataframe:

```
create_df(ratings_df = readers.read_file(data_file, sep="::"))
```

This line will create the user versus item table for the training set and save the dataframe as a `user_item_table_train.pkl` file in your specified directory.

Clustering similar movies

For this section, refer to the `kmean.py` script. This script takes the rating data file as input and returns movies along with their respective clusters.

More technically, the aim of this section is to find similar movies; for example, user 1 liked movie 1, and because movie 1 and movie 2 are similar, the user would like movie 2. Let's get started by importing required packages and modules:

```
import tensorflow as tf
import numpy as np
import pandas as pd
import time
import readers
import matplotlib.pyplot as plt
import seaborn as sns
from sklearn.decomposition import PCA
```

Now let's define the data parameters to be used: the path of the rating data file, number of clusters, K, and maximum number of iterations. Additionally, we also define whether we would like to use a trained user versus item matrix:

```
data_file = "Input/ratings.dat" #Data source for the positive data
K = 5 # Number of clusters
MAX_ITERS =1000 # Maximum number of iterations
TRAINED = False # Use TRAINED user vs item matrix
```

Then the `k_mean_clustering ()` method is defined. It returns the movies along with their respective clusters. It takes the ratings dataset, `ratings_df`, which is a rating data frame. It then stores all the user ratings for respective movies, K is the number of clusters, `MAX_ITERS` is the maximum number of recommendations, and `TRAINED` is a Boolean type that signifies whether to use the trained user versus movie table or the untrained one.

How to find the optimal K value

Here we set the value of K naively. However, to fine-tune the clustering performance, we can use a heuristic approach called Elbow method. We start from K = 2, then, we run the K-means algorithm by increasing K and observe the value of the cost function (CF) using WCSS. At some point, a big drop in CF will happen. Nevertheless, the improvement then becomes marginal with an increasing value of K. In summary, we can pick the K after the last big drop of WCSS as an optimal one.

Finally, the `k_mean_clustering()` function returns a list of movies/items and a list of clusters:

```
def k_mean_clustering(ratings_df,K,MAX_ITERS,TRAINED=False):
    if TRAINED:
        df=pd.read_pickle("user_item_table_train.pkl")
    else:
        df=pd.read_pickle("user_item_table.pkl")
    df = df.T
    start = time.time()
    N=df.shape[0]

    points = tf.Variable(df.as_matrix())
    cluster_assignments = tf.Variable(tf.zeros([N], dtype=tf.int64))
    centroids = tf.Variable(tf.slice(points.initialized_value(),
 [0,0], [K,df.shape[1]]))
    rep_centroids = tf.reshape(tf.tile(centroids, [N, 1]), [N, K, df.shape[1]])
    rep_points = tf.reshape(tf.tile(points, [1, K]), [N, K, df.shape[1]])
    sum_squares = tf.reduce_sum(tf.square(rep_points - rep_centroids),reduction_indices=2)

    best_centroids = tf.argmin(sum_squares, 1)    did_assignments_change = tf.reduce_any(tf.not_equal(best_centroids, cluster_assignments))
```

```
        means = bucket_mean(points, best_centroids, K)

        with tf.control_dependencies([did_assignments_change]):
            do_updates = tf.group(
                centroids.assign(means),
                cluster_assignments.assign(best_centroids))

    init = tf.global_variables_initializer()
    sess = tf.Session()
    sess.run(init)
    changed = True
    iters = 0

    while changed and iters < MAX_ITERS:
        iters += 1
        [changed, _] = sess.run([did_assignments_change, do_updates])
    [centers, assignments] = sess.run([centroids, cluster_
assignments])
    end = time.time()

    print (("Found in %.2f seconds" % (end-start)), iters,
"iterations")
    cluster_df=pd.DataFrame({'movies':df.index.
values,'clusters':assignments})

    cluster_df.to_csv("clusters.csv",index=True)
    return assignments,df.index.values
```

In the preceding code, we have a silly initialization in a sense that we use the first K points as the starting centroids. In the real world, it can be further improved.

In the preceding code block, we replicate N copies of each centroid and K copies of each data point. Then we subtract and compute the sum of squared distances. We then use the argmin to select the lowest-distance point. However, we do not write the assigned clusters variable until after computing whether the assignments have changed, hence with dependencies.

If you look at the preceding code carefully, there is a function called bucket_mean(). It takes the data points, the best centroids, and the number of the tentative cluster, K, and computes the mean to use in cluster computation:

```
def bucket_mean(data, bucket_ids, num_buckets):
    total = tf.unsorted_segment_sum(data, bucket_ids, num_buckets)
    count = tf.unsorted_segment_sum(tf.ones_like(data), bucket_ids,
num_buckets)
    return total / count
```

Once we have trained our K-means model, the next task is to visualize those clusters representing similar movies. For this, we have a function called showClusters(), which takes the user-item table, the clustered data written in a CSV file (clusters.csv), the number of principal components (the default is 2), and the SVD solver (possible values are randomized and full).

The thing is, in a 2D space it would be difficult to plot all the data points representing the movie clusters. For this reason, we have applied Principal Component Analysis (PCA) to reduce the dimensionality without sacrificing the quality much:

```
user_item=pd.read_pickle(user_item_table)
cluster=pd.read_csv(clustered_data, index_col=False)
user_item=user_item.T
pcs = PCA(number_of_PCA_components, svd_solver)
cluster['x']=pcs.fit_transform(user_item)[:,0]
cluster['y']=pcs.fit_transform(user_item)[:,1]
fig = plt.figure()
ax = plt.subplot(111)
ax.scatter(cluster[cluster['clusters']==0]['x'].values,cluster[cluster['clusters']==0]['y'].values,color="r", label='cluster 0')
ax.scatter(cluster[cluster['clusters']==1]['x'].values,cluster[cluster['clusters']==1]['y'].values,color="g", label='cluster 1')
ax.scatter(cluster[cluster['clusters']==2]['x'].values,cluster[cluster['clusters']==2]['y'].values,color="b", label='cluster 2')
ax.scatter(cluster[cluster['clusters']==3]['x'].values,cluster[cluster['clusters']==3]['y'].values,color="k", label='cluster 3')
ax.scatter(cluster[cluster['clusters']==4]['x'].values,cluster[cluster['clusters']==4]['y'].values,color="c", label='cluster 4')
ax.legend()
plt.title("Clusters of similar movies using K-means")
plt.ylabel('PC2')
plt.xlabel('PC1');
plt.show()
```

Well done. We will evaluate our model and plot the clusters in the evaluation step.

Movie rating prediction by users

For this I have written a function called prediction(). It takes the sample input about users and items (in this case, movies), and creates TensorFlow placeholders from the graph by name. It then evaluates those tensors. In the following code, it is to be noted that TensorFlow assumes that the checkpoint directory already exists, so make sure that it already exists. For details on this step refer to the run.py file. Note that this script does not show any result but a function from this script named *prediction* is further invoked in the main.py script for making predictions:

```
def prediction(users=predicted_user, items=predicted_item, allow_soft_placement=allow_soft_placement,\
log_device_placement=log_device_placement, checkpoint_dir=checkpoint_dir):
    rating_prediction=[]
    checkpoint_prefix = os.path.join(checkpoint_dir, "model")
    graph = tf.Graph()
    with graph.as_default():
        session_conf = tf.ConfigProto(allow_soft_placement=allow_soft_placement,log_device_placement=log_device_placement)
        with tf.Session(config = session_conf) as sess:
            new_saver = tf.train.import_meta_graph("{}.meta".format(checkpoint_prefix))
            new_saver.restore(sess, tf.train.latest_checkpoint(checkpoint_dir))
            user_batch = graph.get_operation_by_name("id_user").outputs[0]
            item_batch = graph.get_operation_by_name("id_item").outputs[0]
            predictions = graph.get_operation_by_name("svd_inference").outputs[0]
            pred = sess.run(predictions, feed_dict={user_batch: users, item_batch: items})
            pred = clip(pred)
        sess.close()
    return pred
```

We will see how we could use this method to predict the top k movies and user ratings for movies. In the preceding code segment, clip() is a user-defined function that limits the values in an array. Here is the implementation:

```
def clip(x):
    return np.clip(x, 1.0, 5.0) # rating 1 to 5
```

Now let's see how we could use the `prediction()` method to make a set of movie ratings predictions by a user:

```
def user_rating(users,movies):
    if type(users) is not list: users=np.array([users])
    if type(movies) is not list:
        movies=np.array([movies])
    return prediction(users,movies)
```

The preceding function returns a user rating for respective user. It takes a list of one or more numbers, a list of one or more user IDs, and a list of one or more numbers and a list of one or more movie IDs. Finally, it returns a list of predicted movies.

Finding top k movies

The following method extracts the top k movies that a user has not seen where k is an arbitrary integer such as 10. The name of the function is `top_k_movies()`. It returns the top k movies for a certain user. It takes a list of user IDs and the rating dataframe. It then stores all the user ratings for these movies. The output is a dictionary containing the user ID as the key and the list of the top k movies for that user as the value:

```
def top_k_movies(users,ratings_df,k):
    dicts={}
    if type(users) is not list:
        users = [users]
    for user in users:
        rated_movies = ratings_df[ratings_df['user']==user].drop(['st', 'user'], axis=1)
        rated_movie = list(rated_movies['item'].values)
        total_movies = list(ratings_df.item.unique())
        unseen_movies = list(set(total_movies) - set(rated_movie))
        rated_list = []
        rated_list = prediction(np.full(len(unseen_movies),user),np.array(unseen_movies))
        useen_movies_df = pd.DataFrame({'item': unseen_movies,'rate':rated_list})
        top_k = list(useen_movies_df.sort_values(['rate','item'], ascending=[0, 0])['item'].head(k).values)
        dicts.update({user:top_k})
    result = pd.DataFrame(dicts)
    result.to_csv("user_top_k.csv")
    return dicts
```

[387]

In the preceding code segment, `prediction()` is a user-defined function that we described previously. We will see an example of how to predict the top k movies (see `Test.py` for more or in a later section).

Predicting top k similar movies

I have written a function called `top_k_similar_items()` that computes and returns k movies that are similar to a particular movie. It takes a list of numbers, or number, a list of movie IDs, and the rating dataframe. It stores all user ratings for these movies. It also takes k as a natural number.

The value of TRAINED can be either TRUE or FALSE, and it specifies whether to use the trained user versus movie table or the untrained one. Finally, it returns a list of k movies that are similar to the one passed as input:

```python
def top_k_similar_items(movies,ratings_df,k,TRAINED=False):
    if TRAINED:
        df=pd.read_pickle("user_item_table_train.pkl")
    else:
        df=pd.read_pickle("user_item_table.pkl")
    corr_matrix=item_item_correlation(df,TRAINED)
    if type(movies) is not list:
        return corr_matrix[movies].sort_values(ascending=False).drop(movies).index.values[0:k]
    else:
        dict={}
        for movie in movies:          dict.update({movie:corr_matrix[movie].sort_values(ascending=False).drop(movie).index.values[0:k]})
        pd.DataFrame(dict).to_csv("movie_top_k.csv")
        return dict
```

In the preceding code, the `item_item_correlation()` function is a user-defined function that computes the movie-movie correlation that is used in when predicting the top k similar movies. The method is as follows:

```python
def item_item_correlation(df,TRAINED):
    if TRAINED:
        if os.path.isfile("model/item_item_corr_train.pkl"):
            df_corr=pd.read_pickle("item_item_corr_train.pkl")
        else:
            df_corr=df.corr()
            df_corr.to_pickle("item_item_corr_train.pkl")
    else:
        if os.path.isfile("model/item_item_corr.pkl"):
```

```
        df_corr=pd.read_pickle("item_item_corr.pkl")
    else:
        df_corr=df.corr()
        df_corr.to_pickle("item_item_corr.pkl")
    return df_corr
```

Computing user-user similarity

To compute user-user similarity, I have written the user_similarity() function, which returns the similarity between two users. It takes three parameters: user 1, user 2; the ratings dataframe; and the value of TRAINED can be either TRUE or FALSE and refers to whether the trained user versus movie table or untrained one should be used. Finally, it computes the Pearson coefficient between users (a value between -1 and 1):

```
def user_similarity(user_1,user_2,ratings_df,TRAINED=False):
    corr_matrix=user_user_pearson_corr(ratings_df,TRAINED)
    return corr_matrix[user_1][user_2]
```

In the preceding function, user_user_pearson_corr() is a function that computes the user-user Pearson correlation:

```
def user_user_pearson_corr(ratings_df,TRAINED):
    if TRAINED:
        if os.path.isfile("model/user_user_corr_train.pkl"):
            df_corr=pd.read_pickle("user_user_corr_train.pkl")
        else:
            df =pd.read_pickle("user_item_table_train.pkl")
            df=df.T
            df_corr=df.corr()
            df_corr.to_pickle("user_user_corr_train.pkl")
    else:
        if os.path.isfile("model/user_user_corr.pkl"):
            df_corr=pd.read_pickle("user_user_corr.pkl")
        else:
            df = pd.read_pickle("user_item_table.pkl")
            df=df.T
            df_corr=df.corr()
            df_corr.to_pickle("user_user_corr.pkl")
    return df_corr
```

Evaluating the recommender system

In this sub-section, we will evaluate the clusters by plotting them to see how the movies are spread across different clusters.

We will then see top k movies and see the user-user similarity and other metrics we have previously discussed. Now let's get started by importing required libraries:

```
import tensorflow as tf
import pandas as pd
import readers
import main
import kmean as km
import numpy as np
```

Next, let's define the data parameters to use for the evaluation:

```
DATA_FILE = "Input/ratings.dat" # Data source for the positive data.
K = 5 #Number of clusters
MAX_ITERS = 1000 #Maximum number of iterations
TRAINED = False # Use TRAINED user vs item matrix
USER_ITEM_TABLE = "user_item_table.pkl"
COMPUTED_CLUSTER_CSV = "clusters.csv"
NO_OF_PCA_COMPONENTS = 2 #number of pca components
SVD_SOLVER = "randomized" #svd solver -e.g. randomized, full etc.
```

Let's see load the ratings dataset that will be used in the invoke call to the `k_mean_clustering()` method:

```
ratings_df = readers.read_file("Input/ratings.dat", sep="::")
clusters,movies = km.k_mean_clustering(ratings_df, K, MAX_ITERS,
TRAINED = False)
cluster_df=pd.DataFrame({'movies':movies,'clusters':clusters})
```

Well done! Now let's see some clusters of simple inputs (movies along with respective clusters):

```
print(cluster_df.head(10))
>>>
clusters  movies
0         0       0
1         4       1
2         4       2
3         3       3
4         4       4
5         2       5
```

```
      6              4         6
      7              3         7
      8              3         8
      9              2         9
print(cluster_df[cluster_df['movies']==1721])
>>>
      clusters   movies
1575           2     1721
print(cluster_df[cluster_df['movies']==647])
>>>
     clusters   movies
627           2      647
```

Let's see how the movies are scattered across clusters:

```
km.showClusters(USER_ITEM_TABLE, COMPUTED_CLUSTER_CSV, NO_OF_PCA_
COMPONENTS, SVD_SOLVER)
>>>
```

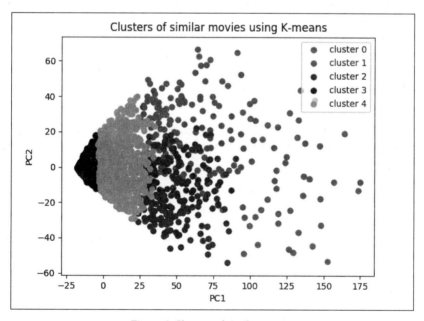

Figure 6: Clusters of similar movies

If we look at the graph, it is clear that the data points are more accurately clustered across clusters 3 and 4. However, clusters 0, 1, and 2 are more scattered and did not cluster well.

Here we did not compute any accuracy metric because train data doesn't have labels. Now let's compute the top k similar movies for a given respective movie name and print them:

```
ratings_df = readers.read_file("Input/ratings.dat", sep="::")
topK = main.top_k_similar_items(9,ratings_df = ratings_df,k = 10,TRAINED = False)
print(topK)
>>>
[1721, 1369, 164, 3081, 732, 348, 647, 2005, 379, 3255]
```

The above result computes Top-K similar movies for the movie `9::Sudden Death (1995)::Action`. Now if you observe the `movies.dat` file, you will see that the following movies are similar to this one:

```
1721::Titanic (1997)::Drama|Romance
1369::I Can't Sleep (J'ai pas sommeil) (1994)::Drama|Thriller
164::Devil in a Blue Dress (1995)::Crime|Film-Noir|Mystery|Thriller
3081::Sleepy Hollow (1999)::Horror|Romance
732::Original Gangstas (1996)::Crime
348::Bullets Over Broadway (1994)::Comedy
647::Courage Under Fire (1996)::Drama|War
2005::Goonies, The (1985)::Adventure|Children's|Fantasy
379::Timecop (1994)::Action|Sci-Fi
3255::League of Their Own, A (1992)::Comedy|Drama
```

Now let's compute the user-user Pearson correlation. When you run this user similarity function, on the first run it will take time to give output but after that, its response is in real time:

```
print(main.user_similarity(1,345,ratings_df))
>>>
0.15045477803357316
Now let's compute the aspect rating given by a user for a movie:
print(main.user_rating(0,1192))
>>>
4.25545645
print(main.user_rating(0,660))
>>>
3.20203304
```

Let's also see the top K movie recommendations for the user:

```
print(main.top_k_movies([768],ratings_df,10))
>>>
{768: [2857, 2570, 607, 109, 1209, 2027, 592, 588, 2761, 479]}
print(main.top_k_movies(1198,ratings_df,10))
>>>
{1198: [2857, 1195, 259, 607, 109, 2027, 592, 857, 295, 479]}
```

So far, we have seen how to develop a simple RE using a movies and rating dataset. However, most recommendation problems assume that we have a consumption/rating dataset formed by a collection of (user, item, rating) tuples. This is the starting point for most variations of collaborative filtering algorithms, and they have proven to yield good results; however, in many applications, we have plenty of item metadata (tags, categories, and genres) that can be used to make better predictions.

This is one of the benefits of using FMs with feature-rich datasets, because there is a natural way in which extra features can be included in the model, and higher-order interactions can be modeled using the dimensionality parameter d (see figure 7 below for more detail).

A few recent types of research show that feature-rich datasets give better predictions: i) Xiangnan He and Tat-Seng Chua, Neural Factorization Machines for Sparse Predictive Analytics. In Proceedings of SIGIR '17, Shinjuku, Tokyo, Japan, August 07-11, 2017. ii) Jun Xiao, Hao Ye, Xiangnan He, Hanwang Zhang, Fei Wu and Tat-Seng Chua (2017) Attentional Factorization Machines: Learning the Weight of Feature Interactions via Attention Networks IJCAI, Melbourne, Australia, August 19-25, 2017.

These papers explain how to make existing data into a feature-rich dataset and how FMs were implemented on the dataset. Therefore, researchers are trying to use FMs to develop more accurate and robust REs. In the next section, we will see some examples of using FMs and some variations.

Factorization machines for recommendation systems

In this section, we will see two examples of developing a more robust recommendation system using FMs. We will start with a brief explanation of FM and their application to the cold-start recommendation problem.

Then we will see a short example of using an FM to developing a real-life recommendation system. After that, we will see an example using an improved version of the FM algorithm called a **Neural Factorization Machine (NFM)**.

Factorization machines

FM-based techniques are at the cutting edge of personalization. They have proven to be extremely powerful with enough expressive capacity to generalize existing models, such as matrix/tensor factorization and polynomial kernel regression. In other words, this type of algorithm is a supervised learning approach that enhances the performance of linear models by incorporating second-order feature interactions that are absent in matrix factorization algorithms.

Existing recommendation algorithms require consumption (product) or rating (movie) dataset in (*user*, *item*, and *rating*) tuples. These types of the dataset are mostly used by variations of Collaborative Filtering (CF) algorithms. CF algorithms have gained wide adoption and have proven to yield good results. However, in many instances, we have plenty of item metadata (tags, categories, and genres) that can be used to make better predictions as well. Unfortunately, CF algorithms do not use these types of metadata.

FMs can make use of these feature-rich (meta) datasets. An FM can consume these extra features to model higher-order interactions specifying the dimensionality parameter d. Most importantly, FMs are also optimized for handling large-scale sparse datasets. Therefore, a second order FM model would suffice because there is not enough information to estimate interactions that are more complex:

	Feature vector **x**																				Target **y**	
x_1	1	0	0	...	1	0	0	0	...	0.3	0.3	0.3	0	...	13	0	0	0	0	...	5	y_1
x_2	1	0	0	...	0	1	0	0	...	0.3	0.3	0.3	0	...	14	1	0	0	0	...	3	y_2
x_3	1	0	0	...	0	0	1	0	...	0.3	0.3	0.3	0	...	16	0	1	0	0	...	1	y_3
x_4	0	1	0	...	0	0	1	0	...	0	0	0.5	0.5	...	5	0	0	0	0	...	4	y_4
x_5	0	1	0	...	0	0	0	1	...	0	0	0.5	0.5	...	8	0	0	1	0	...	5	y_5
x_6	0	0	1	...	1	0	0	0	...	0.5	0	0.5	0	...	9	0	0	0	0	...	1	y_6
x_7	0	0	1	...	0	0	1	0	...	0.5	0	0.5	0	...	12	1	0	0	0	...	5	y_7
	A	B	C	...	TI	NH	SW	ST	...	TI	NH	SW	ST	...		TI	NH	SW	ST	...		
	User				Movie					Other Movie rated					Time	Last Movie rated						

Figure 7: An example training dataset representing a personalization problem with the feature vectors x and the target y. Here rows refer to movies and columns to director, actor and genre info

Let's assume that the dataset of a prediction problem is described by a design matrix $X \in \mathbb{R}^{n \times p}$, as shown in figure 7. In figure 1, the ith row, $x_i \in \mathbb{R}^p$ of X, describes one case, where p is a real-valued variable. On the other hand, y_i is the prediction target of the i_{th} case. Alternatively, we can describe this set as a set S of tuples (**x**, y), where (again) $x \in \mathbb{R}^p$ is a feature vector and y is its corresponding target or label.

In other words, in figure 7, every row represents a feature vector x_i with its corresponding target y_i. For easier interpretation, the features are grouped into indicators for the active user (blue), the active item (red), other movies rated by the same user (orange), the time in months (green), and the last movie rated (brown). Then, the FM algorithm models all the nested interactions (up to order d) between p input variables in **x** using the following factorized interaction parameters:

$$\hat{y}(x) = w_0 + \sum_{i=1}^{n} w_i x_i + \sum_{i=1}^{n} \sum_{j=i+1}^{n} \langle v_i, v_j \rangle x_i x_j$$

In the equation, the v's represent k-dimensional latent vectors associated with each variable (the users and the items), and the bracket operator represents the inner product. This kind of representation with data matrices and feature vectors is common in many machine-learning approaches, for example, in linear regression or support vector machines (SVM).

However, if you are familiar with the Matrix Factorization (MF) models, the preceding equation should look familiar: it contains a global bias as well as user/item-specific biases and includes user-item interactions. Now, if we assume that each x(j) vector is only non-zero at positions u and i, we get the classic MF model:

$$\hat{y}(x) = w_0 + w_i + w_u + \langle v_i, v_j \rangle$$

Nevertheless, MF models for recommendation systems often suffer from the cold-start problem. We will talk about this in the next section.

Cold-start problem and collaborative-filtering approaches

The term cold-start problem sounds funny, but as the name implies, it derives from cars. Suppose you live in Alaska. Due to the cold, your car's engine might not start smoothly, but once it reaches its optimal operating temperature, it will start, run, and operate normally.

In the realm of recommendation engines, the term cold-start simply means a circumstance that is not yet optimal for the engine to provide the best possible results. In e-commerce, there are two distinct categories for a cold start: product cold-start and user cold-start.

Cold-start is a potential problem in computer-based information systems that involve a degree of automated data modeling. Specifically, it concerns the issue that the system cannot draw any inferences on users or items about which it has not yet gathered sufficient information.

The cold-start problem is most prevalent in recommender systems. In the collaborative filtering approach, the recommender system would identify users who share preferences with the active user and propose items that the like-minded users have favored (and the active user has not yet seen). Due to the cold-start problem, this approach would fail to consider items that no one in the community has rated.

The cold-start problem is often reduced by adopting a hybrid approach between content-based matching and collaborative filtering. New items that have not yet received any ratings from users would be assigned a rating automatically, based on the ratings assigned by the community to other similar items. Item similarity would be determined according to the item's content-based characteristics.

Recommendation engines using CF-based approaches recommend each item based on user actions. The more user actions an item has, the easier it is to tell which user would be interested in it and what other items are similar to it. As time progresses, the system will be able to give more and more accurate recommendations. At a certain stage, when new items or users are added to the user-item matrix, this problem occurs:

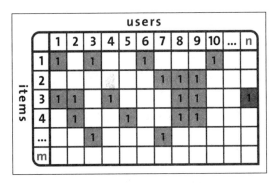

Figure 8: Users versus items matrixes sometimes lead to the cold-start problem

In this case, the RE does not have enough knowledge about this new user or this new item yet. The content-based filtering approach, similar to FM, is a method that can be incorporated to alleviate the cold-start problem.

The main difference between the previous two equations is that FM introduces higher-order interactions in terms of latent vectors that are also affected by categorical or tag data. This means that the models go beyond co-occurrences in order to find stronger relationships between latent representations of each feature.

Problem definition and formulation

Given a sequence of click events performed by a user during a typical session on an e-commerce website, the goal is to predict whether the user is going to buy something or not, and if they are buying, what items they would buy. The task, therefore, could be divided into two sub-goals:

- Is the user going to buy items in this session?
- If yes, what are the items that are going to be bought?

To predict the quantity of an item bought in a session, a robust classifier can help to predict whether a user will buy that item or not. Following the original implementation of FM, the training data should be structured as follows:

	User				Item				Categories				History				Quantity	
x_1	1	0	1	...	0	1	0	...	1	2	3	...	1	1	0	...	3	y_1
x_2	0	0	1	...	1	0	1	...	8	9	6	...	0	1	0	...	7	y_2
x_3	0	1	1	...	1	0	0	...	5	2	7	...	1	1	1	...	9	y_3
...	...	...	...	...	...	...	...	...	...	...	...	...	...	...	...	...	...	...
x_n	1	0	1	...	1	1	1	...	2	4	6	...	0	1	1	...	8	y_n

Figure 9: A user versus item/category/history table can be used to train the recommendation model

To prepare our training set like this, we can use the `get_dummies()` method from pandas to transform all the columns into categorical data, because FM models work with categorical data represented as integers.

We used two functions, `TFFMClassifier` and `TFFMRegressor`, to make a prediction (see `items.py`) and calculate MSE respectively (see `quantity.py` script from the tffm library (under the MIT license)). The tffm is a TensorFlow-based implementation of FM and pandas for pre-processing and structuring data. This TensorFlow-based implementation provides an arbitrary order (>=2) Factorization Machine, which supports:

- Dense and sparse inputs
- Different (gradient-based) optimization methods
- Classification/regression via different loss functions (logistic and mse implemented)
- Logging via TensorBoard

Another good thing is that the inference time is linear with respect to the number of features.

We would like to thank the authors and cite their work as follows: Mikhail Trofimov, Alexander Novikov, TFFM: TensorFlow implementation of an arbitrary order Factorization Machine, GitHub repository, https://github.com/geffy/tffm, 2016.

To use this library, just issue the following command on Terminal:

```
$ sudo pip3 install tffm # For Python3.x
$ sudo pip install tffm # For Python 2.7.x
```

Before we start the implementation, let's take a look at the dataset we will use for this example.

Dataset description

For this example, I will use the RecSys 2015 challenge dataset to illustrate how to fit an FM model to get a personalized recommendation. The data contains click and purchase events for an e-commerce site, with additional item category data. The dataset's size is about 275 MB, and it can be downloaded from https://s3-eu-west-1.amazonaws.com/yc-rdata/yoochoose-data.7z.

There are three files and a readme file; however, we will be using *youchoose-buys.dat* (buy events) and *youchoose-clicks.dat* (click events):

- `youchoose-clicks.dat`: Each record/line in the file has the following fields:
    - **Session ID**: One or many clicks in one session
    - **Timestamp**: The time when the click occurred
    - **Item ID**: The unique identifier of the item
    - **Category**: The category of the item

- `youchoose-buys.dat`: Each record/line in the file has the following fields:
    - **Session ID**: session ID: One or many buying events in a session
    - **Timestamp**: The time when the buy occurred
    - **Item ID**: A unique identifier of items
    - **Price**: The price of the item
    - **Quantity**: How many of this item were bought

The session IDs in youchoose-buys.dat also exist in the youchoose-clicks.dat file. That means the records with the same session ID together form the sequence of click events of a certain user during the session.

The session could be short (a few minutes) or very long (a few hours), and it could have one click or hundreds of clicks. It all depends on the activity of the user.

Workflow of the implementation

Let's develop a recommendation model that predicts and generates a `solution.data` file. Here is the short workflow:

1. Download and load the **RecSys 2015** challenge dataset and copy in the 'data' folder in the code repository of this chapter
2. Buy data contains Session ID, Timestamp, Item ID, Category, and Quantity. In addition, youchoose-clicks.dat contains Session ID, Timestamp, Item ID, and Category. We will not be using Timestamp here. We remove the time stamps, one-hot encode all the columns, and merge the buy and click datasets to make the dataset feature-rich. After pre-processing, the data looks similar to that shown in figure 11.

3. For simplification, we consider only the top 10,000 sessions and split the dataset into training (75%) and testing (25%) sets.
4. We then split the test into normal (keeping historical data) and cold-start (by removing historical data) to differentiate the model for the users/items with history or without history.
5. We then use the **tffm to train our FM model**, which is an implementation of FM in TensorFlow and train the model using the training data.
6. Finally, we evaluate the model on both the normal and cold-start datasets.

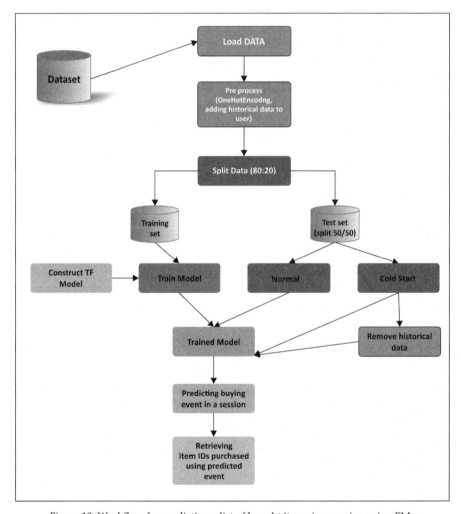

Figure 10: Workflow for predicting a list of bought items in a session using FMs

Preprocessing

If we want to make the full use of the categories and the expanded historical data, we need to load and convert the data into the right format. Thus, some preprocessing is necessary before getting the training set prepared. Let's start by loading the packages and modules:

```
import tensorflow as tf
import pandas as pd
from collections import Counter
from tffm import TFFMClassifier
from sklearn.metrics import mean_squared_error
from sklearn.model_selection import train_test_split
import numpy as np
from sklearn.metrics import accuracy_score
import os
```

I am assuming that you have already downloaded the dataset from the previously mentioned link. Now let's load the dataset:

```
buys = open('data/yoochoose-buys.dat', 'r')
clicks = open('data/yoochoose-clicks.dat', 'r')
```

Now create the pandas dataframe for the click and buys datasets:

```
initial_buys_df = pd.read_csv(buys, names=['Session ID', 'Timestamp',
'Item ID', 'Category', 'Quantity'], dtype={'Session ID': 'float32',
'Timestamp': 'str', 'Item ID': 'float32','Category': 'str'})
initial_buys_df.set_index('Session ID', inplace=True)
initial_clicks_df = pd.read_csv(clicks, names=['Session ID',
'Timestamp', 'Item ID', 'Category'],dtype={'Category': 'str'})
initial_clicks_df.set_index('Session ID', inplace=True)
```

We do not need to use the timestamps in this example, so let's drop them from the dataframe:

```
initial_buys_df = initial_buys_df.drop('Timestamp', 1)
    print(initial_buys_df.head())    # first five records
    print(initial_buys_df.shape)     # shape of the dataframe
>>>
```

	Session ID	Timestamp	Item ID	Category	Quantity
0	420374.0	2014-04-06T18:44:58.314Z	214537888.0	12462	1
1	420374.0	2014-04-06T18:44:58.325Z	214537856.0	10471	1
2	281626.0	2014-04-06T09:40:13.032Z	214535648.0	1883	1
3	420368.0	2014-04-04T06:13:28.848Z	214530576.0	6073	1
4	420368.0	2014-04-04T06:13:28.858Z	214835024.0	2617	1

(1150753, 5)

```
initial_clicks_df = initial_clicks_df.drop('Timestamp', 1)
print(initial_clicks_df.head())
print(initial_clicks_df.shape)
>>>
```

	Session ID	Timestamp	Item ID	Category
0	1	2014-04-07T10:51:09.277Z	214536502	0
1	1	2014-04-07T10:54:09.868Z	214536500	0
2	1	2014-04-07T10:54:46.998Z	214536506	0
3	1	2014-04-07T10:57:00.306Z	214577561	0
4	2	2014-04-07T13:56:37.614Z	214662742	0

(33003944, 4)

Since we won't use timestamps in this example, remove the Timestamp column from dataframe(df):

```
initial_buys_df = initial_buys_df.drop('Timestamp', 1)
print(initial_buys_df.head(n=5))
print(initial_buys_df.shape)
>>>
```

	Item ID	Category	Quantity
Session ID			
420374.0	214537888.0	12462	1
420374.0	214537856.0	10471	1
281626.0	214535648.0	1883	1
420368.0	214530576.0	6073	1
420368.0	214835024.0	2617	1

(1150753, 3)

```
initial_clicks_df = initial_clicks_df.drop('Timestamp', 1)
print(initial_clicks_df.head(n=5))
print(initial_clicks_df.shape)
>>>
```

	Item ID	Category
Session ID		
1	214536502	0
1	214536500	0
1	214536506	0
1	214577561	0
2	214662742	0

(33003944, 2)

Let's take the top 10,000 buying users:

```
x = Counter(initial_buys_df.index).most_common(10000)
top_k = dict(x).keys()
initial_buys_df = initial_buys_df[initial_buys_df.index.isin(top_k)]
    print(initial_buys_df.head())
    print(initial_buys_df.shape)
>>>
```

	Item ID	Category	Quantity
Session ID			
420471.0	214717888.0	2092	1
420471.0	214821024.0	1570	1
420471.0	214829280.0	837	1
420471.0	214819552.0	418	1
420471.0	214746384.0	784	1

(106956, 3)

```
initial_clicks_df = initial_clicks_df[initial_clicks_df.index.isin(top_k)]
    print(initial_clicks_df.head())
    print(initial_clicks_df.shape)
>>>
```

	Item ID	Category
Session ID		
932	214826906	0
932	214826906	0
932	214826906	0
932	214826955	0
932	214826955	0

(209024, 2)

Chapter 9

Now let's create a copy of the index, since we will also apply one-hot encoding to it:

```
initial_buys_df['_Session ID'] = initial_buys_df.index
print(initial_buys_df.head())
print(initial_buys_df.shape)
>>>
```

	Item ID	Category	Quantity	_Session ID
Session ID				
420471.0	214717888.0	2092	1	420471.0
420471.0	214821024.0	1570	1	420471.0
420471.0	214829280.0	837	1	420471.0
420471.0	214819552.0	418	1	420471.0
420471.0	214746384.0	784	1	420471.0

(106956, 4)

As we mentioned earlier, we can introduce historical engagement data into our FM model. We will use some `group_by` magic to generate a history profile of the entire user's engagement. At first, we one-hot encode all columns for clicks and buys:

```
transformed_buys = pd.get_dummies(initial_buys_df)
    print(transformed_buys.shape)
>>>
(106956, 356)
transformed_clicks = pd.get_dummies(initial_clicks_df)
print(transformed_clicks.shape)
>>>
(209024, 56)
```

[405]

Now it's time to aggregate the historical data for items and categories:

```
filtered_buys = transformed_buys.filter(regex="Item.*|Category.*")
    print(filtered_buys.shape)
>>>
(106956, 354)
filtered_clicks = transformed_clicks.filter(regex="Item.*|Category.*")
    print(filtered_clicks.shape)
>>>
(209024, 56)
historical_buy_data = filtered_buys.groupby(filtered_buys.index).sum()
    print(historical_buy_data.shape)
>>>
(10000, 354)
historical_buy_data = historical_buy_data.rename(columns=lambda column_name: 'buy history:' + column_name)
    print(historical_buy_data.shape)
    >>>
    (10000, 354)
historical_click_data = filtered_clicks.groupby(filtered_clicks.index).sum()
    print(historical_click_data.shape)
    >>>
(10000, 56)
historical_click_data = historical_click_data.rename(columns=lambda column_name: 'click history:' + column_name)
```

Then we merge the historical data of every `user_id`:

```
merged1 = pd.merge(transformed_buys, historical_buy_data, left_index=True, right_index=True)
print(merged1.shape)
merged2 = pd.merge(merged1, historical_click_data, left_index=True, right_index=True)
print(merged2.shape)
>>>
(106956, 710)
(106956, 766)
```

Then we take the quantity as the target and convert it into binary:

```
y = np.array(merged2['Quantity'].as_matrix())
```

Now let's convert *y* into binary [if buying happens, 1; else 0]:

```
for i in range(y.shape[0]):
    if y[i]!=0:
        y[i]=1
    else:
        y[i]=0
print(y.shape)
print(y[0:100])
print(y, y.shape[0])
print(y[0])
print(y[0:100])
print(y, y.shape)
>>>
```

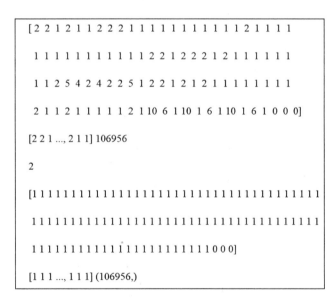

Training the FM model

Since we have prepared the dataset, the next task is to create the MF model. First, though, let's split the data into training and testing sets:

```
X_tr, X_te, y_tr, y_te = train_test_split(merged2, y, test_size=0.25)
```

Then we split the testing data into half, one for normal testing and one for cold-start testing:

```
X_te, X_te_cs, y_te, y_te_cs = train_test_split(X_te, y_te, test_size=0.5)
```

Now let's include the session ID and item ID in the dataframe:

```
test_x = pd.DataFrame(X_te, columns = ['Item ID'])
print(test_x.head())
>>>
```

Session ID	Item ID
2614096	214829888.0
6388687	214845456.0
517818	214837488.0
6498748	214691520.0
2541201	214845104.0

(10696, 1)

```
test_x_cs = pd.DataFrame(X_te_cs, columns = ['Item ID'])
print(test_x_cs.head())
>>>
```

Session ID	Item ID
17929	214827008.0
161673	214826928.0
10914216	214854848.0
9075227	214678368.0
8356289	214716672.0

Then we have to remove the unwanted features from the datasets:

```
X_tr.drop(['Item ID', '_Session ID', 'click history:Item ID', 'buy history:Item ID', 'Quantity'], 1, inplace=True)
X_te.drop(['Item ID', '_Session ID', 'click history:Item ID', 'buy history:Item ID', 'Quantity'], 1, inplace=True)
X_te_cs.drop(['Item ID', '_Session ID', 'click history:Item ID', 'buy history:Item ID', 'Quantity'], 1, inplace=True)
```

Then we need to convert the DataFrames into arrays:

```
ax_tr = np.array(X_tr)
ax_te = np.array(X_te)
ax_te_cs = np.array(X_te_cs)
```

Now that the pandas DataFrames have been converted into NumPy arrays, we need to do some null treatment. We simply replace NaN with zeros:

```
ax_tr = np.nan_to_num(ax_tr)
ax_te = np.nan_to_num(ax_te)
ax_te_cs = np.nan_to_num(ax_te_cs)
```

Then we instantiate the TF model with optimized hyperparameters for classification:

```
model = TFFMClassifier(
        order=2,
        rank=7,
        optimizer=tf.train.AdamOptimizer(learning_rate=0.001),
        n_epochs=100,
        batch_size=1024,
        init_std=0.001,
        reg=0.01,
        input_type='dense',
        log_dir = ' logs/',
        verbose=1,
        seed=12345
    )
```

Before we start training the model, we have to prepare the data for the cold-start:

```
cold_start = pd.DataFrame(ax_te_cs, columns=X_tr.columns)
```

As was mentioned earlier, we are also interested in seeing what happens if we only have access to categories and no historical click/purchase data. Let's delete historical click and purchasing data for the `cold_start` testing set:

```
for column in cold_start.columns:
    if ('buy' in column or 'click' in column) and ('Category' not in column):
        cold_start[column] = 0
```

Now let's train the model:

```
model.fit(ax_tr, y_tr, show_progress=True)
```

One of the most important tasks is predicting the buying events in the sessions:

```
predictions = model.predict(ax_te)
print('accuracy: {}'.format(accuracy_score(y_te, predictions)))
print("predictions:",predictions[:10])
print("actual value:",y_te[:10])
>>>
accuracy: 1.0
predictions: [0 0 1 0 0 1 0 1 1 0]
actual value: [0 0 1 0 0 1 0 1 1 0]

cold_start_predictions = model.predict(ax_te_cs)
print('Cold-start accuracy: {}'.format(accuracy_score(y_te_cs, cold_start_predictions)))
print("cold start predictions:",cold_start_predictions[:10])
print("actual value:",y_te_cs[:10])
>>>
Cold-start accuracy: 1.0
cold start predictions: [1 1 1 1 1 0 1 0 0 1]
actual value: [1 1 1 1 1 0 1 0 0 1]
```

Then let's add the predicted values to the testing data:

```
test_x["Predicted"] = predictions
test_x_cs["Predicted"] = cold_start_predictions
```

Now it's time to find all the buy events for each `session_id` in the testing data and retrieve the respective item IDs:

```
sess = list(set(test_x.index))
fout = open("solution.dat", "w")
print("writing the results into .dat file....")
for i in sess:
    if test_x.loc[i]["Predicted"].any() != 0:
```

```
            fout.write(str(i)+";"+','.join(s for s in str(test_x.loc[i]
["Item ID"].tolist()).strip('[]').split(','))+'\n')
    fout.close()
>>>
writing the results into .dat file....
```

Then we do the same for the cold-start testing data:

```
sess_cs = list(set(test_x_cs.index))
fout = open("solution_cs.dat", "w")
print("writing the cold start results into .dat file....")
for i in sess_cs:
    if test_x_cs.loc[i]["Predicted"].any() != 0:
        fout.write(str(i)+";"+','.join(s for s in str(test_x_cs.loc[i]
["Item ID"].tolist()).strip('[]').split(','))+'\n')
    fout.close()
>>>
writing the cold start results into .dat file....
print("completed..!!")
>>>
completed!!
```

Finally, we destroy the model to free the memory:

```
model.destroy()
```

Additionally, we can see the sample contents of the file:

```
11009963;214853767
10846132;214854343, 214851590
8486841;214848315
10256314;214854125
8912828;214853085
11304897;214567215
9928686;214854300, 214819577
10125303;214567215, 214853852
10223609;214854358
```

The experimental results are good, considering that we have used a relatively small dataset to fit our model. As expected, it is easier to generate predictions if we have access to all of the information set with item purchases and clicks, but we still get a decent predictor for cold-start recommendations using only aggregated category data.

Now that we have seen that the customer will buy in each session, it would be great to compute the mean squared error for both testing sets. The `TFFMRegressor` method can help us with this. For this, use the `quantity.py` script.

First, the question is what happens if we only have access to categories and no historical click/purchase data. Let's delete historical click and purchasing data for the cold_start testing set:

```
for column in cold_start.columns:
    if ('buy' in column or 'click' in column) and ('Category' not in column):
        cold_start[column] = 0
```

Let's create the MF model. You can play around with the hyperparameters:

```
reg_model = TFFMRegressor(
    order=2,
    rank=7,
    optimizer=tf.train.AdamOptimizer(learning_rate=0.1),
    n_epochs=100,
    batch_size=-1,
    init_std=0.001,
    input_type='dense',
    log_dir = ' logs/',
    verbose=1
    )
```

In the preceding code block, feel free to put in your own logging path. Now it is time to train the regression model using the normal and the cold-start training sets:

```
reg_model.fit(X_tr, y_tr, show_progress=True)
```

Then we compute the mean squared error for both testing sets:

```
predictions = reg_model.predict(X_te)
print('MSE: {}'.format(mean_squared_error(y_te, predictions)))
print("predictions:",predictions[:10])
print("actual value:",y_te[:10])
cold_start_predictions = reg_model.predict(X_te_cs)
print('Cold-start MSE: {}'.format(mean_squared_error(y_te_cs, cold_start_predictions)))
print("cold start predictions:",cold_start_predictions[:10])
print("actual value:",y_te_cs[:10])
print("Regression completed..!!")
>>>
MSE: 0.4897467853668941
predictions: [ 1.35086     0.03489107  1.0565269  -0.17359206
 -0.01603088  0.03424695
```

```
    2.29936886   1.65422797   0.01069662   0.02166392]
actual value: [1 0 1 0 0 0 1 1 0 0]
Cold-start MSE: 0.5663486183636738
cold start predictions: [-0.0112379    1.21811676   1.29267406
 0.02357371  -0.39662406   1.06616664
 -0.10646269   0.00861482   1.22619736   0.09728943]
actual value: [0 1 1 0 1 1 0 0 1 0]
Regression completed..!!
```

Finally, we destroy the model to free the memory:

```
reg_model.destroy()
```

So, dropping category columns from the training dataset makes the MSE even smaller, but doing so means that we cannot tackle the cold-start recommendation problem. The experimental results are good given the condition that we have used a relatively small dataset.

As expected, it is easier to generate predictions if we have access to the full information setting with item purchases and clicks, but we still get a decent predictor for cold-start recommendations using only aggregated category data.

Improved factorization machines

Many predictive tasks for web applications need to model categorical variables, such as user IDs, and demographic information, such as genders and occupations. To apply standard ML techniques, these categorical predictors need to be converted to a set of binary features via one-hot encoding (or any other technique). This makes the resultant feature vector highly sparse. To learn effectively from such sparse data, it is important to consider the interactions between features.

In the previous section, we saw that FM could be applied to model second-order feature interactions effectively. However, FM models feature interactions in a linear way, which is insufficient if you want to capture the non-linear and inherently complex structure of real-world data.

Xiangnan He and Jun Xiao et al. have proposed several research initiatives, such as Neural Factorization Machine (NFM) and Attentional Factorization Machine (AFM), in an attempt to overcome this limitation.

For more information, see the following papers:

- Xiangnan He and Tat-Seng Chua, Neural Factorization Machines for Sparse Predictive Analytics. In Proceedings of SIGIR '17, Shinjuku, Tokyo, Japan, August 07-11, 2017.
- Jun Xiao, Hao Ye, Xiangnan He, Hanwang Zhang, Fei Wu and Tat-Seng Chua (2017). Attentional Factorization Machines: Learning the Weight of Feature Interactions via Attention Networks IJCAI, Melbourne, Australia, August 19-25, 2017.

NFMs can be used to make predictions under sparse settings by seamlessly combining the linearity of FM in modeling second-order feature interactions and the non-linearity of the neural network in modeling higher-order feature interactions.

On the other hand, AFMs can be used to model the data even if all the feature interactions have the same weight, as not all feature interactions are equally useful and predictive.

In the next section, we will see an example of using an NFM for the movie recommendation.

Neural factorization machines

Using the original FM algorithm, performance can be hindered by the way it models all feature interactions with the same weight, as not all feature interactions are equally useful and predictive. For example, the interactions with useless features may even introduce noise and adversely degrade the performance.

Recently, Xiangnan H. et al. proposed an improved version of FM algorithm called Neural Factorization Machines (NFMs). NFMs seamlessly combines the linearity of FMs in modeling second-order feature interactions and the non-linearity of the neural network in modeling higher-order feature interactions. Conceptually, NFMs are more expressive than FMs since FMs can be seen as a special case of NFMs without hidden layers.

Dataset description

We used the MovieLens data for personalized tag recommendation. It contains 668,953 tag applications of users on movies. Each tag application (user ID, movie ID, and tag) is converted into a feature vector using one-hot encoding. This leaves 90,445 binary features, called the `ml-tag` dataset.

I have used a Perl script to convert it from .dat to.libfm format. The conversion procedure is described at http://www.libfm.org/libfm-1.42.manual.pdf (section 2.2.1). The converted dataset has files for the training, validation, and the testing, as follows:

- `ml-tag.train.libfm`
- `ml-tag.validation.libfm`
- `ml-tag.test.libfm`

For more information about this file format, see http://www.libfm.org/.

Using NFM for the movie recommendation

We have reused and extended the NFM implementation using TensorFlow from this GitHub, https://github.com/hexiangnan/neural_factorization_machine. This is a deep version of an FM, which is more expressive compared to regular FMs. The repository has three files, namely `NeuralFM.py`, `FM.py`, and `LoadData.py`:

- `FM.py` is used to train the dataset. This is the original implementation of the FMs.
- `NeuralFM.py` is used to train the dataset. This is the original implementation of NFMs but with some improvements and extension.
- `LoadData.py` is used to preprocess and load the dataset in libfm format.

Model training

First, we train the FM model with the following command. The command also includes the parameters needed to perform the training:

```
$ python3 FM.py --dataset ml-tag --epoch 20 --pretrain -1 --batch_size
4096 --lr 0.01 --keep 0.7
>>>
FM: dataset=ml-tag, factors=16, #epoch=20, batch=4096, lr=0.0100,
lambda=0.0e+00, keep=0.70, optimizer=AdagradOptimizer, batch_norm=1
#params: 1537566
Init:       train=1.0000, validation=1.0000 [5.7 s]
Epoch 1  [13.9 s]    train=0.5413, validation=0.6005 [7.8 s]
Epoch 2  [14.2 s]    train=0.4927, validation=0.5779 [8.3 s]
...
Epoch 19 [15.4 s]    train=0.3272, validation=0.5429 [8.1 s]
Epoch 20 [16.6 s]    train=0.3242, validation=0.5425 [7.8 s]
```

Once the training is finished, the trained model will be saved in the `pretrain` folder in your home directory:

>>>

Save model to file as pretrain.

Additionally, I have tried to make training and the validation error visible for the validation and the training loss using the following code:

```
# Plot loss over time
plt.plot(epoch_list, train_err_list, 'r--', label='FM training
loss per epoch', linewidth=4)
plt.title('FM training loss per epoch')
plt.xlabel('Epoch')
plt.ylabel('Training loss')
plt.legend(loc='upper right')
plt.show()

# Plot accuracy over time
plt.plot(epoch_list, valid_err_list, 'r--', label='FM validation
loss per epoch', linewidth=4)
plt.title('FM validation loss per epoch')
plt.xlabel('Epoch')
plt.ylabel('Validation loss')
plt.legend(loc='upper left')
plt.show()
```

The preceding code produces graphs showing the training versus validation loss per iteration in the FM model:

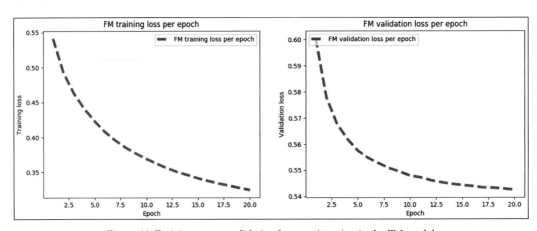

Figure 11: Training versus validation loss per iteration in the FM model

If you look at the preceding output logs, the best training (that is for both validation and training) occurs at the 20th and last iteration. However, you could do more iterations to improve the training, which means a low RMSE value in the evaluation step:

```
Best Iter(validation) = 20    train = 0.3242, valid = 0.5425 [490.9 s]
```

Now let's train the NFM model using the following command (but play with the parameters too):

```
$ python3 NeuralFM.py --dataset ml-tag --hidden_factor 64 --layers
[64] --keep_prob [0.8,0.5] --loss_type square_loss --activation relu
--pretrain 0 --optimizer AdagradOptimizer --lr 0.01 --batch_norm 1
--verbose 1 --early_stop 1 --epoch 20
>>>
Neural FM: dataset=ml-tag, hidden_factor=64, dropout_keep=[0.8,0.5],
layers=[64], loss_type=square_loss, pretrain=0, #epoch=20, batch=128,
lr=0.0100, lambda=0.0000, optimizer=AdagradOptimizer, batch_norm=1,
activation=relu, early_stop=1
#params: 5883150
Init:     train=0.9911, validation=0.9916, test=0.9920 [25.8 s]
Epoch 1 [60.0 s]    train=0.6297, validation=0.6739, test=0.6721 [28.7
s]
Epoch 2 [60.4 s]    train=0.5646, validation=0.6390, test=0.6373 [28.5
s]
...
Epoch 19 [53.4 s]    train=0.3504, validation=0.5607, test=0.5587
[25.7 s]
Epoch 20 [55.1 s]    train=0.3432, validation=0.5577, test=0.5556
[27.5 s]
```

Additionally, I have tried to make the training and the validation error visible for the validation and the training loss using the following code:

```
    # Plot test accuracy over time
    plt.plot(epoch_list, test_err_list, 'r--', label='NFM test loss
per epoch', linewidth=4)
    plt.title('NFM test loss per epoch')
    plt.xlabel('Epoch')
    plt.ylabel('Test loss')
    plt.legend(loc='upper left')
    plt.show()
```

The preceding code produces training versus validation loss per iteration in the NFM model:

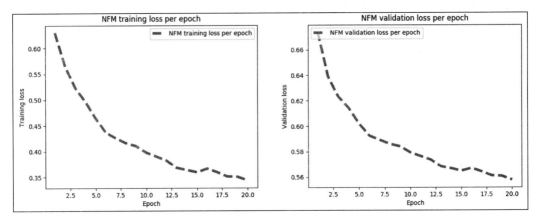

Figure 12: training vs validation loss per iteration in NFM model

For the NFM model, the best training (for both validation and training) occurs at the 20th and last iteration. However, you could do more iterations to improve the training, which means a low RMSE value in the evaluation step:

```
Best Iter (validation) = 20    train = 0.3432, valid = 0.5577, test = 0.5556 [1702.5 s]
```

Model evaluation

Now, to evaluate the original FM model, execute the following command:

```
$ python3 FM.py --dataset ml-tag --epoch 20 --batch_size 4096 --lr 0.01 --keep 0.7 --process evaluate
Test RMSE: 0.5427
```

 For an Attentional Factorization Machines implementation on TensorFlow, interested readers can refer to the GitHub repository at https://github.com/hexiangnan/attentional_factorization_machine. However, note that some codes might not work. I updated them to be TensorFlow v1.6 compatible. Therefore, I would highly recommend using the code provided with this book.

To evaluate the NFM model, just add the following line to the `main()` method in the `NeuralFM.py` script as follows:

```
# Model evaluation
print("RMSE: ")
print(model.evaluate(data.Test_data)) #evaluate on test set
>>>
RMSE: 0.5578330373003925
```

Therefore, the RMSE is almost the same as the FM model. Now let's see how the test errors went per iteration:

```
# Plot test accuracy over time
plt.plot(epoch_list, test_err_list, 'r--', label='NFM test loss per epoch', linewidth=4)
plt.title('NFM test loss per epoch')
plt.xlabel('Epoch')
plt.ylabel('Test loss')
plt.legend(loc='upper left')
plt.show()
```

The preceding code plots the test loss per iteration in NFM model:

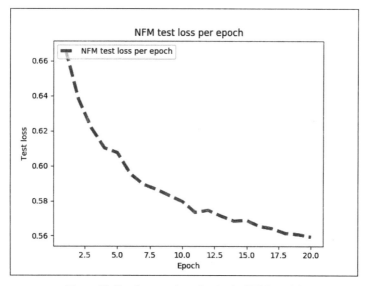

Figure 13: Test loss per iteration in the NFM model

Summary

In this chapter, we have discussed how to develop scalable recommendation systems with TensorFlow. We have seen some of the theoretical backgrounds of recommendation systems and using a collaborative filtering approach in developing recommendation systems. Later in the chapter, we saw how to use SVD, and K-means, to develop a movie recommendation system.

Finally, we saw how to use FMs and a variation called NFM to develop more accurate recommendation systems that can handle large-scale sparse matrixes. We have seen that the best way to handle the cold-start problem is to use a collaborative filtering approach with FMs.

The next chapter is about designing an ML system driven by criticisms and rewards. We will see how to apply RL algorithms to make a predictive model for real-life datasets.

10
Reinforcement Learning

Reinforcement learning (**RL**) is an area of machine learning that studies the science of decision-making processes, in particular trying to understand what the best way is to make decisions in a given context. The learning paradigm of RL algorithms is different from most common methodologies, such as supervised or unsupervised learning.

In RL, an agent is programmed as if he were a human being who must learn through a trial and error mechanism in order to find the best strategy to achieve the best result in terms of long-term reward.

RL has achieved incredible results within games (digital and table) and automated robot control, so it is still widely studied. In the last decade, it has been decided to add a key component to RL: neural networks.

This integration of RL and **deep neural networks** (**DNNs**), called **deep reinforcement learning**, has enabled Google DeepMind researchers to achieve amazing results in previously unexplored areas. In particular, in 2013, the Deep Q-Learning algorithm achieved the performance of experienced human players in the Atari games domain by taking the pixels that represented the game screen as input, placing the agent in the same situation as a human being playing a game.

Another extremely important achievement came in October 2015 when the same research lab, using the same family of algorithms, beat the European Go champion (Go is a Chinese game of great complexity), and finally beat the world champion in March 2016.

Reinforcement Learning

The chapter covers the following topics:

- The RL problem
- Open AI gym
- The Q-Learning algorithm
- Deep Q-Learning

The RL problem

RL differs greatly from supervised learning. In *supervised learning*, each example is a pair consisting of an input object (typically a vector) and a desired output value (also called the supervisory signal). The supervised learning algorithm analyzes the training data and produces an inferred function, which can be used to map new examples.

RL does not provide an association between incoming data and the desired output values, so the learning structure is completely different. The main concept of RL is the presence of two components that interact with one another: an *agent* and an *environment*.

An RL agent learns to make decisions within an unfamiliar environment by performing a series of actions and obtaining the numerical rewards associated with them. By accumulating experience through a *trial and error process*, the agent learns which actions are the best to perform depending on the state it is in, defined by the environment and the set of previously performed actions. The agent has the ability to figure out what the most successful moves are by simply assessing the reward it has earned and adjusting its policy, in order to get the maximum cumulative reward over time.

The RL model is made up of the following:

- A *set of states* $(s_0, s_1, s_2, \ldots, s_n) \in S$, defined by the interaction between the environment and the agent
- A *set of possible actions* $((a_0, a_1, a_2, \ldots a_m) \in A$, suitably selected by the agent according to the input state
- A *reward*, r, associated with each interaction between the environment and the agent
- A *policy* mapping each state into an output action
- A *set of functions* called state-value functions and action-value functions that determine the value of the state of the agent at a given time and the value that the agent performs a specific action on a given moment.

Chapter 10

An RL agent interacts with the environment at a certain time t. At each t, the agent receives a state $S_t \in S$ and a reward r_t as input. Accordingly, the agent determines the action $a_t \in A(s_t)$ to be performed, where $A(s_t)$ represents the set of possible actions in a given state.

The latter is received by the environment, which processes a new S_{t+1} state and a new reward signal, r_{t+1}, corresponding to the next agent input at time $t + 1$. This recursive process is the learning algorithm of the RL agent. The agent's goal is to earn as much as possible in terms of the final cumulative reward. The purpose can be achieved by using different methodologies.

During training, the agent is able to learn appropriate strategies that allow it to gain a more immediate reward or gain a greater long-term reward at the expense of immediate rewards.

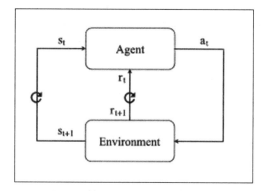

Figure 1: RL model

OpenAI Gym

OpenAI Gym is an open source Python framework developed by OpenAI, a non-profit AI research company, as a toolkit for developing and evaluating RL algorithms. It gives us a set of test problems, known as environments, that we can write RL algorithms to solve. This enables us to dedicate more of our time to implementing and improving the learning algorithm instead of spending a lot of time simulating the environment. In addition, it provides a medium for people to compare and review the algorithms of others.

OpenAI environments

OpenAI Gym has a collection of environments. At the time of writing this book, the following environments are available:

- **Classic control and toy text**: Small-scale tasks from the RL literature.
- **Algorithmic**: Performs computations such as adding multi-digit numbers and reversing sequences. Most of these tasks require memory, and their difficulty can be changed by varying the sequence length.
- **Atari**: Classic Atari games, with screen images or RAM as input, using the Arcade Learning Environment.
- **Board games**: Currently, we have included the game of Go on 9x9 and 19x19 boards, and the Pachi engine [13] serves as an opponent.
- **2D and 3D robots**: Allows you to control a robot in simulation. These tasks use the MuJoCo physics engine, which was designed for fast and accurate robot simulation. A few of the tasks are adapted from RLLab.

The env class

OpenAI Gym allows the use of the `env` class, which encapsulates the environment and any internal dynamics. This class has different methods and attributes that enable you to implement to create a new environment. The most important methods are named `reset`, `step`, and `render`:

- The `reset` method has the task of resetting the environment by initializing it to the initial state. Within the reset method, the definitions of the elements that make up the environment (in this case, the definition of the mechanical arm, the object to be grasped, and its support) must be contained.
- The `step` method is used to advance the environment temporally. It requires the action to be entered and returns the new observation to the agent. Within the method, movement dynamics management, status and reward calculation, and episode completion controls must be defined.
- The last method is `render`, which is used to visualize the current state.

Using the `env` class proposed by the framework as the basis for new environments, it adopts the common interface provided by the toolkit.

This way, built environments can be integrated into the library of the toolkit, and their dynamics can be learned from algorithms that have been made by the users of the OpenAI Gym community.

Installing and running OpenAI Gym

For a more detailed explanation of how to use and run OpenAI Gym, please refer to the official documentation page at (https://gym.openai.com/docs/). A minimal installation of OpenAI Gym can be achieved with the following command:

```
git clone https://github.com/openai/gym
cd gym
pip install -e
```

After OpenAI Gym has been installed, you can instantiate and run an environment in your Python code:

```
import gym
env = gym.make('CartPole-v0')


obs = env.reset()

for step_idx in range(500):
  env.render()
  obs, reward, done, _ = env.step(env.action_space.sample())
```

This code snippet will first import the `gym` library. Then it creates an instance of the Cart-Pole (https://gym.openai.com/envs/CartPole-v0/) environment, which is a classical problem in RL. The Cart-Pole environment simulates an inverted pendulum mounted on a cart. The pendulum is initially vertical, and your goal is to maintain its vertical balance. The only way to control the pendulum is to choose a horizontal direction for the cart to move (either to left or right).

The preceding code runs the environment for 500 time steps, and it chooses a random action to perform at each step. As a result, you see in the video below that the pole is not kept stable for long. The reward is measured by the number of time steps elapsed before the pole becomes more than 15 degrees away from the vertical. The longer you remain within this range, the higher your total reward.

The Q-Learning algorithm

Solving an RL problem requires an estimate, during the learning process, of an evaluation function. This function must be able to assess, through the sum of the rewards, the success of a policy.

The basic idea of Q-Learning is that the algorithm learns the optimal evaluation function for the entire space of states and actions ($S \times A$). This so-called Q-function provides a match in the form $Q: S \times A \rightarrow R$, where R is the expected value of the future rewards of an action $a \in A$ executed in the state, $s \in S$. Once the agent has learned the optimal function, Q, it will be able to recognize what action will lead to the highest future reward in a certain state.

One of the most commonly used examples of implementing the Q-Learning algorithm involves the use of a table. Each cell of the table is a value $Q(s; a) = R$ and it is initialized to 0. The action $a \in A$, performed by the agent, is chosen using a policy which is epsilon-greedy with respect to Q.

The basic idea of the Q-Learning algorithm is the training rule, which updates a table element $Q(s; a)$.

The algorithm follows these basic steps:

1. Initialize $Q(s; a)$ arbitrarily.
2. Repeat the following (for each episode):
    1. Initialize s.
    2. Repeat (for each step of episode):
    3. Choose an action $a \in A$ from $s \in S$ using policy derived from Q.
    4. Take an action a, observe r, s':

    $$Q(s;a) \leftarrow Q(s;a) + a \cdot (r + \gamma \cdot \max Q(s';a) - Q(s;a))$$

    s': s <- s'

    5. Continue until s is terminal.

We have depicted the algorithm in the following diagram:

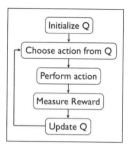

Figure 2: Q-Learning algorithm

Let's summarize the parameters used in the Q-value update process:

- α is the learning rate, which is set between 0 and 1. Setting it to 0 means that the Q-values are never updated, and hence nothing is learned. Setting a high value such as 0.9 means that learning can occur quickly.
- γ is the discount factor, which is set between 0 and 1. This models the fact that future rewards are worth less than immediate rewards. Mathematically, the discount factor needs to be set less than 1 for the algorithm to converge.
- $max\ Q(s';a)$ is the maximum reward that is attainable in the state following the current one, that is, the reward for taking the optimal action thereafter.

The FrozenLake environment

The agent controls the movement of a character in a 4×4 grid world. Some tiles of the grid are walkable, and others lead to the agent falling into the water. Additionally, the movement direction of the agent is uncertain and only partially depends on the chosen direction. The agent is rewarded for finding a walkable path to a goal tile:

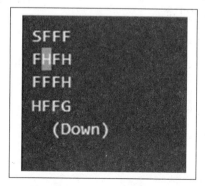

Figure 3: A representation of the Frozen-Lake v0 grid word

Reinforcement Learning

The surface shown above is described using a grid such as the following:

```
SFFF    (S: starting point, safe)
FHFH    (F: frozensurface, safe)
FFFH    (H: hole, fall to yourdoom)
HFFG    (G: goal, where the frisbee islocated)
```

The episode ends when we reach the goal or fall in a hole. We receive a reward of 1 if we reach the goal, and 0 otherwise.

Q-Learning for the FrozenLake problem

Neural networks are exceptionally strong at coming up with good features for highly structured data.

To resolve the FrozenLake problem, we'll build a one-layer network that takes the state encoded in a [1× 16] vector and learns the best move (action), mapping the possible actions in a vector of length four.

The following implementation is based in TensorFlow:

First, we need to import all the libraries:

```
import gym
import numpy as np
import random
import tensorflow as tf
import matplotlib.pyplot as plt
```

Then we load and set the environment to test:

```
env = gym.make('FrozenLake-v0')
```

The input network is a state, encoded in a tensor of shape [1,16]. For this reason, we define the inputs1 placeholder:

```
inputs1 = tf.placeholder(shape=[1,16],dtype=tf.float32)
```

The network weights are initially chosen randomly by the `tf.random_uniform` function:

```
W = tf.Variable(tf.random_uniform([16,4],0,0.01))
```

The network output is given by the product of the `inputs1` placeholder and the weights:

```
Qout = tf.matmul(inputs1,W)
```

The `argmax` evaluated on `Qout` will give the predicted value:

```
predict = tf.argmax(Qout,1)
```

The best move (`nextQ`) is encoded in a tensor of shape [1,4]:

```
nextQ = tf.placeholder(shape=[1,4],dtype=tf.float32)
```

Next, we define a loss function to implement the backpropagation procedure.

The loss function is $loss = \sum (Q-target-Q)^2$, where the difference between the current predicted Q-values and the target value is computed and the gradients are passed through the network:

```
loss = tf.reduce_sum(tf.square(nextQ - Qout))
```

The optimizing function is the well-known `GradientDescentOptimizer`:

```
trainer = tf.train.GradientDescentOptimizer(learning_rate=0.1)
updateModel = trainer.minimize(loss)
```

Reset and initialize the computational graph:

```
tf.reset_default_graph()
init = tf.global_variables_initializer()
```

Then we set the parameter for the Q-Learning training procedure:

```
y = .99
e = 0.1
num_episodes = 6000

jList = []
rList = []
```

We define the session, `sess`, in which the network will have to learn the best possible sequence of moves:

```
with tf.Session() as sess:
    sess.run(init)
    for i in range(num_episodes):
        s = env.reset()
        rAll = 0
        d = False
        j = 0

        while j < 99:
            j+=1
```

Reinforcement Learning

The input state is used here to feed the network:

```
a,allQ = sess.run([predict,Qout],\
                feed_dict=\
                {inputs1:np.identity(16)[s:s+1]})
```

A random state is chosen from the output tensor a:

```
if np.random.rand(1) < e:
    a[0] = env.action_space.sample()
```

Evaluate the `a[0]` action using the `env.step()` function, obtaining the reward, `r`, and the state, `s1`:

```
s1,r,d,_ = env.step(a[0])
```

This new state, `s1`, is used to update the Q-tensor:

```
Q1 = sess.run(Qout,feed_dict=\
             {inputs1:np.identity(16)[s1:s1+1]})
maxQ1 = np.max(Q1)
targetQ = allQ
targetQ[0,a[0]] = r + y*maxQ1
```

Of course, the weights must be updated for the backpropagation procedure:

```
_,W1 = sess.run([updateModel,W],\
              feed_dict=\
              {inputs1:np.identity(16)
[s:s+1],nextQ:targetQ})
```

rAll here defines the total reward that will be gained during the session. Let's recall that the goal of an RL agent is to maximize the total reward that it receives in the long run:

```
rAll += r
```

Update the state of the environment for the next step:

```
s = s1
if d == True:
    e = 1./((i/50) + 10)
    break
jList.append(j)
rList.append(rAll)
```

When the computation ends, the percent of successful episodes will be displayed:

```
print ("Percent of successfulepisodes: " +\
str(sum(rList)/num_episodes) + "%")
```

If we run the model, we should get a result like this, which can be improved by tuning the network parameters:

```
>>>[2017-01-15 16:56:01,048] Making new env: FrozenLake-v0
Percentage of successful episodes: 0.558%
```

Deep Q-learning

Thanks to the recent achievements of Google DeepMind in 2013 and 2016, which succeeded at reaching so-called superhuman levels in Atari games and beat the world champion Go, RL has become very interesting in of the machine learning community. This renewed interest is also due to the advent of **Deep Neural Networks (DNNs)** as approximation functions, bringing the potential value of this type of algorithm to an even higher level. The algorithm that has gained the most interest in recent times is definitely Deep Q-Learning. The following section introduces the Deep Q-Learning algorithm and also discusses some optimization techniques to maximize its performance.

Deep Q neural networks

The Q-learning base algorithm can cause tremendous problems when the number of states and possible actions increases and becomes unmanageable from a matrix point of view. Just think of the input configuration in the case of the structure used by Google to achieve the level of performance in the Atari games. State space is discrete, but the number of states is huge. This is the point where deep learning steps in. Neural networks are exceptionally good at coming up with good features for highly structured data. In fact, we can identify the Q function with a neural network, which takes the state and action as input and outputs the corresponding Q value:

Q (state; action) = value

Reinforcement Learning

The most common implementation of a deep neural network is pictured below:

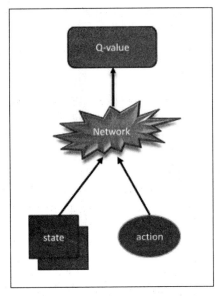

Figure 4: Common implementation of a Deep Q neural network

Alternatively, it can take the state as input and produce the corresponding value for each possible action:

Q (state) = value for each possible action

This optimized implementation can be seen in the following diagram:

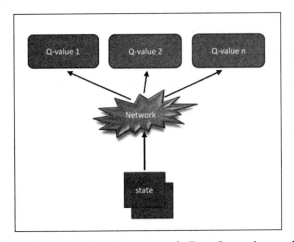

Figure 5: Optimized implementation of a Deep Q neural network

This last approach is computationally advantageous, because to update the Q value (or choose the highest Q value) we just have to take a step forward through the network and immediately we will have all Q values for all available actions.

The Cart-Pole problem

We'll build a deep neural network that can learn to play games through RL. More specifically, we'll use Deep Q-learning to train an agent to play the *Cart-Pole* game.

In this game, a freely swinging pole is attached to a cart. The cart can move to the left and right, and the goal is to keep the pole upright as long as possible:

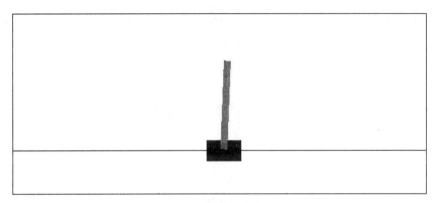

Figure 6: Cart-Pole

We simulate this game using OpenAI Gym. We need to import the required libraries:

```
import gym
import tensorflow as tf
import numpy as np
import time
```

Let's create the Cart-Pole game environment:

```
env = gym.make('CartPole-v0')
```

Initialize the environment, the rewards list, and the starting time:

```
env.reset()
rewards = []
tic = time.time()
```

Reinforcement Learning

The `env.render()` statement is used here to show the window with the running simulation:

```
for _ in range(1000):
    env.render()
```

`env.action_space.sample()` is passed to the `env.step()` statement to build the next step in the simulation:

```
state, reward, done, info = \
    env.step\
    (env.action_space.sample())
```

In the Cart-Pole game, there are two possible actions: moving the cart left or right. So, there are two actions we can take, encoded as 0 and 1.

Here, we take a random action:

```
        rewards.append(reward)
        if done:
            rewards = []
            env.reset()
toc = time.time()
```

After 10 seconds, the simulation ends:

```
if toc-tic > 10:
    env.close()
```

To shut the window showing the simulation, use `env.close()`.

When we run the simulation, we have a list of rewards, as follows:

```
[1.0, 1.0, 1.0, 1.0, 1.0, 1.0, 1.0, 1.0, 1.0, 1.0, 1.0, 1.0, 1.0, 1.0,
1.0, 1.0, 1.0, 1.0, 1.0]
```

The game resets after the pole has fallen past a certain angle. The simulation returns a reward of 1.0 for each frame that it is running. The longer the game runs, the more rewards we get. So, our network's goal is to maximize the reward by keeping the pole vertical. It will do this by moving the cart to the left and the right.

Deep Q-Network for the Cart-Pole problem

We train our Q-learning agent again using the Bellman equation:

$$Q(s,a) = r + \gamma \max_{a'} Q(s',a')$$

Here, s is a state, a is an action, and s' is the next state from state s and action a.

Earlier, we used this equation to learn values for a **Q-table**. However, there are a huge number of states available for this game. The state has four values: the position and velocity of the cart, and the position and velocity of the pole. These are all real-valued numbers, so if we ignore floating point precisions, we practically have infinite states. Instead of using a table, then, we'll replace it with a neural network that will approximate the Q-table lookup function.

The Q value is calculated by passing in a state to the network, while the output will be Q-values for each available action, with fully connected hidden layers:

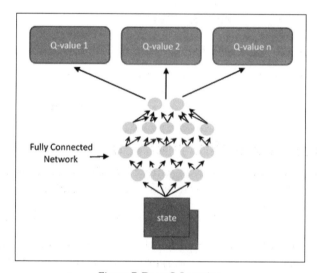

Figure 7: Deep Q-Learning

In this Cart-Pole game, we have four inputs, one for each value in the state; and two outputs, one for each action. The network weights update will take place by choosing an action and simulating the game with the chosen action. This will take us to the next state and then to the reward.

Here is a brief code snippet of the neural network used to solve the Cart-Pole problem:

```
import tensorflow as tf
class DQNetwork:
    def __init__(self,\
                 learning_rate=0.01, \
                 state_size=4,\
                 action_size=2, \
                 hidden_size=10,\
                 name='DQNetwork'):
```

The hidden layers consist of two fully connected layers with ReLU activations:

```
self.fc1 =tf.contrib.layers.fully_connected\
            (self.inputs_,\
             hidden_size)
self.fc2 = tf.contrib.layers.fully_connected\
            (self.fc1,\
             hidden_size)
```

The output layer is a linear output layer:

```
self.output = tf.contrib.layers.fully_connected\
            (self.fc2,\
             action_size,activation_fn=None)
```

The Experience Replay method

The approximation function can suffer greatly from the presence of non-independent and identically distributed and non-stationary data (correlations between states).

This kind of problem can be overcome by using the *Experience Replay* method.

During the interaction between the agent and the environment, all experiences (`state`, `action`, `reward`, and `next_state`) are saved in a replay memory, which is fixed size memory and operates in First In First Out (FIFO).

Here is the implementation of the replay memory class:

```
from collections import deque
import numpy as np

class replayMemory():
    def __init__(self, max_size = 1000):
        self.buffer = \
```

```
            deque(maxlen=max_size)

    def build(self, experience):
        self.buffer.append(experience)

    def sample(self, batch_size):
        idx = np.random.choice\
            (np.arange(len(self.buffer)),
                            size=batch_size,
                            replace=False)
        return [self.buffer[ii] for ii in idx]
```

This will allow the use of mini-batches of experiences taken randomly within the replay memory during the training of the network, instead of using recent experiences one after the other.

Using the experience replay method helps to mitigate the problem of sequential training data that could lead to the algorithm remaining stuck in a local minimum, denying it the chance to reach the optimal solution.

Exploitation and exploration

Whenever the agent has to choose which action to take, it basically has two ways that it can carry out its strategy. The first mode is called **exploitation** and consists of taking the best possible decision given the information obtained so far, that is, the past and stored experiences. This information is always available as a value function, which expresses which of the actions offers the greatest final cumulative return for each state-action pair.

The second mode is called **exploration**, and it is a strategy of making decisions that are different from what is currently considered optimal.

The exploration phase is very important, because it is used to gather information on unexplored states. In fact, it is possible that an agent that only performs the optimal action is limited to always follow the same sequence of actions without ever having had the opportunity to explore and find out that there could be strategies that, in the long run, could lead to much better results, even if this if it means the immediate gain is lower.

The policy most often used to reach the right compromise between exploration and exploitation is the **greedy policy**. It represents a methodology of selection of actions based on the possibility of choosing a random action with uniform probability distribution.

The Deep Q-Learning training algorithm

Let's see how to build a deep Q-Learning algorithm to solve the Cart-Pole problem.

The project is rather complex. For this reason. it has been subdivided into several file modules:

- `DQNetwork.py`: Implements the Deep Neural Network
- `memory.py`: Implements the experience replay method
- `start_simulation.py`: Creates the Cart-Pole environment that we want to resolve
- `solve_cart_pole.py`: Solves the Cart-Pole environment with the trained neural network
- `plot_result_DQN.py`: Plots the final rewards versus the episodes
- `deepQlearning.py`: The main program

The following commands provide a brief description of the implementation of the `deepQlearning.py` file:

```
import tensorflow as tf
import gym
import numpy as np
import time
import os
from create_cart_pole_env import *
from DQNetwork import *
from memory import *
from solve_cart_pole import *
from plot_result_DQN import *
```

The next thing to do is to define the hyperparameters used for this implementation, so we need to define the maximum number of episodes to learn from, the maximum number of steps in an episode, and the future reward discount:

```
train_episodes = 1000
max_steps = 200
gamma = 0.99
```

Exploration parameters are the exploration probability at the start, the minimum exploration probability, and the exponential decay rate for the exploration probability:

```
explore_start = 1.0
explore_stop = 0.01
decay_rate = 0.0001
```

Network parameters are the number of units in each hidden Q-network layer and the Q-network learning rate:

```
hidden_size = 64
learning_rate = 0.0001
```

Define the following memory parameters:

```
memory_size = 10000
batch_size = 20
```

Then we have the number of experiences to use to pretrain the memory:

```
pretrain_length = batch_size
```

Now we can create the environment and start the Cart-Pole simulation:

```
env = gym.make('CartPole-v0')
start_simulation(env)
```

Next, we instantiate the DNN with the `hidden_size` and `learning_rate` hyperparameters:

```
tf.reset_default_graph()
deepQN = DQNetwork(name='main', hidden_size=64, \
                   learning_rate=0.0001)
```

Finally, we re-initialize the simulation:

```
env.reset()
```

Let's take a random step, from which we can get the state and the reward:

```
state, rew, done, _ = env.step(env.action_space.sample())
```

Instantiate the `replayMemory` object to implement the Experience Replay method:

```
memory = replayMemory(max_size=10000)
```

Take a chunk of random actions to store the relative experiences, the state and actions, using the `memory.build` method:

```
pretrain_length= 20

for j in range(pretrain_length):
    action = env.action_space.sample()
    next_state, rew, done, _ = \
                env.step(env.action_space.sample())
```

Reinforcement Learning

```
            if done:
                env.reset()
                memory.build((state,\
                            action,\
                            rew,\
                            np.zeros(state.shape)))
                state, rew, done, _ = \
                        env.step(env.action_space.sample())
            else:
                memory.build((state, action, rew, next_state))
                state = next_state
```

With the new experiences obtained, we can carry out the training of the neural network:

```
rew_list = []
train_episodes = 100
max_steps=200
```

```
with tf.Session() as sess:
    sess.run(tf.global_variables_initializer())
    step = 0
    for ep in range(1, train_episodes):
        tot_rew = 0
        t = 0
        while t < max_steps:
            step += 1
            explore_p = stop_exp + (start_exp - stop_exp)*\
                        np.exp(-decay_rate*step)

            if explore_p > np.random.rand():
                action = env.action_space.sample()

            else:
```

Then we compute the Q state:

```
                Qs = sess.run(deepQN.output, \
                        feed_dict={deepQN.inputs_: \
                                    state.reshape\
                                    ((1, *state.shape))})
```

We can now obtain the action:

```
            action = np.argmax(Qs)

next_state, rew, done, _ = env.step(action)
tot_rew += rew

if done:
    next_state = np.zeros(state.shape)
    t = max_steps

    print('Episode: {}'.format(ep),
          'Total rew: {}'.format(tot_rew),
          'Training loss: {:.4f}'.format(loss),
          'Explore P: {:.4f}'.format(explore_p))

    rew_list.append((ep, tot_rew))
    memory.build((state, action, rew, next_state))
    env.reset()
    state, rew, done, _ = env.step\
                        (env.action_space.sample())

else:
    memory.build((state, action, rew, next_state))
    state = next_state
    t += 1



batch_size = pretrain_length
states = np.array([item[0] for item \
            in memory.sample(batch_size)])
actions = np.array([item[1] for item \
            in memory.sample(batch_size)])
rews = np.array([item[2] for item in \
            memory.sample(batch_size)])
next_states = np.array([item[3] for item\
            in memory.sample(batch_size)])
```

Reinforcement Learning

Finally, we start training the agent. The training is slow because it's rendering the frames slower than the network can train:

```
target_Qs = sess.run(deepQN.output, \
                            feed_dict=\
                    {deepQN.inputs_: next_states})

target_Qs[(next_states == \
            np.zeros(states[0].shape))\
        .all(axis=1)] = (0, 0)

targets = rews + 0.99 * np.max(target_Qs, axis=1)

loss, _ = sess.run([deepQN.loss, deepQN.opt],
                    feed_dict={deepQN.inputs_: states,
                                deepQN.targetQs_: targets,
                                deepQN.actions_: actions})



    env = gym.make('CartPole-v0')
```

To test the model, we call the following function:

```
solve_cart_pole(env,deepQN,state,sess)

plot_result(rew_list)
```

This is the implementation of the `solve_cart_pole function.py`, which is used here to test the neural network on the cart pole problem:

```
import numpy as np

def solve_cart_pole(env,dQN,state,sess):
    test_episodes = 10
    test_max_steps = 400
    env.reset()
    for ep in range(1, test_episodes):
        t = 0
        while t < test_max_steps:
            env.render()
            Qs = sess.run(dQN.output, \
                        feed_dict={dQN.inputs_: state.reshape\
                                    ((1, *state.shape))})
            action = np.argmax(Qs)
```

Chapter 10

```
            next_state, reward, done, _ = env.step(action)

        if done:
            t = test_max_steps
            env.reset()
            state, reward, done, _ =

                            env.step(env.action_space.sample())

        else:
            state = next_state
            t += 1
```

Finally, if we run the `deepQlearning.py` script we should obtain a result like this:

```
[2017-12-03 10:20:43,915] Making new env: CartPole-v0
[]
Episode: 1 Total reward: 7.0 Training loss: 1.1949 Explore P: 0.9993
Episode: 2 Total reward: 21.0 Training loss: 1.1786 Explore P: 0.9972
Episode: 3 Total reward: 38.0 Training loss: 1.1868 Explore P: 0.9935
Episode: 4 Total reward: 8.0 Training loss: 1.3752 Explore P: 0.9927
Episode: 5 Total reward: 9.0 Training loss: 1.6286 Explore P: 0.9918
Episode: 6 Total reward: 32.0 Training loss: 1.4313 Explore P: 0.9887
Episode: 7 Total reward: 19.0 Training loss: 1.2806 Explore P: 0.9868
......
Episode: 581 Total reward: 47.0 Training loss: 0.9959 Explore P:
0.1844
Episode: 582 Total reward: 133.0 Training loss: 21.3187 Explore P:
0.1821
Episode: 583 Total reward: 54.0 Training loss: 42.5041 Explore P:
0.1812
Episode: 584 Total reward: 95.0 Training loss: 1.5211 Explore P:
0.1795
Episode: 585 Total reward: 52.0 Training loss: 1.3615 Explore P:
0.1787
Episode: 586 Total reward: 78.0 Training loss: 1.1606 Explore P:
0.1774
........
Episode: 984 Total reward: 199.0 Training loss: 0.2630 Explore P:
0.0103
Episode: 985 Total reward: 199.0 Training loss: 0.3037 Explore P:
0.0103
Episode: 986 Total reward: 199.0 Training loss: 256.8498 Explore P:
0.0103
Episode: 987 Total reward: 199.0 Training loss: 0.2177 Explore P:
0.0103
```

```
Episode: 988 Total reward: 199.0 Training loss: 0.3051 Explore P: 0.0103
Episode: 989 Total reward: 199.0 Training loss: 218.1568 Explore P: 0.0103
Episode: 990 Total reward: 199.0 Training loss: 0.1679 Explore P: 0.0103
Episode: 991 Total reward: 199.0 Training loss: 0.2048 Explore P: 0.0103
Episode: 992 Total reward: 199.0 Training loss: 0.4215 Explore P: 0.0102
Episode: 993 Total reward: 199.0 Training loss: 0.2133 Explore P: 0.0102
Episode: 994 Total reward: 199.0 Training loss: 0.1836 Explore P: 0.0102
Episode: 995 Total reward: 199.0 Training loss: 0.1656 Explore P: 0.0102
Episode: 996 Total reward: 199.0 Training loss: 0.2620 Explore P: 0.0102
Episode: 997 Total reward: 199.0 Training loss: 0.2358 Explore P: 0.0102
Episode: 998 Total reward: 199.0 Training loss: 0.4601 Explore P: 0.0102
Episode: 999 Total reward: 199.0 Training loss: 0.2845 Explore P: 0.0102
[2017-12-03 10:23:43,770] Making new env: CartPole-v0
>>>
```

The total reward increases as the training loss decreases.

During the test, the cart pole balances perfectly:

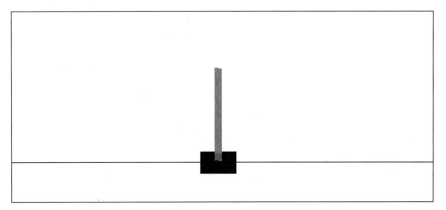

Figure 8: Resolved Cart-Pole problem

To visualize the training, we have used the `plot_result()` function (it is defined in the `plot_result_DQN.py` function).

The `plot_result()` function plots the total reward for each episode:

```
def plot_result(rew_list):
    eps, rews = np.array(rew_list).T
    smoothed_rews = running_mean(rews, 10)
    smoothed_rews = running_mean(rews, 10)
    plt.plot(eps[-len(smoothed_rews):], smoothed_rews)
    plt.plot(eps, rews, color='grey', alpha=0.3)
    plt.xlabel('Episode')
    plt.ylabel('Total Reward')
    plt.show()
```

The following graph shows the total reward per episode increasing as the agent improves its estimate of the value function:

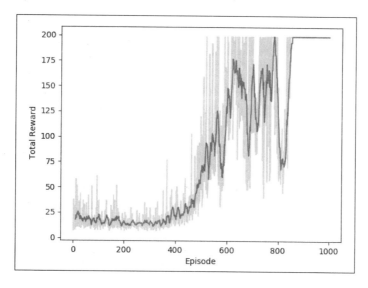

Summary

Many researchers believe that RL is the best shot we have of creating artificial general intelligence. It is an exciting field, with many unsolved challenges and huge potential. Although it can appear challenging at first, getting started in RL is actually not so difficult. In this chapter, we have described some basic principles of RL.

The main thing we have discussed is the Q-Learning algorithm. Its distinctive feature is the capacity to choose between immediate rewards and delayed rewards. Q-learning at its simplest uses tables to store data. This very quickly loses viability when the size of the state/action space of the system it is monitoring/controlling increases.

We can overcome this problem using a neural network as a function approximator that takes the state and action as input and outputs the corresponding Q-value.

Following this idea, we implemented a Q-learning neural network using the TensorFlow framework and the OpenAI Gym toolkit to win at the FrozenLake game.

In the last part of the chapter, we introduced deep reinforcement learning. In traditional RL, the problem spaces were very limited and there were only a few possible states in an environment. This was one of the major limitations of traditional approaches. Over the years, there have been a couple of relatively successful approaches that were able to deal with larger state spaces by approximating the state.

The advances in deep learning algorithms have led to a new wave of successful applications in RL, because it offers the opportunity to efficiently work with high-dimensional input data (such as images). In this context, trained DNNs can be seen as a kind of end-to-end RL approach, where the agent can learn a state abstraction and a policy approximation directly from its input data. Following this approach, we implemented a DNN to solve the Cart-Pole problem.

Our journey in Deep Learning with TensorFlow ends here. Deep learning is a very productive research area; there are many books, courses, and online resources that may help the reader to go deeper into the theory and programming. In addition, TensorFlow provides a rich set of tools for working with deep learning models. I would like the reader of this book to be a part of the TensorFlow community, which is very active and expects enthusiastic people to join them soon.

Other Books You May Enjoy

If you enjoyed this book, you may be interested in these other books by Packt:

Python Machine Learning - Second Edition

Sebastian Raschka, Vahid Mirjalili

ISBN: 978-1-78712-593-3

- Understand the key frameworks in data science, machine learning, and deep learning
- Harness the power of the latest Python open source libraries in machine learning
- Master machine learning techniques using challenging real-world data
- Master deep neural network implementation using the TensorFlow library
- Ask new questions of your data through machine learning models and neural networks
- Learn the mechanics of classification algorithms to implement the best tool for the job
- Predict continuous target outcomes using regression analysis
- Uncover hidden patterns and structures in data with clustering
- Delve deeper into textual and social media data using sentiment analysis

Other Books You May Enjoy

Python Interviews

Mike Driscoll

ISBN: 978-1-78839-908-1

- How successful programmers think
- The history of Python
- Insights into the minds of the Python core team
- Trends in Python programming

Leave a review – let other readers know what you think

Please share your thoughts on this book with others by leaving a review on the site that you bought it from. If you purchased the book from Amazon, please leave us an honest review on this book's Amazon page. This is vital so that other potential readers can see and use your unbiased opinion to make purchasing decisions, we can understand what our customers think about our products, and our authors can see your feedback on the title that they have worked with Packt to create. It will only take a few minutes of your time, but is valuable to other potential customers, our authors, and Packt. Thank you!

Index

A

activation functions
  about 81, 82
  ReLU 84
  sigmoid 84
  softmax 85
  tanh 84
AlexNet
  about 139, 152
  architecture 153
Alternating Least Squares (ALS) algorithm
  about 359
  advantages 359
Amazon Web Services (AWS) 29
AMIs (Amazon Machine Images) 30
Area Under the Precision-Recall
    Curve (AUPRC) 218
Artificial Neural Networks (ANNs)
  about 9, 11
  backpropagation algorithm 15, 16
  learning 15
artificial neuron 13, 14
artistic style learning, with VGG-19 162
autoencoders
  about 18, 24, 25
  as unsupervised feature learning
    algorithm 224-229
  implementing, with TensorFlow 195-200
  robustness, improving 200, 201
  working 192-194
axon 12

B

backpropagation algorithm 77-79
backward pass 322
Basic Linear Algebra Subroutines
    (BLAS) 36
basic RNNs
  implementing, in TensorFlow 239-243
Berkeley Vision and Learning Center
    (BVLC) 28
bias neuron 95
Bi-directional RNN (BRNN) 244, 245
biological neurons 12

C

Caffe 28
Cart-Pole problem
  about 433, 434
  Deep Q-Learning training
    algorithm 438-445
  Deep Q-Network, for 435, 436
  experience replay method, using 436, 437
  exploitation 437
  exploration 437
  reference 425
classification 5
clustering 7
code structure, TensorFlow 41-44
cold-start problem 396
collaborative filtering approaches
  about 356, 357
  content-based filtering approaches 358
  hybrid recommender systems 358

issues 357
model-based collaborative filtering 358
components, TensorFlow graph
  placeholders 40
  session 40
  tensors 40
  variables 40
computational graph, TensorFlow 37-40
computations
  visualizing, through TensorBoard 57
content-based filtering approaches 358
contrastive divergence 114
convolutional autoencoders
  decoder, working 208-216
  encoder, working 207
  implementing 207
Convolutional AutoEncoders (CAEs) 27
convolutional layer 142
Convolutional Neural Network (CNNs)
  about **18-23**, 140, 141, **217**
  emotion recognition with 173-185
  in action 142
convolution matrix 141
cross-entropy 149
CUDA architecture 311

D

data model, TensorFlow
  about 46
  data type 50-52
  feeds 55, 56
  fetches 54
  placeholders 55, 56
  rank 48, 49
  shape 48, 49
  tensor 46-48
  variables 53
data parallelism
  about 321
  asynchronous training 323
  synchronous training 323
dataset preparation 156
decoders
  working, in convolutional
    autoencoders 208-216
deconvolution 211

Deep Belief Networks (DBNs)
  about **18**, 20, 111
  implementing, with TensorFlow for
    client-subscription assessment 118-125
  Restricted Boltzmann Machines
    (RBMs) 112-114
  simple DBN, constructing 115
  supervised fine-tuning 117
  unsupervised pre-training 116
deep learning 8, 10
deep learning frameworks
  about 27
  Caffe 28
  Keras 27
  MXNet 28
  Neon 28
  TensorFlow 27
  Theano 28
  Torch 28
Deep Neural Networks
    (DNNs)
  about 18, **140**, 217, **270, 431**
  Deep Belief Networks (DBNs) 20
  multilayer perceptron 19
Deep Q-Learning
  about 431
  Cart-Pole problem 433, 434
  Deep Q neural networks 431-433
Deep Q-Learning training
    algorithm 438-445
deep reinforcement learning 421
Deep SpatioTemporal Neural
    Networks (DST-NNs) 27
dendrites 12
denoising autoencoder
  implementing 201-206
design principles, Keras
  extensibility 343
  minimalism 343
  modularity 343
development set 3
deviations
  reference 226
distributed computing
  about 320
  data parallelism 321, 323
  model parallelism 320, 321

distributed TensorFlow setup 323, 324
dropout operator 149
dropout optimization 133-136

E

eager execution, TensorFlow 44, 45
edges, TensorFlow
  normal 38
  special 38
Elastic Compute Cloud (EC2) 30
Emergent Architectures (EAs) 18
emotion recognition, with CNNs
  about 174-185
  model, testing on own image 185-187
  source code 187, 188
encoders
  working, in convolutional
    autoencoders 207
env class
  about 424
  render method 424
  reset method 424
  step method 424
errors
  reference 226
estimator 328
ETL (Extraction, Transformation,
    and Load) 4
expected value
  reference 226
exploitation 437
exploration 437
exploration, versus exploitation example 8

F

Factorization Machines (FMs)
  about 355
  cold-start problem 395
  collaborative-filtering approaches 396, 397
  dataset description 398
  FM model, training 408-413
  formulation 397
  for recommendation systems 393-395
  improved factorization machines 413
  preprocessing 401-406

problem definition 397
workflow of implementation 399
Factorization Matrix (FM) 358
Fast Fourier Transformation (FFT) 35
feature map 142
feed-forward neural networks (FFNNs)
  about 76, 77
  activation functions 81, 82
  backpropagation algorithm 77-79
  biases 79-81
  implementing 85, 86
  MNIST dataset, exploring 86, 87
  weights 79, 80
FFNN hyperparameters
  biases initialization 128
  GridSearch, for hyperparameter tuning 130
  number of hidden layers 126, 127
  number of neurons per hidden
    layer 127, 128
  randomized search, for hyperparameter
    tuning 130
  suitable optimizer, selecting 129
  tuning 126
  weight initialization 128
fine-tuning implementation
  about 157-159
  artistic style learning, with VGG-19 162
  content extractor 164-166
  content loss 167
  input images 163
  merger 168
  style extractor 167
  style loss 168
  total loss 168
  training 168, 170
  VGG 160, 161
forward pass 322
fraud analytics, with autoencoders
  about 217
  dataset, description 217
  exploratory data analysis 219-223
  model evaluation 229-233
  normalization 224
  problem description 218, 219
  testing set preparation 223
  training set preparation 223
  validation set preparation 223

FrozenLake environment 427
FrozenLake problem
  resolving, with Q-Learning 428-431

G

Gated Recurrent Unit (GRU) cell 252
Gated Recurrent units (GRUs) 26
GoogLeNet 171
GPGPU 310
GPGPU computing
  about 310
  CUDA architecture 311
  GPGPU history 310
  GPU programming model 312
GPU programming model
  working 312
Gradient Descent (GD) 16
Gram matrix 167
greedy policy 437
GridSearchCV
  reference 130

H

Human Activity Recognition (HAR),
    with LSTM model
  about 294
  dataset description 294, 295
  implementation 297-307
  workflow 296
hybrid recommender systems 358
hyperparameters 3

I

ImageNet dataset 139
ImageNet Large-Scale Visual Recognition
    Challenge (ILSVRC) 139
improved factorization machines
  about 413, 414
  Neural Factorization Machines (NFMs) 414
Inception-v3
  about 171, 172
  exploring, with TensorFlow 172, 173
Inception-v3 model 139
Inception vN 171
input neurons 95

IPython Notebook
  reference 52
issues, collaborative filtering approaches
  cold start 357
  scalability 357
  sparsity 357

K

Kaggle platform
  about 139
  reference 173
Keras
  about 27, 342
  design principles 343
Keras implementation, of SqueezeNet
  reference 351
Keras programming models
  functional APIs 343, 348
  sequential model 343
Keras v2.1.4
  reference 348
K-means 9

L

latent factors (LFs) 358
LeNet5
  about 139, 143, 144
  AlexNet 152
  implementing 144-152
  pre-trained AlexNet 154
  transfer learning 154
linear combination 14
linear regression
  about 59-66
  for real dataset 67-72
Long Short-Term Memory (LSTM) 131
loss 157
LSTM networks 249-251
LSTM predictive model, for sentiment
    analysis
  about 270
  LSTM model evaluation 291-293
  LSTM model training 271-289
  network design 270
  visualization, through
      TensorBoard 289, 290

M

machine learning
  about 2
  reinforcement learning 8
  supervised learning 4, 5
  unsupervised learning 6, 7
Markov Chain Monte Carlo (MCMC) 114
Markov Random Fields (MRF) 111
Mean Squared Error (MSE) 226
Microsoft Cognitive Toolkit (CNTK)
  about 29
  reference 30
MNIST digit classification system 139
model-based collaborative filtering 358
model inferencing 3
model parallelism 320
MovieLens dataset
  about 362
  exploratory analysis 364-370
  movies data 362
  ratings data 362
  users data 363
MovieLens website
  reference 362
movie recommendation, with
    collaborative filtering
  dataset 362
  developing 359
  evaluating 390-393
  exploratory analysis, of MovieLens
      dataset 364-370
  implementing 370
  utility matrix 359
movie RE implementation
  model, training with available
      ratings 371-379
  movie rating prediction, by users 386, 387
  performing 370
  saved model, inferencing 380
  similar movies, clustering 382-385
  top k movies, finding 387, 388
  top k similar movies, predicting 388
  user-item table, generating 380-382
  user-user similarity, computing 389

Multi-Dimensional Recurrent Neural
    Networks (MD-RNNs) 27
multilayer perceptron (MLP)
  about 18, 19
  dataset description 99, 100
  implementing 95, 96
  preprocessing 101-103
  TensorFlow implementation, for
      client-subscription assessment 103-111
  training 96-98
  using 98
MXNet 28

N

Natural Language Processing (NLP) 35
Neon 28
Neural Factorization Machines (NFMs)
  about 414
  dataset description 414
  FM model, evaluating 418
  FM model training 415, 416
  NFM model, evaluating 419
  NFM model training 417, 418
  using, for movie recommendation 415
neural network architectures
  about 18
  autoencoders 24, 25
  Convolutional Neural Networks
      (CNNs) 22, 23
  Deep Neural Networks (DNNs) 18, 19
  Recurrent Neural Networks (RNNs) 26
neuron 12, 13
normalization
  min-max scaling 224
  Z-score 224
NVIDIA CUDA toolkit
  reference 36
NVIDIA cuDNN
  reference 36
NVIDIA GPU Cloud (NGC) 29
NVIDIA Graph Analytics Library 37

O

OpenAI environments
  2D and 3D robots 424
  algorithmic 424

Atari 424
board games 424
classic control and toy text 424
OpenAI Gym
about 423
env class 424
installing 425
OpenAI environments 424
reference 425
running 425
Open Neural Network Exchange (ONNX) 28
overfitting 2

P

parameter server 322
PIL (Pillow) 154
pixels 140
pixel shaders 310
placeholders 43
predictive model, for time series
dataset description 260, 261
developing 260
exploratory analysis 262, 263
LSTM predictive model 264-266
model evaluation 267-269
pre-processing 262, 263
pre-trained AlexNet 154-156
pre-trained VGG-19 neural network 164
PrettyTensor
about 337
branch method 338
chaining layers 337
digit classifier 338-342
join method 338
normal mode 337
sequential mode 338

Q

Q-Learning algorithm
about 426, 427
for FrozenLake problem 428
FrozenLake environment 427
Q-table 435

R

RandomizedSearchCV
reference 130
receptive field 142
recommendation systems
about 356
collaborative filtering approaches, using 356
Recurrent Neural Networks (RNNs)
about 18, 26
Bi-directional RNNs 244, 245
Gated Recurrent Unit (GRU) cell 252, 253
gradient vanishing-exploding problem 246-248
long-term dependency problem 243, 244
LSTM networks 249-251
working principles 236-239
Recurrent Neural Networks (RNNs), for spam prediction
data description 253-260
data preprocessing 253-260
implementing 253
regression 5
regularization
L1 regularization 130
L2 regularization 130
max-norm constraints 131
reinforcement learning (RL) 8, 421
ReLU
using 84
ReLU operator 149
residuals
reference 226
Restricted Boltzmann Machines (RBMs) 20, 21, 112-114
RL problem 422, 423
RMSPropOptimizer 150

S

sequential model, Keras
sentiment classification, of movie reviews 344-347
shared memory 311
sigmoid
about 14
using 84

Singular Value Decomposition (SVD) 358
softmax activation function 143
softmax classifier 88-94
softmax function 83
  using 85
SqueezeNet 348-350
Staked Auto-Encoders (SAEs) 18, 77
step function 14
Stochastic Gradient Descent (SGD) 17, 228
streaming multiprocessor (SM) 311
supervised learning 4, 5
synapses 12
synaptic terminals 12

T

tanh
  using 84
telodendria 12
TensorBoard
  computations, visualizing 57
  working 57, 58
TensorFlow
  about 27, 157
  autoencoder, implementing 195-200
  basic RNNs, implementing in 239-243
  code structure 41-44
  computational graph 37-40
  configuring 36
  data model 46
  edges 38
  features, by latest release 32, 33
  Inception, exploring with 172, 173
  installing 36
  overview 32
  reference 33
TensorFlow GPU setup
  about 313
  GPU memory management 316
  GPU representation 314
  GPU, using 314, 315
  multiple GPUs, using 318
  single GPU, assigning on multi-GPU system 316, 317
  source code, for GPU 317
  TensorFlow, updating 313

TensorFlow graph
  components 40
  tf.Operation objects 38
  tf.Tensor objects 38
TensorFlow Lite 34, 35
TensorFlow v1.6
  about 33
  eager execution 35
  Nvidia GPU support optimized 34
  optimized accelerated linear algebra (XLA) 35
tensors
  about 46-48
  reference 43, 46
test set 3
tf.estimator
  about 327
  flower predictions 329-332
  graph actions 328
  resources, parsing 328
TFLearn
  about 333
  estimator 333
  graph_actions 333
  installing 334
  layers 333
  Titanic survival predictor 334-336
Theano 28
TITO (tensor-in-tensor-out) 38
Torch 28
training set 3
transfer learning 154

U

unbalanced data 6
unsupervised learning 6
utility matrix 359, 360

V

validation set 3
vector space model
  reference 102
VGG
VGG-n 160
Visual Geometry Group (VGG) 139, 160, 161

W

weight optimization 16
workers 322

X

Xavier initialization 81